Minerals and Rocks 15

J. Barry Dawson

Kimberlites and Their Xenoliths

With 84 Figures

Springer-Verlag
Berlin Heidelberg New York 1980

Professor J. BARRY DAWSON
Department of Geology, University of Sheffield
Mappin Street
Sheffield S1 3JD, Great Britain

Volumes 1 to 9 in this series appeared under the title
Minerals, Rocks and Inorganic Materials

ISBN 3-540-10208-6 Springer-Verlag Berlin Heidelberg NewYork
ISBN 0-387-10208-6 Springer-Verlag NewYork Heidelberg Berlin

Library of Congress Cataloging in Publication Data. Dawson, John Barry 1932–. Kimberlites and their xenoliths. (Minerals and rocks; 15). Includes bibliographical references and index. 1. Kimberlites. 2. Kimberlites – Inclusions. I. Title. II. Series. QE462.K5D38 552'.3 80-23462.

Typesetting, printing and binding: Brühlsche Universitätsdruckerei, Gießen.
2132/3130-543210

To my wife, Christine

Preface

In these days of information explosion and high-cost publishing, it is perhaps only reasonable for an author to convince the reading public that it is getting something worth reading. After all, intense research into the upper mantle over the past two decades has already resulted in a number of volumes on kimberlites and their xenoliths. So why yet another one? First, in this book I have concentrated on kimberlite as an individual rock-type and a sampler of the upper mantle, in the hope of complementing such monographs as *Deep seated inclusions in kimberlites and the problem of the composition of the upper mantle* by N.V. Sobolev and *Géologie du Diamant* by M.G. Bardet, which have concentrated more on specific aspects of kimberlite and upper mantle geology; as a result I have not attempted to describe diamond prospecting and mining and, although I have attempted to give some of the up-to-date views on xenolith petrology, the confines of space have not permitted me to explore the complexities of the upper mantle as deeply as has Sobolev. Second, the literature is vast and I have tried to pull together for the reader the wide range of views and information presented both in the recent geological literature and in recent multi-author volumes such as the Proceedings Volumes of the two International Conferences on Kimberlite held at Capetown in 1973 and Santa Fe in 1977. To cut down on the vast literature list, I have, where possible, drawn upon recent review articles; for example, that by Meyer on American kimberlites. As a result, some worthy author may not find his name in the reference list; I trust he will forgive me and understand it is in the interest of space.

This book is a blend of my own experience and the efforts of many other people, and I should like to pay special tribute to two late colleagues and friends, W.Q. Kennedy and C.F. Davidson, who did much to stimulate my early interest in kimberlite. I acknowledge also the following people for preprints, valuable discussion and analytical assistance: J.R. Baldwin, F.R. Boyd, D.A. Carswell, T.N. Clifford, P. Deines, C.H. Emeleus, J.J. Gurney, S.E. Haggerty, J.W. Harris, B. Harte, J.B. Hawthorne, P. Kresten, M.E. McCallum, R.H. Mitchell, A. Nicolas, D.J. Powell, A.M. Reid, J.V. Smith, N.V. Sobolev, W.E. Stephens, and P.J. Wyllie. In particular, I should like to thank John Gurney and Joe Smith with whose research groups at the universities of Cape Town and Chicago I have had long and fruitful collaboration. Much of my fieldwork in southern Africa has been carried out with the cooperation and assistance of the geologists of De Beers Consolidated Mines Ltd., notably L.G. Murray, C.D. Hallam and J.B. Hawthorne.

The Natural Environment Research Council, the Royal Society, the Carnegie Trust for the Universities of Scotland and the Travel Fund of the University of St. Andrews are acknowledged for financial assistance in connection with my own research. In addition, I have enjoyed the cheerful and unstinting assistance of the secretarial and technical staff in the departments of Geology in the universities of St. Andrews and Sheffield; I particularly thank S. Fell and M. Townsend who typed the manuscript with the help of S. Forster, P. Mellor and B. Wilson, and J. Allen and M. Cooper for photography and draughting.

Finally, I pay tribute to my wife, Christine, whose constant cheerfulness has done much to maintain my equanimity during the writing of this book.

Sheffield, September 1980 J. BARRY DAWSON

Contents

Chapter 1 Introduction

Kimberlite is a rare and volumentrically insignificant igneous rock. What, then, are the reasons for studying it? First, kimberlite and some rare xenoliths within it are the principal primary terrestrial source of the strategic mineral, diamond, and the study of kimberlite may eventually outline the conditions for the formation of diamond in nature. Second, it can now be demonstrated that the ultrabasic and basic xenoliths within kimberlite have originated within the upper mantle. Kimberlite has carried out a more extensive sampling of the upper mantle than any other type of magmatic activity, and it seems highly unlikely that the abandoned Mohole project would have provided even a small proportion of the information about the petrology and mineralogy of the upper mantle that has been gained from the study of xenoliths in kimberlite. In addition, the four-phase garnet lherzolite xenoliths have received considerable attention from experimental petrologists in connection with the origin of basaltic magmas. The third principal reason for studying kimberlite is its chemistry. It contains high concentrations of lithophile elements in a rock that is essentially ultrabasic, and it also contains a much higher concentration of the so-called "incompatible" trace elements than other ultrabasic rocks. These features focus attention upon a source of these elements within the upper mantle and the processes concentrating them in kimberlite.

The study of kimberlites may be said to have begun in 1866 when the children of a farmer playing on the banks of the Orange River near Hopetown, Cape Province, South Africa, found a beautiful, white pebble. This curiosity passed through several hands until it came into the hands of the mineralogist W. Guybon Atherston of Grahamstown who identified it as a fine diamond. Although exhibited at the Paris Exhibition that year, little attention was paid to the find until in 1867 another magnificent diamond, later named "the Star of South Africa", was found on the banks of the Orange River at Zendfontein. Following this, prospectors swarmed into the area, gradually following scattered alluvial diamond finds up the Vaal River to a point where a rich alluvial deposit was found at Klipdrift (now Barkly West) in 1870. A different kind of deposit was discovered in August of the same year when De Klerk, the overseer at the farm Jagersfontein on the arid plains to the south-east of the alluvial diggings on the Vaal, noticed garnets in a dry river-bed. Knowing that garnets often accompany alluvial diamond, he had the curiosity to investigate further and was rewarded by finding a diamond weighing 50 carats. In the following month diamonds were found in the same area at Dutoitspan on the farm Dorstfontein, and then on the neighbouring farm Bultfontein. Early in May 1871 diamonds were found at two localities on the nearby farm Vooruitzigt belonging to Jan de Beer. One of the deposits, originally called Colesberg Kopje, was an exceptionally rich find and soon a large, prosperous town, named Kimberley after the Secretary of State for the Colonies at that time, sprang up around the diggings. The mines around Kimberley were known as the "dry diggings", by contrast with the "river diggings" on the Vaal. The river diggings were similar to the alluvial diamond deposits of Brazil and India, but the deposits in the vicinity of Kimberley were an entirely new phenomenon in diamond mining. The diamonds were found in loose superficial deposits underlain by a yellowish clay rock which contained as many diamonds as the superficial gravels. This passed downwards into a hard bluish-green-rock – the "blue-ground" of the prospectors – which was also diamondiferous. By the end of 1872 the diamondiferous rock at Kimberley had been sufficiently mined to reveal that the host-rock of the diamond occupied a roughly circular depression and that the rock persisted in depth. It was soon realised that the diamondiferous rock was an igneous rock occupying an old volcanic vent but it was not until 1887 that Lewis gave the first petrographic description of the rock. He described the rock as a porphyritic mica-bearing peridotite that in some parts of the mine contained so many xenoliths of carbonaceous shale as to resemble a volcanic breccia. (He also believed that diamond resulted from interaction between the igneous rock and carbon derived from the shales). However it was in a subsequent paper that Lewis (1888) pointed out the unique nature of the rock and proposed that it should be called "kimberlite". Although Lewis originally specifically included diamond as an essential constituent in kimberlite, the term was later extended to include rocks of similar petrographic character which, although not containing diamonds, occurred in the vicinity of diamondiferous kimberlite. Much later, Dawson (1967a) proposed that diamond should not be regarded as an essential constituent of kimberlite, thus broadening the term to include many of the so-called mica-peridotites that, except for lacking diamond, resemble the kimberlites of the type locality in their tectonic setting, mode of occurrence, petrographic character, mineralogy, geochemistry and xenolith content.

Although it was first found in diatremes, kimberlite was subsequently found in dykes and sills, but in no case were the igneous bodies found to be of any great size, and kimberlite must be regarded as a volumetrically insignificant and very rare rock-type, particularly when compared with major rock-types such as basalt and granite.

Following the discovery of kimberlite in the diatremes and dykes in the Kimberley area, more kimberlite intrusions were found in other parts of South Africa, South-West Africa, Rhodesia, and the (then) Belgian Congo. In addition it was recognized that mica-peridotites occurring in various parts of the U.S.A. were also kimberlites, and it is worthy of note that the first kimberlites in the world, although not recognized as such at the time, were discovered in the earlier part of the nineteenth century in the vicinity of Ludlowville, near Ithaca, New York State

(Vanuxem 1837). In the years between 1870 and 1930 a large amount of data was obtained on the geology and mineralogy of kimberlite and its associated xenoliths and this information was largely collated in the now-classical monographs of Wagner (1914) and Williams (1932). Then from the late 1930's to the late 1950's there was almost no new published work on the geology of kimberlite. In 1954 a team of Russian geologists discovered a kimberlite diatreme (the Zarnitsa pipe) in the headwaters of the Markha River in Siberia. It had been a desire to obtain a source of diamond independent of Western control, combined with a geological analysis by Academician V.S. Sobolev in which he compared the tectonics and geological history of the Siberian and South African cratons, that had caused the Russians to mount extensive prospecting operations on the eastern side of the Anabarsk shield from which rivers drained eastwards into the Lena River where small alluvial diamonds had been found in gold placers. Many more kimberlite intrusions were found in the area and the first monograph on the Siberian kimberlites by Bobrievich et al. (1957) heralded a new era in the study of kimberlite both in Russia and in the West.

In the past 20 years the older literature has been supplemented by a wealth of new data. Not only have there been the finds of entirely new kimberlite provinces, but the application of isotope geochemistry, geochronology and high pressure/ high temperature experimentation has clarified many problems of kimberlite geology that could not be solved by the earlier workers. In addition electron probe microanalysis and more sophisticated modern analytical techniques have given us much more detailed information on the chemistry and mineralogy of kimberlite and its associated xenoliths, and of their individual phases. It is the main purpose of this book to review these later contributions to kimberlite geology.

Chapter 2 Distribution and Tectonic Setting of Kimberlites

The general world-wide distribution of kimberlites is shown in Fig. 1 and in greater detail for the main kimberlite provinces in Figs. 2–6. From the aspect of the tectonic setting of kimberlite, Dawson (1970) established that, in the case of the kimberlites of Africa, most kimberlites are confined to the ancient cratons (or areas underlain by the cratons), and that relatively few are to be found in the circum-cratonic fold-belts. This general principle is true for other kimberlite provinces and, in the following review, the kimberlite distribution will be discussed within their tectonic framework. In addition, it is apparent that in many areas, and particularly in the circum-cratonic fold-belts, the intrusion of kimberlite is part of a phase of very limited igneous activity manifested by intrusion of other rare rock-types; brief summaries of this contemporary activity are included below.

In the past, various types of lamprophyric dyke rocks found within carbonatite complexes have been referred to as "kimberlites" or "kimberlitic rocks". To distinguish between these rocks and the products of true, regional kimberlite magmatism, Dawson (1967a, 1971) suggested the term "central complex kimberlite". More recently Mitchell (1979) has shown, mainly on the basis of groundmass spinel composition, that some, such as those at Alnö (Sweden), Fen (Norway) and Île Bizard (Canada) are not kimberlites; he also suggests that other kimberlitic rocks occurring in regional swarms such as Coral Rapids, Bachelor Lake and Saguenay Valley (all in Canada) are calcite-rich lamprophyres, not kimberlites. Although these, and other "kimberlitic" rocks are included in the present review, the reader should be aware that subsequent detailed mineralogical studies may prove that some of these are not kimberlites.

2.1 Africa

It is perhaps appropriate that kimberlite should have its type locality in Africa as it is within this landmass that this rare rock-type is most widespread, being found in a region extending from well north of the equator to almost the southern tip of the continent. The more important areas of kimberlite intrusion are West Africa, Angola, the central plateau of Tanzania and the upland interior of South Africa. In the other countries – Gabon, Zaire, Kenya, Zambia and Zimbabwe – kimberlite occurs on a relatively minor scale.

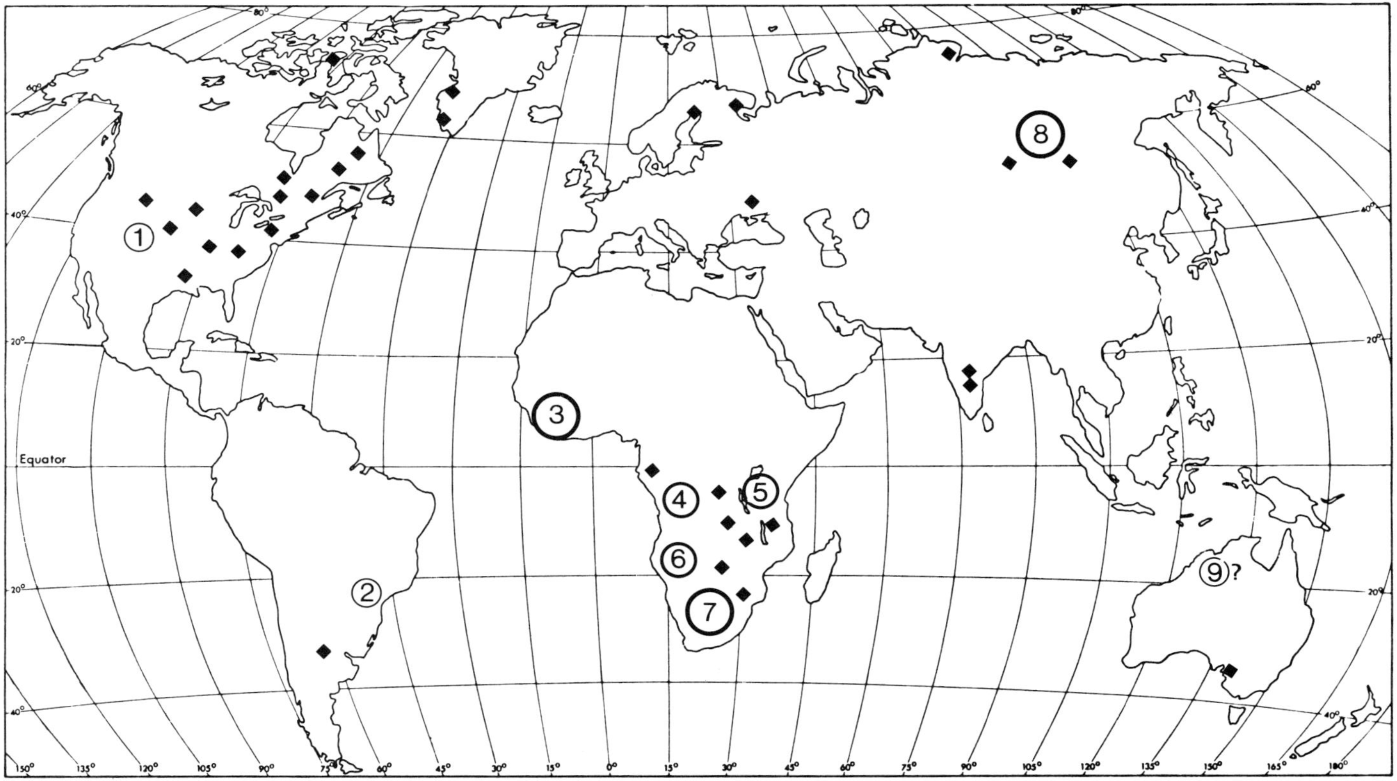

Fig. 1. Map showing the world-wide distribution of kimberlite provinces *(numbered)* and more isolated occurrences *(solid black triangles)*. The numbered provinces are: *1* Navajo-Hopi Province, South-West U.S.A.; *2* Brazil; *3* West-Africa; *4* Angola; *5* Tanzania; *6* Namibia; *7* Southern Africa; *8* Yakutia, U.S.S.R.; *9* North-West Australia. Of these provinces, *1, 2* and *9* are either relatively insignificant or, in the case of *9,* its potential as a kimberlite province is not yet known

The tectonic setting of African kimberlites has been analysed by Dawson (1970). The primary tectonic units are the older cratons – the West African craton, the Angola-Kasai craton, the Tanzania craton and the Transvaal-Rhodesia craton – which have not suffered orogeny since at least 1500 m.y. ago. After the Kibaran orogenesis about 1100 ± 200 m.y. ago, the consolidated Kibarides linked up the Angola-Kasai and Tanzania shield to form, with the accreted Irumide belt, the Congo craton; in the south of the continent, the Orange River (or Natal-Namaqualand) Belt was fused to the Transvaal-Rhodesia shield to form the Kalahari craton. The two cratons, together with the West African craton, acted as stable blocks during the late Precambrian – early Palaeozoic Katangan orogeny when the Damaride, Mozambique and West Congo fold-belts were formed. The distribution of kimberlites in Africa and their relation to the major cratons and fold-belts is shown in Fig. 2, from which it is apparent that kimberlite is most extensively developed on the major cratons.

Again looking at the African kimberlites on a continental scale Hawthorne (1975) has pointed out that the kimberlites of Tanzania, Angola, Botswana and Namaqualand (N.W. Cape Province, S. Africa) are much less eroded than those of South Africa; this may be interpreted as

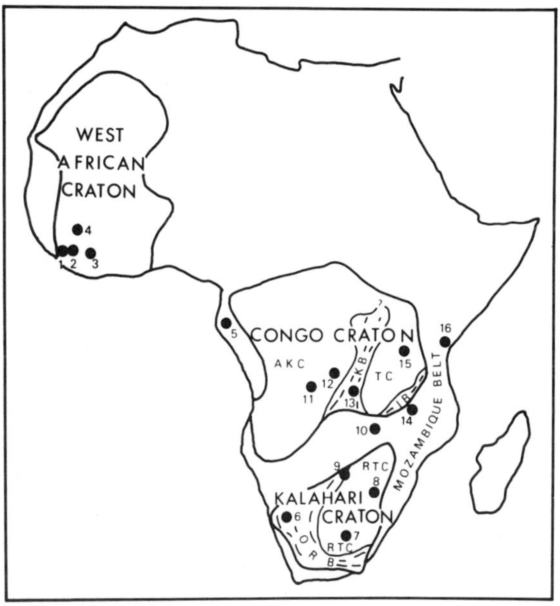

Fig. 2. Map of Africa showing the major cratons and distribution of kimberlites on the cratons. The numbers are as follows: *1* Liberia; *2* Sierra Leone; *3* Ivory Coast; *4* Mali; *5* Gabon; *6* Namibia; *7* South Africa; *8* Swaziland; *9* Botswana; *10* Zambia; *11* Angola; *12* Bakwanga area, Zaire; *13* Kundelungu Plateau, Zaire; *14* Southern Tanzania; *15* Tanzania Main Province. Abbreviations: *AKC* Angola-Kasai Craton; *KB* Kibaride Belt; *TC* Tanzania Craton; *IB* Irumide Belt; *RTC* Rhodesia-Transvaal Craton; *ORB* Orange River Belt

a) either the non-eroded kimberlites are of a substantially younger age, or
b) the regions of the crust into which they are intruded have not been subject to the same amount of uplift and erosion.

Radiometric ages of 93 m.y. and 189 m.y. (Table 2) on zircons from the Orapa and Mwadui pipes respectively indicate that suggestion (b) is the more likely.

2.1.1 West Africa

In West Africa kimberlites are confined to the area underlain by the 2000 m.y. Eburnean craton, although the most diamondiferous kimberlites are situated close to even older cratonic nuclei of Liberian age (Knopf 1970); in Liberia one of these older nuclei has now been dated as 2700 m.y. old (White and Leo 1970). According to Bardet and Vachette (1966), kimberlites of four ages are known in West Africa: 2100–2300, 1150, 700 and 80–100 m.y. old. On the West African craton, kimberlite dykes occupy fractures that have three dominant strikes – N55°E, N85°E and E40°S, and these trends link up groups of kimberlite intrusions (Bardet 1963).

In *Mali* kimberlite pipes intrude Precambrian schists and Ordovician sediments in the vicinity of Keneiba. They are intruded into an area that has been domed in post-Ordovician times and are believed to be of Cretaceous age (Bureau d'Études Géologiques et Minières1958).

In *Guinea* dykes and pipes are found in the south-east of the country, in a triangular area the apices of which are the towns of Keruane, Beyla and Macenta. Some pipes are in the vicinity of the Bonankoro alluvial diamond deposits. The dykes mainly strike NE-SW or EW, across the trend of the basement gneisses and earlier dolerite dykes (Kozlov 1966).

Kimberlite pipes and dykes occur in the Yengema-Koidu and Tongo areas of *Sierra Leone* (Grantham and Allen 1960; Fairbairn and Robertson 1966). The country-rocks are granites and NNW-SSE-trending schist belts which are intruded by two sets of dolerite dykes, one set trending E-W, the other NNE-SSW. The kimberlite dykes strike N55°E and their intrusion was controlled by shear-belts (Andrews-Jones 1966) that cut across all the earlier regional structures. In addition a small ring-dyke of micaceous kimberlite, 170 m in diameter, is known in the Yengema area (E. Gerryts, personal communication 1969).

In *Liberia,* several small discontinuous dykes occur in the west of the country, in the vicinity of Zoi, near the Sierra Leone border. They strike NNE and, as in the case of the Sierra Leone dykes, cut obliquely across the enclosing gneissic rocks and approximately normal to a zone of earlier dolerite dykes; they apparently define another fracture system (White and Leo 1969).

In the *Ivory Coast* kimberlite dykes are known in the Seguela and Haut-Nzi areas, and kimberlitic geochemical and mineralogical anomalies have been found in the Batie-Bouna area in the north-east of the country. In the Seguela field there are, in addition to the normal kimberlites, some that are sufficiently different that they have been termed "metakimberlites" (Bardet 1973). The Toubabouko and Bobi dykes are sheared and have the appearance and mineralogy of talc schists; although diamondiferous, they lack pyrope and ilmenite but contain abundant chromite. In addition the Toubabouko dyke has caused potassium fenitisation of the granite wall-rock (now syenite), and the Bobi dykes are intimately associated with dykes of alnoite and fitzroyite (Knopf 1970; Bardet 1973). Generally in the Ivory Coast, the kimberlite dykes follow the strike of the older trends and fracture systems.

The presence of alluvial diamonds in *Ghana* and *Upper Volta* may be an indication of undiscovered kimberlites in these countries.

2.1.2 Central Africa

In *Gabon* kimberlite dykes are exposed in a tributary of the River Ikoy. The dykes crop out in a elongate dome of Archaean gneisses, flanked by Proterozoic sediments (Choubert 1946). In N.W. Gabon in the Mitzic area are dykes of Precambrian talcose, diamondiferous "metakimberlites" similar to those of the Ivory Coast (Bardet 1973).

In *Angola* at least 94 kimberlites are now known (Reis 1972). They occur along a NW-SE zone, the most important provinces from NE to SW being (1) north-eastern Lunda; (2) the headwaters of the Cucumbi, Cacuilo and Cuago Rivers; (3) the Cuanza region; (4) the headwaters of the Cunene, Queve and Catumbela Rivers in a region extending from Nova Lisboa to Andullo; (5) in the headwaters of the Longa River. There are also outlying occurrences in the Cassinga and Artura de Paiva regions. A NE-SW chain of post-Jurassic alkalic ring-complexes, some of them containing carbonatite, are associated with the kimberlites in the Nova Lisboa area. The kimberlites are of post-Jurassic age, and the main provinces are aligned along a major NE-SW-trending fracture system that is cut by smaller NW-SE-trending cross-faults; this zone is also an area of high magnetic anomaly. A line extended NE from the faulted graben-area of the Luachimo-Chicapa area of Lunda (Real 1959) trends directly towards the Mesozoic kimberlites of *Zaire* (formerly the Belgian Congo) in the Bakwanga-Kalonji area (Meyer de Stadelhofen 1963; Fièremans 1966). It appears that the Angola-Zaire kimberlites were intruded along this major 1600-km-long zone of weakness that fractured the Angola-Kasai shield in Mesozoic times; in north-eastern Angola, this fracture zone is at the southern hinge of the downwarped Congo Basin. In south-eastern Zaire, kimberlite diatremes, some very large, intrude horizontal Precambrian sediments in the western and eastern escarpments of the Kundelungu Plateau (Verhoogen 1938). The area is underlain at depth by rocks of the Kibaran orogenic belt.

In *Zambia* a group of vents and dykes intrude Jurassic sediments in the Luangwa graben (L.G. Murray, personal communication, in Dawson 1970). The trend of the vent zone and dykes is NW-SE, in contrast to the NNE-SSW trend of the Luangwa graben, and to the NE-SW or ENE-WSW trend of basement rocks outside the graben.

2.1.3 East Africa

The central plateau of Tanzania is an area of abundant kimberlite intrusion. More than 120 intrusions are known in the granitic craton south of Lake Victoria. They are largely concentrated in the Mabuki, Shinyanga, Nzega, Kahama and Kimali areas, and in the Iramba plateau west of Singida; only the Mwadui Mine at Shinyanga and one near Kahama have been sufficiently diamondiferous to warrant exploitation. The earlier mines in the Shinyanga-Nzega area were described by Williams (1939), but a more recent account of the whole province is that by Edwards and Howkins (1966), who also give a detailed account of the Mwadui diatreme, the largest known kimberlite (see Table 5); this diatreme has been little eroded and contains original crater-like sediments. More detailed accounts of the kimberlites of the Iramba-Singida area are given by Teale (1932) and Mannard (1962).

Compared with their relative concentration on the Tanzania craton, kimberlites are scarce in the circum-cratonic Mozambique Belt. Kimberlites intrude Mesozoic sediments overlying the Mozambique Belt in the Ruhuhu graben of S.W. Tanzania (McKinlay 1955).

The late Tertiary-Recent rift valleys of East Africa, which are bounded by very fundamental fractures that follow the older fold-belts around the Tanzania craton, contain a very few rare occurrence of rocks that have been interpreted as young kimberlite extrusives. Nixon (1973a) interprets calcareous tuffs at Losorogoi, Kenya, and pyroclastics containing ultramafic xenoliths near Moroto, Uganda, as having kimberlitic affinites. However, another occurrence, the Lashaine volcano, Tanzania, which was also believed on the basis of its xenolith content to be the surface expression of a modern kimberlite diatreme (Dawson 1964), has been proved to have more affinities with the regional nephelinites (Dawson et al. 1970). The Igwisi Hills, west of Tabora, Tanzania, have also been interpreted as the surface expression of Recent kimberlite activity (Sampson 1953; Reid et al. 1975a). At this locality highly carbonated lava has been extruded; the lava contains microxenolith of ultramafic rocks and resembles the carbonate-rich kimberlite of the Benfontein sills, South Africa (Dawson and Hawthorne 1973). By contrast with the recent volcanics of similar age which erupted in the rift valleys encircling the Tanzania craton, the Igwisi volcanoes lie in the centre of the craton.

2.1.4 Southern Africa

Zimbabwe, Botswana and Swaziland. Four kimberlites, one sill and three pipes, intrude Archaean granite in the Lochard area, 30 miles NE of Bulawayo (MacGregor et al. 1937). In *Botswana* a group of diamondiferous kimberlites that have undergone very little erosion has been found in the Orapa area. This group is apparently close to the edge of the Kalahari craton, where rocks of the Limpopo Belt (a 2100–1950 m.y. orogenic belt that lies between the old Rhodesian and

Transvaal nuclei and forms part of the Kalahari craton) intersect the circum-cratonic Damaran Belt, i.e. at the junction of two extensive zones of crustal weakness (Crockett and Mason 1968). This group contains the AK/1 pipe, the second-largest known kimberlite. Others are reported in S.W. Botswana (Baldock et al. 1976).

A kimberlite of Permian age, the Dokolwayo kimberlite, intrudes late Archaean gneisses near the margin of the Kaapvaal craton in *Swaziland;* the age of this pipe is unique in Southern Africa (Hawthorne et al. 1979).

Republic of South Africa, South-West Africa and Lesotho. The inland plateau of South Africa is the region of most concentrated kimberlite intrusion in the world and, together with the Yakutia province of Siberia, is the most thoroughly investigated. The kimberlites are spread over a huge, roughly oval area north of the Cape Fold Mountains, and cover most of the interior plateau except in the south-east where they are conspicuously absent (Fig. 3). They have been well described by Wagner (1914) and Williams (1932), but there are more recent accounts of the kimberlites of Lesotho (Nixon 1973b); the kimberlites of S.W. Africa (Janse 1975); the kimberlite dyke system in the Harts River Valley (Bosch 1971); of kimberlite sills in the Kimberley area (Hawthorne 1968; Dawson and Hawthorne 1973); and intrusions in Namaqualand (Frick 1974) and Bushmanland (Cornelissen and Verwoerd 1975). A list of the main kimberlite areas is given in Table 1.

Kimberlites occur both on the older Transvaal craton and within the accreted Orange River Belt (Fig. 3). In both tectonic settings the kimberlites tend to occur in clusters as, for example, in the Kimberley, Winburg and Bushmanland areas, and the diatremes are lined up along deep-seated fractures which are infilled by kimberlite dykes. On the Transvaal craton these kimberlite-filled fractures run E-W, N-S, WNW-ESE and NE-SW (Fig. 3), and they appear to be independent of pre-existing structures, cutting across the trend of the Basement gneisses. One of these directions – the one striking WNW-ESE – has been termed the Lesotho trend (Crockett and Mason 1968); it is a dominant trend of kimberlite dykes over a zone 600 km long and 280 km wide from Lesotho to the Kimberley and Posmasburg areas. This is also the area of most intense seismic activity on the craton. In Lesotho itself, the strike of the kimberlite dyke (WNW-ESE) is parallel to the dominant strike of the earlier Karroo (U. Jurassic) dolerite dykes and it is apparent that the kimberlites were intruded up the same zone of fundamental fractures already exploited by the dolerites (Dawson 1960). In addition, the Kimberley area, which is the site of some of the most intense diamondiferous kimberlite intrusion, lies at the intersection of three fundamental fracture trends. This has a parallel in the Yakutia province where the diatremes of the Olenek-Markha area are situated at the intersection of two major fracture zones. By contrast, in the areas underlain by the accreted Orange River Belt, groups of kimberlites line up with the trend in the Basement rocks, as in S.W. Africa and the Britstown area.

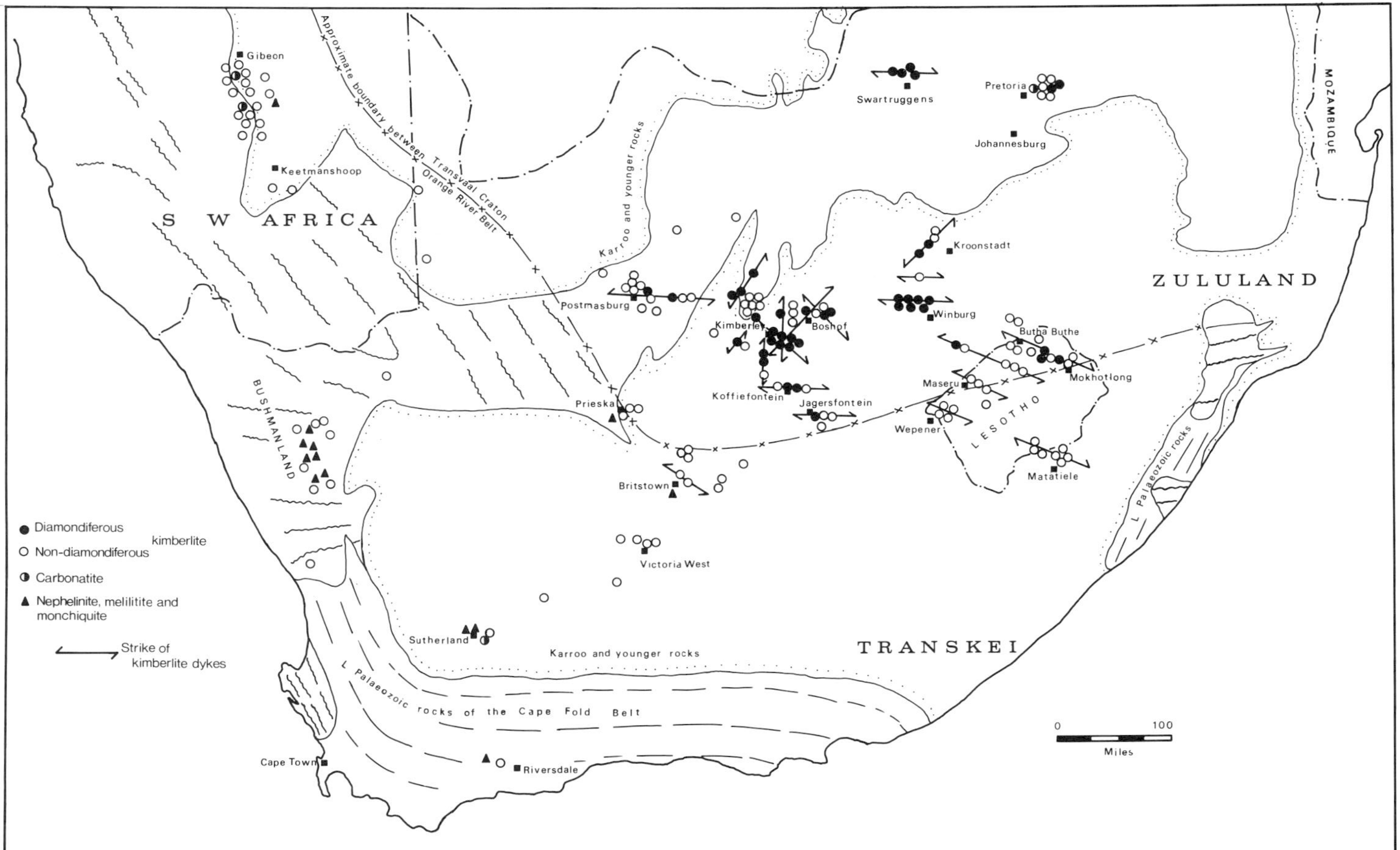

Fig. 3. Distribution of kimberlites and associated alkalic rocks in Namibia, South Africa and Lesotho (Dawson 1970)

Table 1. Kimberlites in Southern Africa

Area	Number of occurrences	Intrusions of note	References [a]
Rhodesia			
Lochard area	4	Colossus Mine, Wessels Sill	MacGregor et al. (1937)
Republic of South Africa			
Cap Province			
Kimberley	> 23	Kimberley, Bultfontein, De Beers, Dutoitspan, Wesselton, Kamfersdam, Benfontein	Hawthorne (1968) Dawson and Hawthorne (1973)
Postmasburg	> 14	Postmas, West End, Finsch, Peiser	
Barkly West-Harts River Valley	> 17	Frank Smith, Newlands, Sonop, Bellsbank, Mitchemanskraal-Sover dykes (Excelsior, Du Plessis Mines)	Bosch (1971)
Sutherland	1		Keyser (1940)
Britstown-De Aar-Phillipstown	9		
Prieska	3		
Victoria West	5	Uintjes Berg	
Riversdale-Robertson	2		
Beaufort West-Frazerburg	> 2		
Kuruman	4		
Matatiele	12		
Bushmanland	> 16		Frick (1974) Cornelissen and Verwoerd (1975)
Orange Free State			
Jagersfontein	5	Jagersfontein, Kalkfontein, Koppiesfontein	
Koffiefontein	6	Koffiefontein, Klipfontein	
Jacobsdal	2	Brakdam, Saltpeterpan	
Clocholan	> 3	Monastery, Lovedale	
Winburg	15	Lion Hill, Gordonia, Theron, Star, Monteleo	

Kroonstad	5	Crown (or Lace), Voorspoed	
Boshof	11	Roberts Victor, New Elands, Zout-en Zuur	
Transvaal			
Swartruggens	9	Helam Mine	Fourie (1958)
Pretoria	10	Premier	Robinson (1975)
Rietfontein	1	Rietfontein	
S.W. Africa			
Gibeon-Keetmanshoop	>40		Wagner (1914) Heath and Toerien (1962) Janse (1975) Hawthorne et al. (1979)
Swaziland			
Dokolwayo			
Botswana			
Orapa (Letlhekane)	> 6	AK/1	
Lesotho	18 diatremes + numerous dykes	Sekhameng, Lipelaneng, Marakabei Kao, Malibamatso, Thaba Putsoa, Letseng-la-terae (Qaqa)	Nixon (1973)

a The references given provide information additional to that given by Wagner (1914) and Williams (1932)

Table 2. Radiometric dates on kimberlites, kimberlite phases, and xenoliths from southern Africa [a]

Locality	Geological age	Radiometric age	Material	Method	References
Premier	Precambrian	1750 ± 100	Galena in kimberlite	Pb-Pb	1
Premier	Precambrian	1115 ± 15	Post-kimberlite sill	Rb-Sr	1
Premier	Precambrian	1200 ± 1250	Bulk kimberlite	Rb-Sr	2
National	Precambrian	1180 ± 30	Kimberlite micas	Rb-Sr	2
Swartruggens	Post-Precambrian	150 ± 3	Kimberlite	Rb-Sr	2, 3
Swartruggens	Post-Precambrian	142 ± 4	Kimberlite mica	K-Ar	4
Wesselton, Kimberley	Post-Jurassic	118 ± 6	Phlogopite nodules	Rb-Sr	3
Wesselton, Kimberley	Post-Jurassic	80 ± 4	Kimberlite mica	Rb-Sr	3
Wesselton, Kimberley	Post-Jurassic	77 ± 2	Mica (?secondary), peridotite nodule	Rb-Sr	3
Wesselton, Kimberley	Post-Jurassic	90.3	Zircon	U-Pb	5
Wesselton, Kimberley	Post-Jurassic	90 ± 4	Kimberlite	U-Pb	3
Dutoitspan, Kimberley	Post-Jurassic	80	Mica, peridotite nodule	Rb-Sr	3
Dutoitspan, Kimberley	Post-Jurassic	77 ± 6	Kimberlite	U-Pb	2
De Beers, Kimberley	Post-Jurassic	92.0	Zircon	U-Pb	5
De Beers, Kimberley	Post-Jurassic	92 ± 4	Kimberlite	U-Pb	2
De Beers, Kimberley	Post-Jurassic	120 ± 100	Kimberlite	U-Pb	2
Kimberley "Pool", Kimberley	Post-Jurassic	95.4	Zircon	U-Pb	5
Kamfersdam	Post-Jurassic	86.9	Zircon	U-Pb	5
Bultfontein, Kimberley	Post-Jurassic	91.2	Zircon	U-Pb	5
Bultfontein, Kimberley	Post-Jurassic	81.7 (duplicate)	Zircon	U-Pb	5
Bultfontein, Kimberley	Post-Jurassic	148 ± 7	K-richterite and mica in peridotite	Rb-Sr	6
Roberts Victor, Boshof Dt.	Post-Jurassic	105 ± 20	Mica (?secondary) in eclogite	Rb-Sr	3
Roberts Victor, Boshof Dt.	Post-Jurassic	125	Mica (?secondary) in eclogite	Rb-Sr	3
Roberts Victor, Boshof Dt.	Post-Jurassic	92.2	Zircon	U-Pb	3
New Elands (dyke), Boshof Dt.	Post-Jurassic	114 ± 1	Groundmass mica	K-Ar	4
New Elands (pipe), Boshof Dt.	Post-Jurassic	127 ± 6	Groundmass mica	K-Ar	4
Zout en Zuur, Boshof Dt.	Post-Jurassic	125	Groundmass mica	K-Ar	4
Leicester, Windsorton Dt.	Post-Jurassic	93.6	Zircon	U-Pb	5

Finsch	Post-Jurassic	130–170	Bulk kimberlite	Rb-Sr	2
Finsch	Post-Jurassic	94.1	Zircon	U-Pb	5
Monastery, Clocholan Dt.	Post-Jurassic	90 ± 1	Groundmass mica	Rb-Sr	3
Monastery, Clocholan Dt.	Post-Jurassic	90	Groundmass mica	K-Ar	4
Monastery, Clocholan Dt.	Post-Jurassic	90.4	Zircon	U-Pb	5
Monastery, Clocholan Dt.	Post-Jurassic	83 ± 5	Bulk kimberlite	U-Pb	2
Star, Winburg Dt.	Post-Jurassic		Groundmass mica	K-Ar	4
Rietfontein (N. Cape)	Post-Jurassic	71.7–71.9	Zircon	U-Pb	5
Bokputs, Namaqualand, N. Cape [b]	Post-Precambrian	67.9	Zircon	U-Pb	5
Platbakbies, Namaqualand, N. Cape [b]	Post-Cambrian	66.7	Zircon	U-Pb	5
Brakfontein, Namaqualand, N. Cape [b]	Post-Cambrian	54.1	Zircon	U-Pb	5
Lushof, Victoria West [b]	Post-Jurassic	67.7	Zircon	U-Pb	5
Mckenzies Post [b]	Post-Jurassic	79–80	Zircon	U-Pb	5
Melkfontein, E. Griqualand [b]	Post-Jurassic	63.4	Zircon in carbonate tuff	U-Pb	5
Orapa, Botswana	Post-Jurassic	93.1	Zircon	U-Pb	5
Mothae, Lesotho	Post-Jurassic	87.1	Zircon	U-Pb	5
Bakwanga, Zaire	?	71.3	Zircon	U-Pb	5
Mwadui, Tanzania	Post-Precambrian	189.3	Zircon	U-Pb	5
Nzega, Tanzania	Post-Precambrian	53 ± 1 (duplicated)	Zircon	U-Pb	5

1. Allsopp et al. (1966); 2. Allsopp and Kramers (1977); 3. Allsopp and Barrett (1975); 4. McIntyre and Dawson (1976); 5. Davis (1977); 6. Erlank and Shimizu (1977)

a Unless otherwise stated, localities are in S. Africa

b Localities in the circum-cratonic Namaquaride fold-belt

Kimberlites of at least three ages are known. The Premier Mine is Precambrian in age (Allsopp et al. 1967), thus linking it with the nepheline syenites, trachytes and carbonatites of the Franspoort Line, a major NNW-SSE Precambrian fracture system (Shand 1922; Verwoerd 1967). The Dokolwayo kimberlite, Swaziland, is Permian in age (Hawthorne et al. 1979). Most South African kimberlites, however, intrude the Carboniferous-Jurassic sediments and lavas of the Karroo Basin and are therefore post-Jurassic. Radiometric ages (Table 2) suggest a protracted period of kimberlite intrusion from ~ 140 m.y. to ~ 80 m.y. b.p. with the most intensive periods at 140, 120 and 80–90 m.y. b.p. The protracted period of intrusion coincides with the period in the Cretaceous when there was strong uplift of the continental landmass with attendant downwarping or faulting around the periphery with a deepening of the contiguous ocean basins (Dingle 1973). From Zululand to the Transkei, successive strips of land were bent down on the seaward side, and along the south coast faults running E-W have downthrows of 5000 m towards the sea (Du Toit 1954). The faults then swing NW following the west coast, and in S.W. Africa there has been severe faulting with a NNW strike, and seaward downthrows.

During this period, the intrusions on the Transvaal craton were exclusively of kimberlite, many of which are diamondiferous. In the area underlain by the circum-cratonic Orange River Belt, rock types of comparable age are basalts, gabbros and comendite in the Damaraland province (123–136 m.y. old), foyaite of the Granitberg, Lüderitz (see summary by Marsh 1973) and the kimberlite/carbonatite volcano of Gross Brukkaros (Ferguson et al 1975; Janse 1975), all these localitites being in S.W. Africa. Non-diamondiferous kimberlites in the circum-cratonic belt are younger than those on the craton (Table 2). It had previously been assumed that these kimberlites, together with the olivine-melilitites and olivine-nephelinites of the Sutherland area and Namaqualand are contemporaneous with the cratonic kimberlites (Holmes 1936; Taljaard 1936; Dawson 1970). Moreover, a radiometric date of 38.5 m.y. on one olivine-melilitite from Namaqualand, and of 35–37 m.y. on phonolite from the coastal belt of S.W. Africa (Kröner 1973) indicates that some, at least, of the post-Jurassic igneous activity in the circum-cratonic belt is protracted and by no means coeval with the kimberlite activity on the craton. There is, therefore, not sufficient evidence of contemporaneous activity to draw a parallel between the broad regional zoning seen in southern Africa, and the Yakutia kimberlite province in Siberia, where the diamondiferous kimberlites in the centre of the province are sourrounded by contemporaneous barren kimberlites and alkalic intrusions (Milashev 1965).

However, in Yakutia the relationship of the differing rock-types to the different tectonic regimes is by no means as apparent as in southern Africa.

2.2 U.S.S.R.

2.2.1 Siberia (Yakutia)

In the U.S.S.R., kimberlite was first reported as a dyke on the Taimyr peninsula, Siberia (Moor 1941). The discovery, however, has been overshadowed by the finding since 1954 of several hundreds of kimberlite pipes and dykes in Eastern Siberia between the Yenesei and Lena Rivers (Fig. 4), in the valleys of the Markha, Muna, Olenek and Vilyui rivers. The greatest concentration of kimberlites lies within the Yakut A.S.S.R., so the area is generally referred to as the "Yakutia kimberlite province". These intrusions have been described in a vast amount of Russian literature, the more important being the works of Bobrievich et al. (1959, 1964), Milashev et al. (1963), Kovalski (1963), Milashev (1965) and Frantsesson (1968). The most comprehensive reviews readily available to most Western readers are those of Lebedev (1964), Bardet (1965) and Davidson (1967), and the English translation of Frantsesson (1970). The following brief account draws upon all these sources and also upon post-1968 literature.

With the exception of the kimberlite of the Taimyr peninsula and the Oka area (Pri-Sayan), the Siberian kimberlites are confined within the borders of the Siberian Platform. Their distribution is shown on Fig. 4, and the location of some of the more important kimberlites is summarized in Table 3. The kimberlite groups have been proved to be of differing ages, also shown on Fig. 4. Within the confines of the Siberian Platform, the crystalline basement is overlain by Proterozoic,

Table 3. List of Yakutia kimberlite groups (shown in Fig. 4) and the more important kimberlite intrusions

1. Malo-Botuobinsk group – Amakinsk, Mir
2. Daldyn-Alakyt group – Zagadochnaya, Dolgozhdannaya, Daldyn, Geofizicheskaya, Dalnaya, Udachnaya, Leningradskaya, Aeromagnetiya, Aykhal, N.I.I.G.A., Moskvicha
3. Muna group – Komsomolskaya, Malaya, Novinka
4. Middle Olenek group – Aerogeologicheskaya, Chomur, Flogopitovaya, Ilmenitovaya, Leningrad, Omonos
5. Lower Olenek group – Obnazhennaya, Slyudyanka
6. Pri-Lena (Molodin) group
7. Udshinsk group
8. Kuonamsk (Anabar) group – Bargydamalakh, Chita, Universitetskaya
9. Meimecha-Kotui group
10. Kamensk group
11. Chadobets group
12. Pri-Sayan (Oka) group – Medvezkaya, Yuzhanaya
13. Chomnolinsk group
14. Tobusk (Aldan) group
15. Arbarastakh group
16. Ingili group

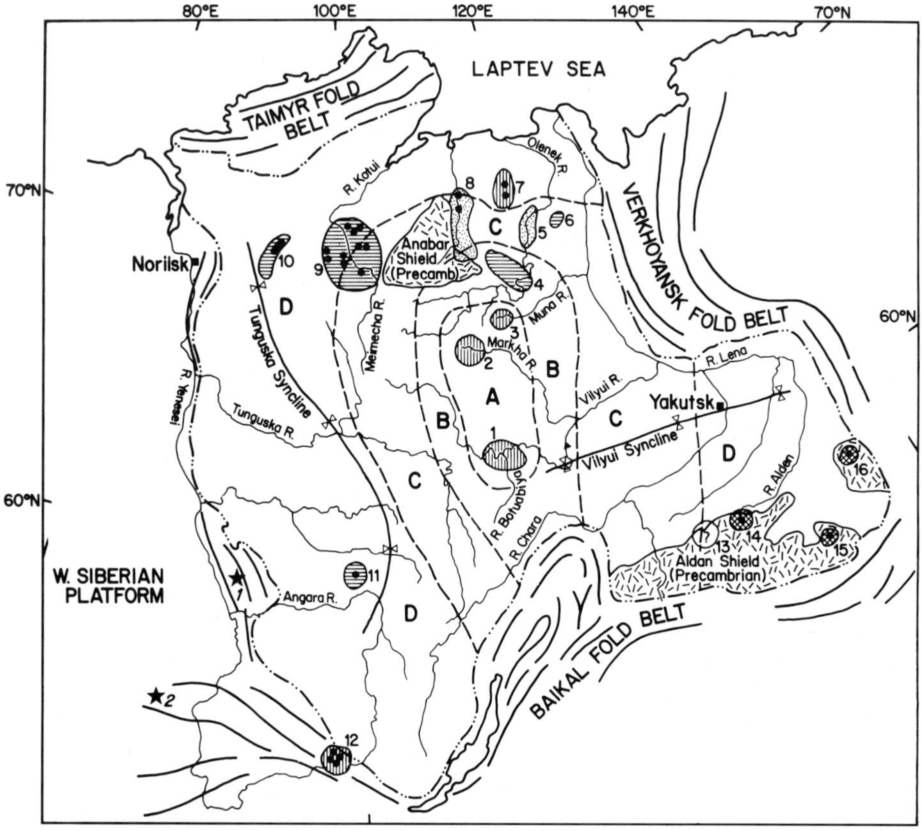

Fig. 4. Distribution of kimberlites on the East Siberian Platform. (Based on the Tectonic Map of the U.S.S.R.; Kaminsky 1972, Fig. 1)

Kimberlite groups: *1* Malo-Botuibinsk; *2* Daldyn-Alakit; *3* Muna; *4* Middle Olenek; *5* Lower Olenek; *6* Pri-Lena; *7* Udschinsk; *8* Kuonamsk (Anabar); *9* Meimecha-Kotui; *10* Kamensk; *11* Chadobets; *12* Pri-Sayan (Oka); *13* Chomiolinsk; *14* Tobusk (Aldan); *15* Arbarastakh; *16* Ingili.

Zones of kimberlites of differing mineralogical sub-facies: *A* Zone of diamond sub-facies kimberlites; *B* Zone of diamond-pyrope subfacies kimberlites; *C* Zone of pyrope sub-facies kimberlites; *D* Zone of rocks of kimberlitic aspect and central-complex kimberlites.

Group of alkalic basaltic diatreme intrusions: *(1)* Yenesei Mountains, *(2)* Minusinsk intermontane basin

Ages of Kimberlite groups
Precambrian
Palaeozoic
Lower Mesozoic
Upper Mesozoic
? Age unknown
• Alkalic ultrabasic ring complexes
4 Kimberlite groups
—··— Boundary of East Siberian Platform
— — — Boundaries between kimberlite zones
★ Basaltic diatreme groups

Palaeozoic and Mesozoic strata, including Lower Mesozoic flood-basalts. The Platform was subjected to Late-Mesozoic flexturing which has given rise to broad epeirogenic anticlinal and synclinal structures with the crystalline basement rocks emerging on the Anabar dome, and in the Aldan Shield in the south-east; there are secondary basin-and-swell structures within the major structures. The borders of the Platform are downwarped on the northern and eastern margins against the Taimyr (Hercynian) and Verkhoyansk (U. Mesozoic) fold-belts; and along its Western and south-western margins near the Baikal (Devonian) fold-belt, the rocks of the platform are more intensely folded. The distribution of the Palaeozoic and Mesozoic strata is controlled by the major swell structures, the Mesozoic strata having been removed around the Anabar and Aldan uplifts. Again reflecting the degree of erosion, only pipes are found in the Malo-Buotobinsk, Muna and Daldyn-Alakyt groups, whereas further north, although of younger geological age, the intrusives of the Meimeicha-Kotui, Kuomansk and Olenek groups are well-dissected and comprise a high proportion of dykes. The platform is cut by a system of deep fundamental fractures often at the hinge-line between uplifted and depressed structural areas; although probably of very great age, the faults have been reactivated at different times, coinciding with the various epochs of kimberlite intrusion. Some of the groups of kimberlite on the Anabar uplift appear to be sited at the intersection of two of these major fracture zones (Arsenyev 1962).

As in the case of South Africa, there appears to be a regional zoning of the intrusions, according to the presence or absence of diamond and pyrope garnet; a central zone of diamondiferous kimberlites is surrounded by a zone in which both pyrope and scarce diamond are present (diamond-pyrope subfacies), then by a zone in which diamond is absent but pyrope is present (pyrope subfacies), and finally by an outer area in which the kimberlites contain neither of these high-pressure minerals (Milashev 1965; Kaminski 1972). Again as in South Africa, the kimberlites in the main diamond-bearing diatreme area of the central zone are the only intrusives, but in the outer zones the kimberlites are accompanied by other alkalic-ultrabasic intrusives. For example in the Lower and Middle Olenek groups, the diamond-pyrope facies kimberlites are accompanied by intrusives of olivine leucitite, nepheline-, melilite- and monticellite-bearing picrite porphyries[1] (Milashev

1 The picrite porphyries differ from kimberlites in the following ways:

a) they formed as the result of a single stage of crystallization;
b) they are massive and lack clastic textures;
c) they have a relatively simple mineralogy, lacking the minerals derived from garnet peridotites, and
d) they tend to contain higher iron, titanium, calcium, alkalies, with decreased magnesium and chromium.

These differences in chemistry reflect the presence of diopside, monticellite, melilite and nepheline (Frantsesson 1970). Some picrite porphyries appear similar to the meimechites of the Meimecha-Kotui area, although Nikishov et al. (1972) refer to these porphyries as varieties of alnöite. Meimechite is an extrusive ultramafic rock first found as both lavas et al. 1963), and in the Kuonamsk group by olivine melilitites and carbonatites (Frantsesson 1969; Marshintsev 1970a) (reminiscent of the Sutherland area, South Africa); in this latter group is an example of olivine melilitite forming a composite diatreme intrusion with kimberlite (Ukhanov 1965). In other kimberlite areas, further away from the main central group of diamondiferous kimberlites, the kimberlites are found as minor intrusives in areas characterized by major ring-complexes of alkalic-ultrabasic plutonic intrusives, as, for example, in the Meimecha-Kotui area, one of the world's major ultrabasic-alkalic provinces. Here, Late-Permian – Early-Triassic extrusion of olivine melilitite, olivine nephelinite, augitite, limburgite and meimechite tuffs and lavas was followed by the intrusion of major ultrabasic-alkalic ring-complexes in which olivinites, melilite-bearing intrusives, jacupirangite, urtite, ijolite, nepheline syenite and carbonatite are present (though not all are present in each complex). The whole area is also intruded by dykes, sills and minor plugs of alnöite, tinguaite, monchiquite and fourchite, in addition to the hypabyssal equivalents of the effusive suite. The dykes occur either in regional swarms trending WNW-ESE and NNW-SSR, or else as radial swarms fanning out from the central ring-complexes. For a detailed account in English the reader is referred to Egorov (1970). Kimberlite pipes and dykes occur as the latest intrusive phase in the vicinity of the Dalbykha complex; the kimberlite contains xenoliths of the earlier intrusives, and in the Krasnoyarsk diatreme forms a multiple intrusion with carbonatised alkalic syenite. The kimberlites contain pyrope and chrome-diopside, but diamond has not yet been found in situ, although occurring in nearby alluvial deposits. The kimberlites of this area are claimed to illustrate more clearly than in any other group the intimate connection between kimberlites and alkalic ultrabasic central ring-complexes with carbonatites (Makhilaev and Surina 1965; Frantsesson 1970).

In the circum-cratonic fold-belts, the kimberlite dykes of the Taimyr peninsula are accompanied by dykes of olivine melilitite (Moor 1941). The ?Devonian kimberlites of the Oka (Pri-Sayan) group are closely associated with the Caledonian central complexes of nepheline syenite, alkalic-ultrabasics and carbonatite (Pecherski 1965). Alkalic basalt breccia pipes, containing blocks of upper mantle peridotite, occur in the Yenesei and Minusinsk intermontane basins a short distance outside the south-western margin of the Siberian Platform.

(Footnote 1 continued)

and pyroclastics in the Meimecha-Kotui area of N.E. Siberia (Moor and Sheinmann (1946). In its richness in olivine and perovskite it resembles kimberlite, but in other respects is fundamentally different (see Dawson 1971; Malkov 1972)

2.2.2 Western Russia

Kimberlite pipes and dykes of probably Devonian age are found in the south of the Russian platform (Gonshakova et al. 1969). The main group is in the area between Kharkov and the Sea of Azov, in the vicinity of one of the many alluvial diamond areas around the Ukranian Precambrian massif. The kimberlites lie on the major-fault zone on the southern boundary of the Dnieper-Donetz (Dombas) depression, and are associated with minor intrusions of limbugite, augitite, monchiquite, leucite porphyries and pyroxenite. Another isolated occurrence is in the Don Valley some 150 miles east of Kharkov on the south-eastern flank side of the Voronezh anteclise. This block structure, which trends NW-SE, is also intruded by peridotites and pyroxenites which, like kimberlite, ascended along deep fractures bounding the structure. The overall tectonic environment of these kimberlites of the Russian platform, where epeirogenic earth movements affecting the Lower Palaeozoic strata overlying the block-faulted Precambrian basement have resulted in the formation of broad anticlinal and synclinal structures that are often separated by deep fractures, is strongly reminiscent of the Siberian Platform. Gonshakova et al. (1968) also report a kimberlite pipe on Yelovy Island in the Gulf of Kandalaksha (a branch of the White Sea), where dykes of Palaeozoic alkali basalt, alnöite, picrite porphyry and monchiquite also occur, and a map in this paper shows another kimberlite pipe near the Russo-Finnish border 100 miles WSW of Murmansk, in an area where a single alluvial diamond has been found. In addition, Popov (1967) records that ultrabasic lavas and tuffs of the composition of "kimberlite", picrite, augitite, melilitite and nephelinite are preserved in the central down-faulted area of the U. Devonian-L. Carboniferous ring-complex of Kontozero, Kola peninsula. Finally, it may be noted that intrusions of picrite porphyry, similar to those associated with the kimberlites in the Olenek group of Yakutia, have been found in the alluvial diamond areas on the western slopes of the Urals; one of these intrusions contains moissanite which has also been found in kimberlite in Yakutia (Starkov 1968).

2.3 India

Although India has been a world source of diamond for over 2000 years, most diamonds are alluvial and kimberlites are known in only three areas. All are within the Precambrian shield area of peninsular India. One is near Majhgawan in Panna State (24°38′N., 80°04′E.) where three diatremes penetrate Lower Proterozoic sandstones (Dubey and Merh 1949; Mathur 1962; Grantham 1964). On the basis of the similarity of the diamonds in the main pipe with those in the nearby Upper Proterozoic conglomerates, Merh (1952) suggests that the pipe is itself Proterozoic in age; phlogopite from the Majhgawan pipe yields a Rb-Sr age of 1140 ± 12 m.y. and a K-Ar date of 1056–1120 m.y. (Crawford and Compston 1970; Paul et al.

1975b). A second area of kimberlite intrusion is near Wajrakurar, Anantapur District, Andhra Pradesh (15°0.1'N., 77°23'E.) where four diatremes intrude Archaean gneisses (Rao and Phadtare 1966). One is dated at 840 ± 33 m.y. (Paul et al. 1975b). Finally, dykes of calcite-rich kimberlite intrude rocks of the Lower Cumbum formation (Upper Precambrian) in the Chelima area of the central Cuddapah Basin, Andhra Pradesh (15°27'N., 78°42'E.); contemporaneous intrusions in the western part of the Cuddapah Basin are of picrite, gabbro and dolerite (Sen and Rao 1967, 1970). According to Crawford (1969) the age of one of the Chelima dykes is "not much less than 1340 m.y. which is also almost certainly the age of similar dykes and pipes cutting the older Precambrian west of the Cuddapah Basin in the Wajrakurar area, one of which has been analysed".

2.4 Australasia

2.4.1 Borneo

Several intrusive peridotite-breccia pipes intrude the Bobaris peridotite massif in the Pamali River area of south-east Borneo. One of these Lower Cretaceous pipes is reported to yield very low amounts of diamond, and diamonds are found in the Upper Cretaceous and Eocene conglomerates and sandstones which lie north and south-east of the Bobaris massif (Van Bemmeln 1949). However, A.J.A. Janse (personal communication) believes that the Pamali breccias are not kimberlite, and certainly their tectonic setting in a geosynclinal area and their intrusion during an orogenic cycle are not typical for kimberlites.

2.4.2 Solomon Islands

Intrusions of alnöite and ankaratrite, the former containing picroilmenite, diopside and pyrope, occur on the island of Malaita (Allen and Deans 1965). One alnöite breccia contains spinel and garnet peridotite xenoliths and a pyroxene/ilmenite intergrowth similar to those found in kimberlites. Although resembling kimberlite to this extent, these intrusions (like those of Borneo) are in an atypical environment for kimberlite. Recent detailed mineralogical studies by Nixon and Boyd (1979) and Dawson et al. (1978a) underline the differences between these rocks and continental kimberlites.

2.4.3 Australia

Although there are numerous cases where alluvial diamonds occur there are as yet only a few authenticated reports of kimberlite in Australia. Fourteen areas of kimberlite and kimberlitic rocks are known in New South Wales, Victoria and south-western S. Australia (Ferguson and Sheraton 1979; Stracke et al. 1979).

There are unconfirmed reports of several groups of kimberlite in the Fitzroy Basin of N.W. Australia in the vicinity of the well-known intrusions of leucite-bearing lamprophyres. The alleged occurrence of diamond in its host rock at Oakey Creek (Inverell), New South Wales has not been confirmed, the host rock proving to be dolerite; this area is also an orogenic area.

2.5 Western Europe and Greenland

Dykes of kimberlite with alnöitic affinities cut Precambrian gneisses in the Kalix-Lulea area of northern Sweden (Geijer 1928; Åhman 1950). This locality is some 400 km north of Alnö Island where kimberlitic and kimberlitic alnöite dykes are among the latest intrusives in the world-famous carbonatite complex of Lower Cambrian age (Eckermann 1948, 1963a,b); the tentative preliminary identification of diamond in one of these kimberlites has not been confirmed (Eckermann 1967).

The pyrope-bearing diatremes of the Ceske Stredhori Mountains, Czechoslovakia, were formerly regarded as having kimberlitic affinities and were believed to be the source of alluvial diamonds found in the area (Kopecky 1960). Further intensive work has now convinced the Czech workers that these rocks are not kimberlites, but are more in keeping with the regional alkali basalts. The pyrope is now believed to originate in the tectonic slices of pyrope peridotite that are known to exist in the crystalline basement of the area (Kopecky et al. 1967).

On the west coast of Greenland, kimberlite dykes and sills of Early Mesozoic age intrude Precambrian gneisses and granite in the area east and south-east of Fredrikashab; these intrusions, which contain xenoliths of spinel- and garnet-bearing peridotite, are part of the Mesozoic basaltic and minor carbonatitic activity that is tentatively linked with the rifting between Greenland and Labrador that led to the opening of the Labrador Sea. Still further north, kimberlitic dykes of Precambrian age are part of the NE-SW swarm of monchiquite and basaltic dykes on the coast east and south of Holsteinborg (Andrews and Emeleus 1975, 1976; Scott 1979).

2.6 North America

The distribution of kimberlites is shown on Figs. 5 and 6, and details of them, together with their known ages, are given in Table 4. The most recent general account is given by Meyer (1976) for those of the U.S.A., and Canadian ones were reviewed by Watson (1967a, 1973). Within these reviews, it is appropriate to acknowledge the numerous recent publications by D. Brookins and co-workers (Kansas kimberlites), McGetchin (Colorado Plateau intrusions) and McCallum and Eggler (N. Colorado). Since these reviews, new kimberlites have been described

Table 4. Locality and age of North American kimberlites

Locality	Form of intrusions [a]	Age [b]	References [c]
Somerset Island, North-West Territories	P	Post-Ordovician	Mitchell and Fritz (1973)
Saglak, Labrador	D	Post-Archaean	Collerson and Malpas (1977)
Allik Bay, Labrador	D	?	
Gauthier Township, Ontario	D	$151 \pm 8^{++}$	Lee and Lawrence (1968)
Coral Rapids, Ontario	D	Post Middle-Devonian	Brown et al. (1967)
Michaud Township, Ontario	D	?	
Keith Township, Ontario	D	?	Watson (1973)
Bachelor Lake, Quebec	D	1100^{+}	
?Castignon Lake, Quebec	D	Post-Upper Proterozoic	Dimroth (1970)
Saguenay Valley, Quebec	D	?	Gittins et al. (1975)
Ile Bizard, Quebec	P	U. Cretaceous	Marchand (1970)
Central Finger Lakes area, New York	D	?	
Ithaca area, New York	D	$145–493^{+}$; $136 \pm 8^{++}$	
Syracuse area, New York	D	$150–306^{+}$; $118–146^{++}$	
Masonville, Fayette Cty., Pa.	D	Post-L Carboniferous; $368–408^{+}$	
Dixonville, Pa.	D	Post-L Carboniferous	
Shenandoah Valley, Virginia	D	?	
Morris, Union City, Tennessee	P	Post-Devonian	
Elliot Cty., N.E. Kentucky	D, P	$270–279^{+}$; $257 \pm 22^{++}$	
W. Kentucky and S. Illinois	D, P	$252–281^{+}$; $260–279^{++}$	
Avon area, Missouri	P, S	$377–388^{+}$; $396–399^{++}$	
Woodson Cty., Kansas	D, P	$88–91^{+}$	
Riley Cty., Kansas	P	$115–123^{FT}$	
Pike Cty., Arkansas	P	Mid-Cretaceous	
Colorado Plateau (Mule Ear, Moses Rock, Garnet Ridge, Buell Park, Green Knobs, Cane Valley)	P	$29.7–33.5^{FT}$	
Colorado (Boulder area)	P	?	Meyer and Kridelbaugh (1977)
Wyoming-Colorado Border	P	U. Silurian-L. Devonian	
N. Montana	P	U. Eocene	

a D = dyke; P = pipe (diatreme); S = sill

b + = K-Ar age; ++ = Rb-Sr age; FT = Fission track age (Radiometric ages in m.y. B.P.). K-Ar and Rb-Sr ages from Zartman et al. (1967); Fission track ages from Naeser (1971)

c References are additional to those in the reviews by Meyer (1976) and Watson (1967a,b; 1973)

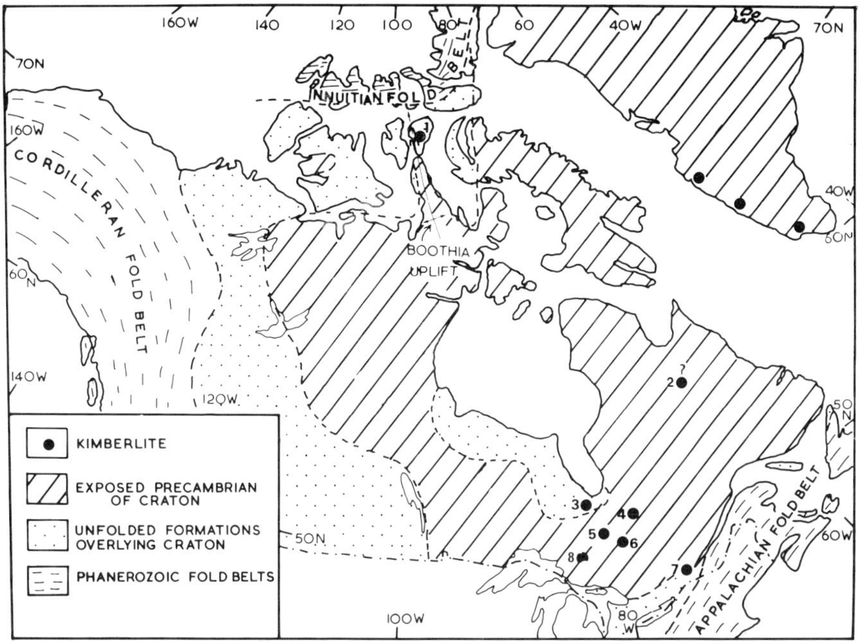

Fig. 5. Distribution of kimberlites in Canada: *1* Somerset Island; *2* Castignon Lake, Labrador; *3* Coral Rapids; *4* Bachelor Lake; *5* Michaud Township; *6* Gauthier Township; *7* Ile Bizard; *8* Keith Township

from Somerset Island; from Coral Rapids in the James Bay Lowlands, and in the Upper Canada Mine (both in Ontario); at Ile Bizard, Montreal, and in the Saguenay River Valley; and on the coast of Labrador. Also more detailed accounts are given of the Green Mountain kimberlite, Colorado (Meyer and Kridelbaugh 1977), and the Murfreesboro kimberlite (Steele and Wagner 1979).

The Somerset Island kimberlites intrude Ordovician sediments but do not contain xenoliths of Silurian or Devonian rocks (Mitchell and Fritz 1973). The group lies near the axis of the Boothia Uplift, a N-S elongate structure composed of a Precambrian core flanked by Lower Palaeozoic sediments; most of the tectonic activity took place in the Devonian, but faulting also occurred in the Caenozoic (Thorsteinsson and Tozer 1969). The Bachelor Lake kimberlites together with spessartite dykes intrude the Algonkian shield; the K-Ar age is 1100 m.y. but in view of the discrepancy between many K-Ar ages and actual ages in rocks of this type in eastern North America (Zartman et al. 1967) this age cannot be regarded as definitive. In the Castignon Lake area of northern Quebec, garnetiferous mica peridotite (which appears to have strong kimberlite affinities) together with carbonatite and meimechite intrude folded Lower Proterozoic sediments of the

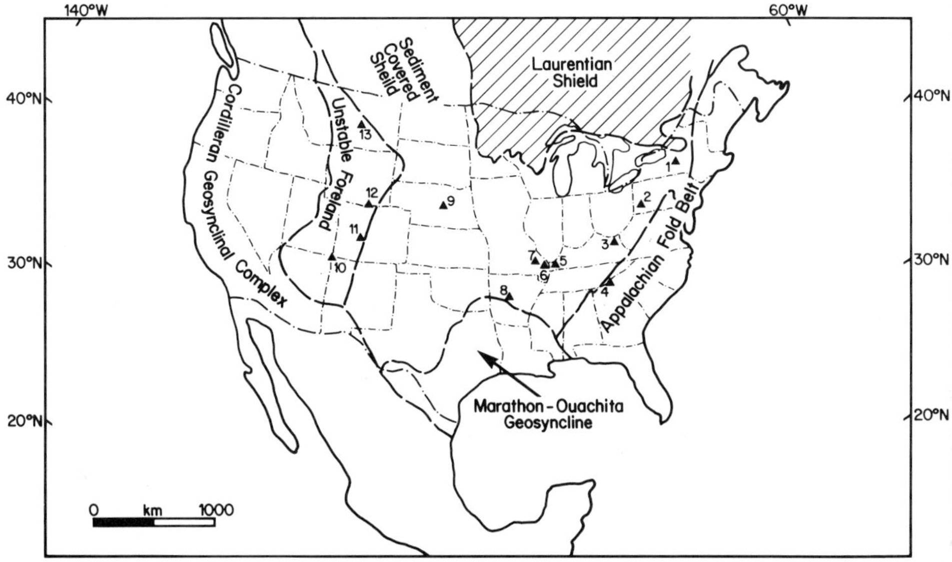

Fig. 6. Distribution of kimberlites in the U.S.A.: *1* Syracuse District, New York State; *2* Fayette County, Pennsylvania; *3* Elliot County, Kentucky; *4* Norris, Tennessee; *5* Crittenden County, West Kentucky; *6* Southern Illinois; *7* Avon, Missouri; *8* Pike County, Arkansas; *9* Riley County, Kansas; *10* Navajo-Hopi Province; *11* Green Mountain, Colorado; *12* Wyoming-Colorado State Line; *13* Montana

Labrador Trough. Most of the kimberlites in eastern North America occur in the interior stable lowlands, along a zone parallel to the Appalachian fold belt. Most intrude Palaeozoic sediments and radiometric dates indicate various ages from Permian to Cretaceous (Table 4). The kimberlites and associated basic dykes in W. Kentucky and S. Illinois intrude the Rough Creek Fault Zone – a zone of discontinuous fractures of small displacement, but to the south-west is the St. Genevieve fault – a fracture of large displacement lying on the boundary between the Ozark uplift and the Illinois Basin. The alnöite intrusions in the Avon area of S.E. Missouri are Upper Devonian, and occur on an E-W-trending axis, extending from the W. Kentucky-S. Illinois intrusion zone through Missouri into eastern Kansas. This axis is delineated by faults, minor ultramafic intrusions (including the kimberlites) and cryptovolcanic structures (Snyder and Gerdemann 1965), and both stratigraphic evidence and radiometric datings on the intrusions indicate that it has been tectonically active and subject to igneous intrusion at various times over a long period (Zartmann et al. 1967). In the southern United States, there are kimberlites, one diamondiferous, at Murfreesboro, Pike County, Arkansas, and another non-diamondiferous one in nearby Cleveland County. They occur at the margin of the major mid-Mesozoic downwarp of the Mississippi embayment, and the Pike County kimberlite is accurately dated as mid-Cretaceous in age. It has been suggested that the intrusion of these and other alkalic rocks of the Arkansas province took place into zones of tensional weakness where the SE-striking Ouachita Mountains were transected by the NE-striking hinge line which forms the approximate border of the Gulf Costal Plain. To the north-west, kimberlites intrude the warped Palaeozoic and Mesozoic sediments of the Colorado Plateau – a relatively stable block (underlain by Algonkian gneisses at depth) among the fold belts of the southern Rocky Mountains. In the Garnet Ridge – Mule Ear – Moses Rock area, the main structural element is the Monument upwarp, a broad flattened anticline whose crest is wrinkled by corrugations. The Colorado Plateau kimberlites are of Middle-Oligocene age. Further north, older, Siluro-Devonian kimberlites near the Colorado-Wyoming border intrude the Precambrian core of the northern Front Range and the southern Laramide Range. In N. Montana, kimberlitic diatremes of mid-Tertiary age penetrate flat-lying Cretaceous sediments to the east of the Rocky Mountain front.

Whereas all the above occurrences are in non-orogenic environments or in areas subject only to broad epeirogenic movements, kimberlites are also found within the Appalachian fold-belt in the Shenandoah Valley area of Virginia and in the Norris area of Tennessee; also the kimberlites in Scott and Logan Counties, Arkansas, intrude folded Palaeozoic rocks of the Ouachita Mountains. However, in these cases, there is no evidence to suggest that the intrusions were other than post-orogenic.

Regarding other igneous activity coeval with the kimberlite emplacement, the Coral Rapids kimberlite forms a composite dyke with monchiquite, and there are other monchiquite dykes in the vicinity. Spessartite dykes occur with the kimberlite at Bachelor Lake and the ?kimberlite of Castignon Lake is one of a complex of carbonatite and meimechite intrusions.

In New York State, the kimberlite dykes are part of a swarm containing alnöite also and some kimberlite dykes grade along their length into alnöite by the additional presence of melilite, and the kimberlites of W. Kentucky and S. Illinois are associated with minor intrusions of kersantite and alnöite. The Avon intrusions also include monticellite-bearing alnöite (Singewald and Milton 1930). The kimberlites of Arkansas are part of the Arkansas alkalic igneous province. This includes the alkalic syenite complexes of Fourche Mountain, Carlsbad and Saline Counties and Magnet Cove, where the rock types are nepheline syenite, tinguaite, foyaite, juvite, monchiquite and, at Magnet Cove, carbonatite. In addition there are many exposed dykes of ouachitite, fourchite, tinguaite and monchiquite (Tolman and Landes 1937), and nepheline syenite and pyroxenite have been encountered in deep drill-holes in the area (Kidwell 1951). To the south-east, numerous small unexposed intrusions of undersaturated lamprophyre, alkalic basalt, trachyte, phonolite and syenite have been found in deep drill-holes in the area of the Monroe Uplift and the Jackson Dome (Kidwell 1951). This confirms the surface evidence that the massive Cretaceous flexturing in the Mississippi embayment was accompanied by wide-spread intrusion of undersaturated igneous rocks.

The Colorado Plateau kimberlites are associated geographically with flows of alkali basalts, and intrusive monchiquites, minettes and vogesites of the Navajo-Hopi alkalic province, and there is carbonatite associated with the Cane Valley diatreme (McGetchin et al. 1973). It was formerly believed that these all belonged to one phase of activity, but recent radiometric work (summarized by Naeser 1971) indicates two periods of volcanic activity, one of Middle-to-Late Pleistocene age when monchiquites dyke and diatreme injection took place with alkalic basalt extrusion in the Hopi Buttes area, and an older phase of activity in the middle Oligocene when the kimberlites were intruded together with minettes. In addition, the Buell Park kimberlite has been intruded by a ring-dyke of minette. Although no other intrusions are reported in association with the N. Colorado-S. Wyoming kimberlites, those in Montana lie within the C. Montana alkalic province that includes the Bearpaw, Highwood, Little Rocky and Judith Mountains (Larsen 1940).

2.7 South America

2.7.1 Guyana

Diamonds in the alluvial deposits of Guyana have not been traced back to a primary host-rock, many being derived from conglomerates and sandstone of the Precambrian Roraima formation. Earlier reports of diamond in a gabbro host-rock have not been confirmed (Pollard et al. 1957).

2.7.2 Brazil

Although Brazil was the foremost diamond-producing country in the world prior to the discovery of the South African fields, most diamonds are alluvial in origin, and only very recently has authenticated diamondiferous kimberlite been reported. Earlier, Draper (1911) had proposed that the localized diamond deposits at Agua Suja on the Bagagem River represent a reworked kimberlite, and reported the occurrence of kimberlite at Patos and Coromandel; he also recorded an intrusion of "porphyritic peridotite" in the Abaete district. Rimann (1931) records the presence of pipes of volcanic breccia in the Agua Suja area and has compared the pipe-filling – "bebedourite" – with kimberlite. The nature of the diamond deposits at Diamantina was in dispute; Draper (1923) and Thompson (1928) believed that these localized conglomerates to be kimberlite diatremes, but Barbosa (1951) contends that they are detrital deposits derived from Precambrian conglomerates. In support of his contention Draper (1923) stated that two of the deposits are linked by a micaceous dyke (?kimberlite) and, whilst not disputing that some diamonds are weathered out of the Algonkian Carains Conglomerate in the vicinity, these conglomerate-derived diamonds are considerably worn, unlike the unabraded diamonds from the Diamantina deposits. Rimann (1915) also records kimberlites, together with dykes of alnöite, limburgite and picrite porphyry in the Sierra dos Lages, 100 km west of Rio de Janeiro.

A recent brief account reports diatremes in south-west Piaui State, and in Minas Gerais in the vicinity of Coromandel; several are stated to be diamondiferous (Svisero et al. 1979) and have zircon ages of 80–120 m.y. (Davis 1977).

2.7.3 Argentine

A single dyke of kimberlite, in which diamond has tentatively been identified, occurs in the vicinity of a carbonatite-intruded ultramafic complex in the Andean foothills north of Mendoza (Villar 1969).

2.8 Indirect Evidence for Kimberlite Intrusions

To the best of our current knowlege, kimberlite is the only major primary rock containing diamond, though there are some rare cases of diamond in Alpine-type ultramafic intrusions and in peridotite and eclogite inclusions in kimberlite (Dawson 1979a). From this, it is tempting to infer that the presence of diamond in either modern sediments or in more ancient clastic sedimentary rocks indicates the presence of earlier kimberlites that were eroded and shed their diamonds into the existing drainage systems. Diamonds are known in Precambrian sedimentary rocks in Brazil (Carains Conglomerate), in South Africa (Witwatersrand conglomerate), in India (Vindhayan conglomerates) and in the Algonkian auriferous conglomerates in the Pilbara goldfields of W. Australia. Diamonds are found in the late Palaeozoic Karroo conglomerates of the Somabula Forest, Rhodesia (MacGregor 1921), in the Tertiary auriferous gravels in Queensland and New South Wales (David 1950) and in the late-Tertiary–Recent coastal deposits of S.W. Africa. The above deposits are mature, well-worked sediments overlying Archaean basement terrains. In addition, many present-day diamondiferous alluvial sediments are derived from Archaean terrains, as are the diamondiferous tills of the northern U.S.A. In many cases, no kimberlite has yet been found in the sediment source areas (which may mean that the diamonds are recycled from older sedimentary formations), but in some areas, as in Angola, the Congo and W. Africa, kimberlites have been found in the source areas. It is a matter of record that it was the presence of diamond in the gold gravels of the Lena River that stimulated the search resulting in the discovery of the Yakutia kimberlite province.

The close connection between placer diamond deposits (of widely varying age) and Archaean gneiss/greenstone terrains is relevant to the discussion of possible diamond sources. Although these placer diamonds *could* have been derived from kimberlites that have yet to be discovered, or which may have been obliterated by erosion, it is *possible* that some at least may have been derived from Archaean greenstone belts; there is no reason why the Birrimian greenstone belt of

Ghana (that contains considerable concentrations of diamond in meta-greywackes – Junner 1943; Grantham 1967) should be unique, and Dawson (1979a) has reviewed the evidence, scanty at present, for igneous rock-types other than kimberlite being the parent of diamond. In short, the diamonds in these placers cannot be unambiguously attributed to a kimberlite source.

2.9 Summary

1. Most kimberlite intrusions are found on the stable cratons that have not been deformed since the Precambrian. The kimberlites are intruded into areas that have undergone epeirogenic flexturing prior to their intrusion, and may be associated with local swell or dome structures. They are intruded up fundamental fractures that transact the trends of the deepest known basement terrains.

2. Kimberlites are also found in the circum-cratonic fold-belts but there is no evidence to suggest that they were intruded during the orogenic cycle.

3. On the cratons kimberlites occur as the only manifestations of very limited magmatic activity. In the areas underlain by the circum-cratonic orogenic belts, the limited contemporaneous magmatic activity gives rise to intrusions of non-diamondiferous kimberlite, and other rare ultra-alkaline rock-types of very limited volume, containing high amounts of volatiles. However, in a few rare examples, as for example in Arkansas, U.S.A., diamondiferous kimberlite may occur as part of a magmatic episode dominated by nepheline syenitic or lamprophyric activity.

4. Kimberlite magmatism has occurred at various times in the earth's history, one major phase being in the Cretaceous. If the presence of diamond in the Precambrian sedimentary rocks can be taken as evidence of Precambrian kimberlite activity, then it follows that areas such as West Africa and South Africa have been the site of at least three (possibly four) periods of kimberlite magmatism. Combined with the evidence of well-established epochs of kimberlite intrusion in the Phanerozoic in the central U.S.A. and Yakutia, it may be said that certain geographical areas, usually coinciding with the old cratons, are "kimberlite-prone", having been subjected to spasmodic intrusion over long periods of geological time.

Chapter 3 Geology of Kimberlite Intrusions

Kimberlite occurs in diatremes (pipes), dykes and sills of very limited size, and one example of a kimberlite ring dyke is known in Sierra Leone. (The words diatreme and pipe will be used synonymously.)

3.1 Diatremes

Diatremes tend to occur in clusters or scattered along elongate zones. The diatremes are of small surface area, varying from several hectares to only a few square metres. The surface area of some of the better-known diatremes is given in Table 5; the areas for Mwadui and AK/1 represent the closest approximation to true surface

Table 5. Surface area of some larger kimberlite diatremes

	Hectares	Square miles
Southern Africa		
Premier	31.5	0.121
Kimberley	4.1	0.015
Bultfontein	9.5	0.036
Dutoitspan	12.3	0.047
AK/1 (Orapa, Botswana)	112.3	0.433
Kao (Lesotho)	17.4	0.067
Zaire (former Belgian Congo)		
Talala	50.0	0.193
Kambeli	26.0	0.100
M'Bo	23.5	0.090
Tanzania		
Mwadui	141.4	0.545
U.S.S.R. (approximate areas)		
Mir	13	0.050
Zarnitsa	28	0.109
Udachnaya	52	0.200

areas as these diatremes have been eroded very little since their emplacement, but other values are for diatremes at various levels of erosion. The small surface of even the largest underline the volumetric insignificance of kimberlite. In plan the diatremes vary from rounded or oval to a variety of bizarre, amoeboid shapes (Fig. 7), and in some cases it can be shown that the surface plan is strongly controlled by the joint-pattern in the surrounding rocks (Malde 1954; Dawson 1962). Other diatremes such as the Moses Rock intrusion, Utah, are little more than elongate dyke enlargements, and the lobate shape of some, such as Kao, Udachnaya and M'Bo, is due to the coalescence of two or more diatremes; Ilupin (1972) has estimated that over 90% of the diatremes in the Daldyn group of Yakutia are double or paired intrusions. Although some pairs may coalesce at depth to form one intrusion, in other cases the two segments or diatremes appear to be the result of different phases of kimberlite intrusion that have given rise to quite different kimberlite types. For example, the kimberlite in the western lobe of the Udachnaya pipe is quite distinct from that in the eastern lobe (Bobrievich et al. 1959), as are the kimberlites in the main and satellite pipes at Letseng, Lesotho (author's own observation). In section most of the diatremes are the shape of a simple inverted cone, tapering downwards (Figs. 8 and 10). There are exceptions, however, as some have a corkscrew-like twist, and in several cases the surface opening of the diatremes does not lie directly above the apex of the inverted cone (e.g. Kimberley Mine). Also, in the case of the Wesselton Mine, the diatreme bifurcates below the 2500' level en route to the surface; one branch terminates upwards in a "blind" intrusion, whereas the other main branch penetrates upwards to the surface.

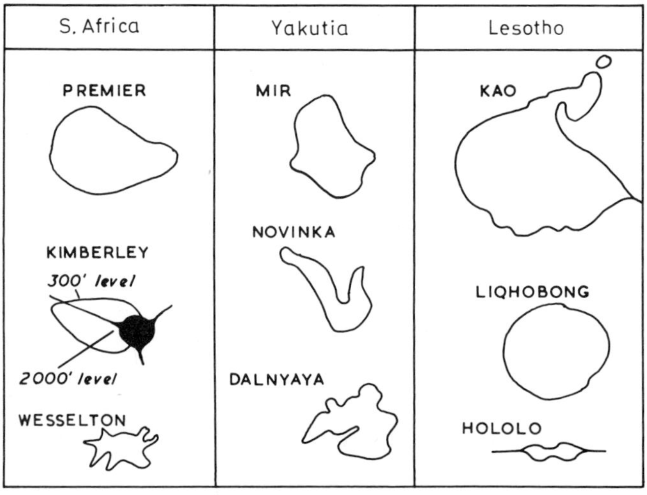

Fig. 7. Outcrop plans of some kimberlite intrusions

Fig. 8. The Big Hole, Kimberley – a mined-out kimberlite diatreme

In general, contacts with the country rocks are very steep – in the order of 75°–85° – and are remarkably constant; only in the superficial parts of some high-level diatremes, such as Mwadui, are relatively shallow contact dips recorded (Hawthorne 1975).

Many larger diatremes are infilled by a number of recognizable columns of petrographically distinctive kimberlite, later columns exhibiting intrusive relationships with earlier ones. Fifteen columns are recorded from the Kimberley Mine (Wagner 1914) but most diatremes contain no more than three or four (Fig. 9). These distinctive types of kimberlite are interpreted as resulting from a number of successive eruptions.

3.2 Dykes

These may occur singly or in swarms of parallel dykes; in the De Beers Mine there is a multiple dyke resulting from the injection of several pulses of kimberlite magma (Donaldson and Reid 1977). A feature is the remarkable persistence of some, although they are generally not more than 1–2 m wide (rarely up to 10 m). For example, one dyke in the Winburg District, S. Africa, has a proven length of

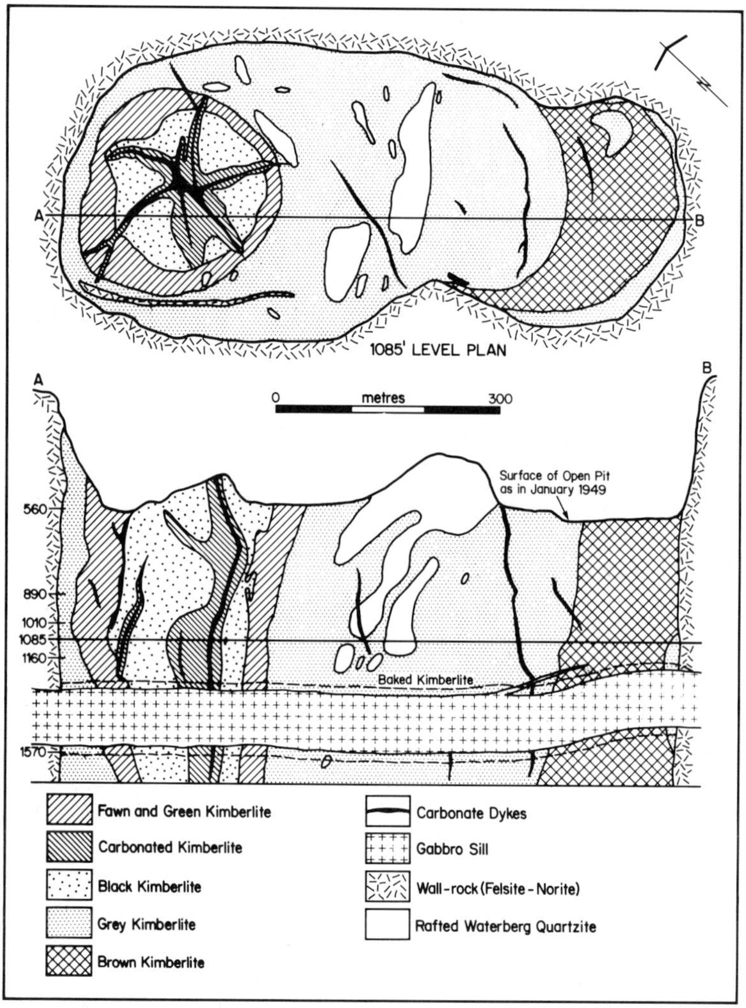

Fig. 9. Plan and section through the Premier Mine, South Africa. From the *Excursion Guide to the First International Conference on Kimberlites, 1973*

65 km and another in the area south-west of Kimberley, although never greater than 1 m wide, persists for over 30 km. The tabular form of the intrusions is maintained in depth as well as along strike, and Wagner (1914) records a 1 m-wide dyke in the Bultfontein Mine whose width showed little variation when followed downwards for 350 m.

Along the dykes, there are sometimes wider bodies of kimberlite referred to as dyke enlargements, chonoliths or, in the expressive language of the prospector, "blows". These were never connected with the surface, being pod-like enlargements on the dykes. They are distinguished from diatremes in being lenticular in section, as well as in plan; they do not contain breccia material, nor do they contain fragments of overlying formations that may have descended into a true diatreme.

3.3 Sills

Kimberlite sills are much rarer than diatremes or dykes, the comparative rarity being due, in part, to the fact that the present-day erosion level must coincide with the injection level of these thin horizontal bodies. Nonetheless, kimberlite sills are known in the Kimberley area of South Africa, in Rhodesia, in the Singida and Kimali areas of Tanzania (Hawthorne 1968); and in S.W. Greenland (Andrews and Emeleus 1975). In addition, the sections through some of the kimberlite diatremes in the Bakwanga area of Zaire show flat-lying tabular bodies, up to several hundred metres thick, extending radially several thousand metres beyond the rims of the true diatremes; these particular bodies developed at the contact between Basement rocks and overlying unconsolidated Mesozoic sediments, and are regarded by Meyer de Stadelhofen (1963) as effusive rather than true intrusive rocks. In some cases there is evidence for localized control of kimberlite sills, as for example in S.W. Greenland and the Singida area of Tanzania where the kimberlite sheets occupy horizontal joints in well-jointed granite; or in the Kimberley area where a massive dolerite sheet appears to have acted as a "cap-rock". Kimberlite sills are very variable in thickness, even within a small area, thinning towards the margin of the sheet. For instance in the Kimberley area, the Wesselton Floors sill varies from 45 m in its central point but dies out marginally, continuing into the country rock as thin veinlets (Hawthorne 1968).

3.4 Diatreme-Dyke-Sill Relationships

Some diatremes (e.g. the Kimberley Mine) have been excavated sufficiently deeply to demonstrate that they contract downwards into dykes or groups of dykes, the intersection of which is the cause of the location of the diatreme. With the further evidence of the diatreme-dyke relationships of the kimberlites in the deeply dissected mountains of Lesotho (Dawson 1960), it is safe to infer that zones of kimberlite diatremes are the surface expression of dykes at depth, and the apparent random distribution within the zone is the result of near-surface weakness. In addition, it can be said that whilst the diatremes penetrate high above the general level of dyke emplacement, the dykes are locally elevated in the vicinity of the roots of the diatremes (Fig. 10). One fairly constant feature is that the diatremes often exhibit cross-cutting relationships to dykes, the so-called antecedent dykes of Wagner (1914). It has hitherto been assumed that the diatremes lie on, and are genetically connected to, these dykes. However, it if is accepted that diatremes

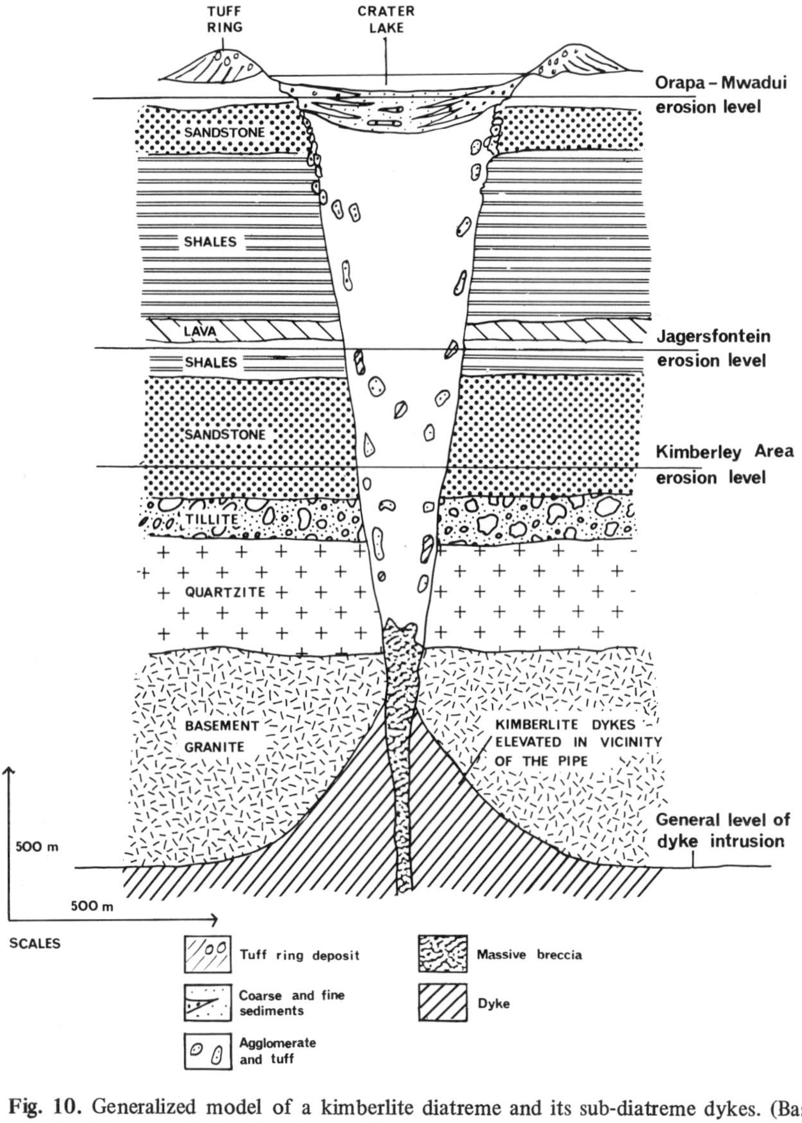

Fig. 10. Generalized model of a kimberlite diatreme and its sub-diatreme dykes. (Based on models of Dawson 1967a; Hawthorne 1975)

flare *above the general level of intrusion of their feeder dykes* (see below), it follows that the diatremes cannot be directly related to the relatively high-level antecedent dykes, despite their geographical proximity; moreover, the diatremes intrude the dykes, and generally contain quite different types of kimberlite (Dawson 1960; Kharkiv 1975). The implication is that the antecedent dykes represent an earlier episode of dyke intrusion that penetrated to a higher level than those feeding the diatremes; the close geographical proximity can be explained by both sets of dykes occupying the same or adjacent fractures. The time difference between the time of emplacement of the two different dyke systems is open to speculation, though if analogies are drawn with the differing columns of kimberlite within individual diatremes (above), a close time relationship might be inferred. However, Clement et al. (1979) report Precambrian (1800 m.y.) ultramafic dykes, some with kimberlitic affinities, in the Kimberley area, and one is cut by the Wesselton pipe. The fragmented material within the diatreme may also be cut by later dykes of either massive kimberlite or kimberlite tuffisite which may also extend out into the country-rock. The later dykes may be simple planar bodies or may wind sinuously through the main diatreme filling; in the case of the Premier Mine (Fig. 9) later calcite-rich dykes radiate out from a vertical central column in the western end of the mine.

In the case of most kimberlite sills, the feeder dykes have not been located. However, Wagner (1914) records a kimberlite dyke sending out minor sills into well-stratified shale wall-rocks in the Kamferdam Mine, Kimberley; and the Saltpeterpan sill, south of Kimberley, has been shown to stand on a N-S striking feeder dyke (Hawthorne 1968).

Regarding diatreme/sill relationships in the Kimberley area, sills exposed in the walls of the Kamferdam Mine diatreme (Wagner 1914) demonstrated an earlier phase of relatively high-level minor injection of kimberlite. In addition, Hawthorne (1968) reports a number of sills exposed on the present-day land-surface, below which level major diatremes such as the Kimberley and Wesselton Mines taper downwards for depths in excess of 700 m before their feeder dykes are encountered. It follows that the feeder dykes for the sills must have penetrated to a much higher level than those feeding the diatremes, and the similarities postulated earlier for the relationships between diatremes and antecedent dykes are apparent.

3.5 Extrusive Kimberlites

These are extremely rare. The only known example of a kimberlite tuff-ring is near Kasma, 140 km south of Kayes, Mali, W. Africa, where a small pipe is surrounded on three sides by lateritized tuffs up to 4 m thick (E. Gerryts, personal communication, in Hawthorne 1975). The Igwisi Hills, Tanzania, comprise three small tuff-cones and a small lava flow consisting of olivine phenocrysts and microxenoliths of peridotite (some containing pyrope, enstatite, chrome-diopside and phlogopite) in a matrix of carbonate, serpentine, spinel and perovskite; trachytic texture due to alignment of calcite plates is prevalent in the lava flow, which is petrographically very similar to the kimberlite of the Benfontein sills (Reid et al. 1975a). In South-West Africa, the Gross Brukkaros volcano is stated to have kimberlite affinities (Janse 1975; Ferguson et al. 1975), and a crypto-volcanic dome on the farm Hatzium is cut by kimberlite dykes (Heath and Toerien 1962). Rocks having affinities with kimberlite occur in small satellite vents of the Saltpetrekop volcano, Cape Province, S. Africa (McIver and Ferguson 1979).

Tuffaceous rocks at Moroto (Uganda) and Lesorogoi (Kenya) have been interpreted as having kimberlite affinities on the basis of their containing pyrope, ilmenite and ultramafic xenoliths (Nixon 1973a). However, on the same basis, Nixon (1973b) also classified the Lashaine volcano as kimberlitic; the extrusive rocks at Lashaine are, in fact, ankaramite and carbonatitic tuffs (Dawson et al. 1970), so the recognition of "kimberlites" on the above criteria must be treated with caution. (An analogy may be drawn with the initial identification of the Malaita alnöite as "kimberlitic" on very similar criteria, see earlier.)

Another rock-type, often claimed to be the extrusive equivalent of kimberlite, is meimechite, an ultrabasic lava first found in north-western Siberia interbedded with picrites, nephelinites, melilitites, limburgites and augites (Butakova and Egorov 1962); their similarity to kimberlite has been discredited on petrographic and chemical grounds (Egorov 1970; Dawson 1971; Malkov 1972). Moreover, whereas the tectonic position of the Siberian meimechites on the Anabar craton is at least consistent with that for most kimberlites, meimichites discovered more recently in the Kamchatka peninsula (Markovski and Rotman 1975) are in an atypical tectonic setting for kimberlite magmatism; meimechites are best regarded as alkali-poor, olivine-rich picrites (Egorov 1970).

3.6 Effects of Kimberlite Intrusions on Wall Rocks

3.6.1 Physical Effects

In most cases, the contacts between the kimberlite and the wall-rock are sharp, though in the case of badly fragmented wall-rocks at the margins of some diatremes, there is a blurred zone in which invading clastic kimberlite material is intimately intermixed with wall-rock material.

As noted above, the surface plan of some kimberlite diatremes may be strongly controlled by the joint-patterns in the country-rock and, there is no arching or doming of the country-rocks. Note, however, that kimberlite is intruded into the crypto-volcanic dome at Hatzium, S.W. Africa, and in their modelling of diatreme emplacement Woolsey et al. (1975) found development of surface doming of overlying sediment, and associated ring and radial fractures, in the initial stages of intrusion. The absence of doming and radial fractures associated with known kimberlites may be a reflection of their present level of erosion. In some instances, soft sediments, particularly shales, may be upturned at their contact with kimberlites, though no disturbance is found beyond a distance of 5–10 m from the contact. This is a purely superficial effect caused by the expansion of the upper part of the kimberlite due to surface-weathering; the country-rocks resume their normal attitude at depths of some 10–15 m. This superficial deformation is not observed where the country-rocks are tough, competent rock-types such as sandstone, basalt or granite. However, on some hard wall-rocks vertical and inclined striae have been observed which may be due to gouging of the wall-rocks by hard xenoliths in kimberlite expanding in the zone of weathering.

3.6.2 Metasomatic Effects

As might be expected with an intrusive rock as rich in volatile materials as kimberlite, metasomatism of the country-rocks is quite common. Zeolitisation and hydrothermal alteration of wall-rock basalts is recorded round some kimberlite diatremes in Lesotho (Dawson 1960; Ferguson et al. 1973a); and silicification and calcification causing induration of wall rocks is known in many instances. Instances are known also of water metasomatism alone, one particularly good example resulting in a peripheral permeation breccia (locally known as McKeown's breccia), at the contact of the Mwadui diatreme, Tanzania; poorly sorted, angular fragments of fresh or altered wall-rock granite-gneiss are set in a matrix of bright-green fine-grained mixture of altered feldspar, kaolin, halloysite, sericite and perovskite (Edwards and Howkins 1966). Whereas the above types of metasomatism may be regarded as hydrothermal effects, there are rare instances of higher-temperature metasomatism. Alkalic metasomatism (fenitisation) of country-rock granite to give a microcline-arfvedsonite rock is recorded from the contact of the Toubabouko dyke, Côte d'Ivoire (Knopf 1970) and replacement of metavolcanic rocks by riebeckite is also known at the contact of calcite-rich kimberlite in Lesueur Township, Quebec (Watson 1973).

The effects of metasomatism on xenoliths are discussed in the section on xenoliths.

3.6.3 Thermal Effects

In a historical context, the hypotheses about the nature and emplacement of kimberlite were dominated by observations made on the kimberlites infilling the major diamond-producing diatremes of South Africa. Together with the observation that kimberlite is essentially a brecciated rock, the other principal conclusion was that kimberlite was emplaced as a cold intrusion. This latter observation was based on the absence of thermal effects upon wall-rocks and xenoliths within the diatremes; even bituminous shales had not been sufficiently heated to expel the hydrocarbons. These observations were correct as far as they went, but they did overshadow less spectacular, but nonetheless highly significant, observations suggesting the existence of hot kimberlite (Wagner 1914). These earlier views, together with more recent observations, have been summarized by Dawson and Hawthorne (1970). In effect, it can be convincingly demonstrated that temperature-sensitive wall-rocks show definite thermal effects at their contact with kimberlite dykes and sills, particularly in the case of kimberlite dykes containing high amounts of mica and calcite. Shales and siltstones are converted to hornstone, and calcite-rich sediments and sandstone recrystallize to marble and quartzite. The contrast between the presence of these thermal effects in the case of the massive dyke and sill kimberlites, and their absence at the contact of the breccia-filled diatremes is an important feature when considering the emplacement of kimberlite. We shall return to this point later. In addition to the thermal effects on the wall rocks, there is evidence of thermal metamorphism of xenoliths within diatremes. Wagner (1911) records thermal metamorphism of shale and dolomite inclusions in the kimberlite of the Premier Mine; augite and garnet develop in the shale, and grossular garnet develops in the dolomite xenoliths which have recrystallized to coarse marble (Fig. 11). Furthermore, Kharkiv (1967) has described metamorphosed marly sediments in the Novinka, Komsomolskaya and Zimnayaya diatremes in Yakutia. Some of these xenoliths are concentrically zoned with marl in the centre, separated from an andradite-cuspidine rim by an intermediate andratite-cuspidine-monticellite zone. Other xenoliths have a pronounced banding in which bands of garnet-cuspidine and garnet-cuspidine-monticillite alternate with bands of unaltered marl; however, the parts of the xenoliths directly in contact with the kimberlite do not show banding but are composed of garnet-cuspidine aggregates. It should, however, be noted that there is no evidence that the metamorphism of the xenoliths in either Yakutia or the Premier Mine took place *within the diatreme;* it may have been effected in ths sub-diatreme dykes before the xenoliths were emplaced with the kimberlite in the diatremes.

Fig. 11. Block of thermally metamorphosed dolomite in kimberlite from the Premier Mine. Note the concentric zonation. The *dark inner zone* consists of grossular and diopside

3.7 Emplacement of Kimberlite Diatremes

For many years, concepts of diatreme emplacement were influenced mainly by two observations made on the kimberlites infilling the major diamond-producing diatremes in South Africa – (a) kimberlite is an essentially fragmental rock; and (b) it is essentially a "cold" rock, since no thermal effects were imposed on even very sensitive wall-rocks and xenoliths. These two observations led some geologists to propose that kimberlite was intruded either as a cold mush, akin to Alpine-type peridotites (Mikheyenko and Nenashev 1962), or as cold, gas-charged mushes ascending to the surface and then breaking through explosively to the surface, with the fragmental material falling back to infill the diatremes (Davidson 1964; Kennedy and Nordlie 1968). These hypotheses ignored the evidence of hot kimberlite magma in dykes and sills. In the kimberlite diatremes of Lesotho, Dawson (1962) observed various features that are inconsistent with an explosive boring or stoping mode of emplacement; the vent shape was found to be largely joint-controlled and there was partial detachment of wall-rock fragments. Large masses of rock ("floating reefs") that have descended into the diatreme are generally undisturbed relative to their oroginal orientation (Fig. 9), and two further features (a) xenoliths from specific horizons tend to be concentrated within certain areas, and (b) many diatremes consist of several distinct columns of kimberlite, are all inconsistent with material falling back with random orientation into the diatreme following explosive eruptions. Dawson (1962) concluded that the features are more consistent with fluidization, a process in which fragmental material is transported in a fast-moving gas stream; this had originally been postulated by Cloos (1941) to explain similar features in the tuff-pipes of Schwabia. Analogous features have now been produced in fluidization models by Woolsey et al. (1975). However, whereas both Cloos and Dawson believed that the fluidization emplacement was effected by solids suspended in gas (CO_2 and H_2O), Clement (1975) has put forward good evidence that at least some of the fluid is calcite-rich liquid that separated as an immiscible fluid from the ascending kimberlite.

A composite model of kimberlite emplacement would be that calcite-rich kimberlite magma, containing megacrysts and xenoliths, is intruded rapidly up a system of deep-seated fractures, these fractures possibly being the result of stress corrosion (Anderson and Grew 1977); speed of ascent must have been sufficiently rapid to preclude sedimentation of high-density peridotite and eclogite xenoliths, and also to prevent total resorption or inversion of diamonds. Then, at preferred points where access to the surface is easiest (intersection of fracture zones, major lithological contacts), and possibly where ground-water percolation is deepest, kimberlites break through to the surface from depths of 2–3 km. This depth is deduced from two independent aspects – (a) the fact that, regardless of temperature, the volume per unit weight of both water and CO_2 increases enormously at pressures of 0.4 0.8 kb, approximating to depths of 1.5–3 km (Dawson 1971); and (b) the mined-out Kimberley Mine diatreme expands at a present-day depth of 800 m which, combined with the estimate of 1600 m of erosion since emplacement (Hawthorne 1975), indicates flaring at a point 2.4 km below the original land-surface. Adiabatic expansion of exsolving CO_2 from the magma and heat absorption by groundwater would cause rapid cooling (accounting for the absence of thermal metamorphic effects), and rapid upward transport of deep-seated material; McGetchin (1968) has calculated that the speed of the fluidized material at the surface was in the order of 400 m/s. The initial explosion vent would then be enlarged by fluidized fragmental material drilling upwards and taking advantage of the country-rock jointing; major marginal expansion could be caused by major wall-rock fragments (the "floating reefs") slumping into the vent; and wall-rock spalling in the initial explosive stage is a possibility. Also some type of pseudo-stoping, perhaps due to hydraulic fracturing, must be invoked to account for "blind" diatremes, such as one of the lobes of the Wesselton Mine. In some cases, later surges emplace later distinctive tuff columns or tuffisite dykes, whilst cavities or consolidation fractures in the vent kimberlite may be infilled by quietly upwelling kimberlite magma to form plugs or dykes of hypabyssal facies kimberlite and kimberlite breccia.

The overall picture is of a diatreme rising above the general level of dyke intrusion, the surface expression being a kimberlite tuff ring surrounding a depression or crater-lake (Fig. 10).

In the emplacement of kimberlite diatremes by fluidization, the role of volatiles is very important. Whereas exsolution of high-pressure gases from the ascending kimberlite has been reconized for some time (Dawson 1962), Clement (1975) has elaborated upon this by proposing a pre-eruption stage when carbonate-rich liquid forms a separate immiscible highly volatile zone at the top of the ascending kimberlite; this is a source of highly concentrated explosive fluid. Whilst much of the carbonate is probably of magmatic origin (although later modified by contact with high-level water), the hydrogen and oxygen isotope ratios of diatreme-facies serpentines strongly suggest that much of the "hydrothermal" water in diatreme serpentines and ground-mass micas is high-level groundwater, rather than of deep, mantle origin (Sheppard and Dawson 1975); a high-level input for this water also explains the extensive serpentinization of olivine in diatreme-facies as opposed to hypabyssal-facies kimberlites, and the relatively low water content in bulk-rock analyses of many hypabyssal-facies kimberlites. In short, there is an intimate relationship between diatreme formation, extensive hydration and involvement of water with high-level (rather than mantle) isotopic characteristics.

Finally, there is some evidence that after the intrusive activity, the vent material settled and compacted in the diatremes, causing minor internal faulting and slumping. In Bushmanland, S. Africa, there are diatremes where > 250 m of lacustrine sediments overlie the kimberlite in some small diatremes (Cornelissen and Verwoerd 1975); it appears unlikely that such narrow pipes would have remained open to such depths for a protracted period, and gradual, continuous subsidence of the kimberlite to provide long-lived sedimentation basins is a more likely explanation.

Chapter 4 Petrography of Kimberlite

4.1 Introduction

Kimberlite is a petrographically complex, hybrid rock, since it contains not only the phases crystallizing from the liquid magma but also polycrystalline fragments or single crystals derived from various types of fragmented xenolith. It is appropriate here to refer the reader to Chapter 7 where the xenolith-derived minerals are described. One of the main characteristics of kimberlite is the inequigranular texture, due to included xenoliths and megacrysts of various minerals set within a finer-grained matrix containing microphenocrysts. The term megacryst is used here deliberately to avoid any genetic connotation since detailed mineralogical studies of kimberlite have shown that these larger crystals consist of both xenocrysts and high-pressure phenocrysts. The commoner megacrysts, that may exceptionally be up to 15 cm in size (though 0.5–3 cm is much more common), are olivine (or serpentine), picro-ilmenite, mica, pyroxene and garnet; these are set in a matrix of serpentine-group minerals and carbonate, containing microphenocrysts of mica, spinel, perovskite and rare diopside and monticellite in varying proportions; the matrix may vary from light yellow to dark green or black, depending upon the type and proportion of phases present. Only the commoner phases are mentioned and a more comprehensive list of kimberlite phases is given in the following chapter. Diamond may or may not be present as a rare accessory mineral; although in his original definition of kimberlite, Lewis (1887) specifically included diamond as an essential constituent of kimberlite, Dawson (1967a) proposed that diamond is not essential and thus extended the name to include many so-called mica-peridotites that in every other respect (mode of occurrence, tectonic setting, mineralogy, petrography and geochemistry) resemble kimberlite.

The megacrysts are generally rounded in shape and show reaction relationships with the matrix; for example, olivine may be partially or wholly replaced by serpentine-group minerals or carbonates, picro-ilmenite is rimmed by perovskite, garnet by rinds of "kelyphite" (fine-grained aggregates of phyllosillicates and spinel), and phlogopite by hydromica. Within these broad limits, kimberlites from certain intrusions are petrographically quite distinctive, by virtue of their matrix colour; their density; the type, size, relative proportions and degree of alteration of the megacrysts; and the type and size of the xenoliths. A striking example is the kimberlite of the Monastery Mine, S. Africa, in which large (> 5 cm) megacrysts of picro-ilmenite, pyroxene, brown olivine and red-brown garnet are set in a light green highly micaceous matrix. The essential feature of kimberlite is that it consists of a series of highly refractory megacrysts set in a matrix of lower temperature minerals with which the megacrysts have reacted to various degrees.

The occurrence of high amounts of phlogopite in the matrix of some kimberlites was originally used by Wagner (1914) so subdivide kimberlite into (1) basaltic, and (2) micaceous or "lamprophyric" varieties, although he never stated the modal proportion of phlogopite required to distinguish between the two types. Later, Rabhkin et al. (1962) proposed that phlogopite should comprise > 5% of the megacrysts or a "considerable proportion of the matrix" for the rock to be termed micaceous. In fact there is a complete gradation from kimberlites bearing virtually no phlogopite to ones, such as those in the Boshof area (S. Africa), that contain > 50% of groundmass phlogopite. Whilst any firm figure would of necessity be highly artibrary, due to (for instance) differing proportions of xenolith, megacrysts and groundmass in any particular kimberlite, in practice the term "micaceous" may be applied to kimberlites in which the eye is immediately attracted to the shiny appearance of the matrix in the hand-specimen. Furthermore, the term "basaltic" is one that is outdated. Whilst adequately describing for field purposes the dark, inequigranular appearance of some kimberlites in the field, the term "basaltic" implies the presence of plagioclase which (apart from very rare xenocrysts) is absent from kimberlite. It is here suggested that the word "basaltic" be dropped, and the single word "kimberlite" be used to designate the kimberlite variety not noticeably rich in mica. A further type of kimberlite, "carbonatitic kimberlite", has been proposed by Malkov (1975); this proposal has been attendant upon the recognition of sometimes considerable quantities of magmatic (rather than hydrothermal or secondary) calcite in some kimberlite intrusions. Again, in practice, there are wide modal variations in this primary calcite from intrusion to intrusion, precluding any arbitrary figure for the calcite content of "carbonatitic kimberlite"; it is really the morphology and carbon and oxygen isotope ratios of the carbonate, rather than its amount, that characterize this type (Dawson and Hawthorne 1973). Skinner and Clement (1979) have proposed five basic subdivisions based on the dominance of some of the five dominant matrix minerals – phlogopite, calcite, serpentine, diopside and monticellite; the presence or absence of some of these proposed "key" minerals will depend upon the mode of emplacement (see above) and the partial pressures of volatiles. For example "monticellite kimberlite" could be transformed into "serpentine kimberlite" or "calcitic kimberlite" depending upon conditions at the time of emplacement, but the use of these proposed terms is useful so long as they are used purely descriptively and the fact is recognized that they may result from an alteration of an earlier now-occult type. The fact that "carbonatitic kimberlite" is confined mainly to dykes and sills, and not found in diatremes, suggests that hypabyssal emplacement (with retention of primary carbonate) is the reason for this type of kimberlite, rather than it being an intrinsically different type of kimberlite.

It may be noted that there appear to be petrographically distinct kimberlite provinces. For example, in South Africa, micaceous kimberlites predominate in the Barkly West, Boshof and Theron areas, whereas less-micaceous kimberlites predominate in the Kimberley area and Lesotho. In Yakutia, non-micaceous kimberlites occupy the Alakit-Daldyn-Vilyui area on the southern side of the Siberian Platform, whilst those in the Anabar-Olenek area on the north-eastern side are mainly micaceous (Milashev et al. 1963).

In addition to its intrinsic mineralogy, the appearance of a given kimberlite can be affected by two further features: (1) its degree of weathering, and (2) its texture, which is a direct reflection of its mode of emplacement. Weathering can give rise to three broad types of kimberlite within an intrusion. At the surface the kimberlite may be oxidized and decomposed and is known, in the expressive terms of the miner, as "yellow-ground". This is succeeded in depth by "blue-ground" which is the same rock in a better state of preservation. Below this is hard unaltered kimberlite, referred to as "hardebank"; often this has a higher calcite content. The distribution of these three broad rock-types is generally controlled by the depth of surface-weathering and fracture-systems controlling the percolation of groundwater, both factors being highly variable. For example in a hot, humid terrain, such as West Africa, yellow-ground may persist to a depth of 100 m (Fairbairn and Robertson 1966), whereas in Yakutia, where the climate is by no means as conducive to chemical weathering, the yellow-ground zone is virtually absent. In addition, it is possible that yellow-ground in the marginal zones of some dykes and sills may be an original feature due to hydration and oxidation when the kimberlite intruded and inter-acted with water-rich sediments.

Regarding the textures of kimberlite, it is a matter of history that deductions as to the nature of kimberlite and concepts about its mode of emplacement were for many years strongly influenced by observations made on the fragmental kimberlites infilling the major diamond-producing diatremes in South Africa, thereby creating the impression that all kimberlite is a fragmental rock. (These features have been noted in the preceding chapter.) A more balanced picture is one in which two main textural types can be recognized (Dawson and Hawthorne 1970):

4.2 Diatreme Facies Kimberlite

This is essentially fragmental and consists of fragments of kimberlite, country-rock xenoliths, deep-seated xenoliths, and single crystals or cleavage fragments cemented by low-temperature, hydrothermal-type minerals (Figs.12 and 13). In the classification of Rabhkin et al. (1962), *kimberlite-tuffs* consist of 60%–90% of kimberlite and country-rock fragments cemented by hydrothermal minerals, with xenolithic material comprizing $< 5\%$ of the bulk; they are named according to the dominant type of kimberlite in the fragments (e.g. micaceous kimberlite tuff). *Tuff-breccia* contains variable amounts of country-rock and, if this predominates the breccia is

Fig. 12. Photograph of kimberlite in field, showing its conglomeratic appearance due to the presence of xenoliths of different types. From the Malibamato pipe, Lesotho

Fig. 13. A hand-specimen of kimberlite from the Jagersfontein Mine, showing its typical inequigranular appearance; this is due mainly, in the hand specimen illustrated, to megacrysts of olivine

termed xeno-tuff-breccia. Both tuffs and tuff-breccias are classified as fine, medium or coarse (< 1, 1–5, > 5 cm). In the tuffs and tuff-breccias, there is virtually no preferred orientation of inequidimensional clasts or minerals; groundmass micas, in particular, are randomly oriented.

Here it should be reiterated that within some kimberlite diatremes, several types of fragmental kimberlite may be recognized due to differences in their colour, mineralogy, xenolith type and content, and type and tenor of diamond.

A particular variety of kimberlite, principally confined to diatremes, is *autolithic kimberlite.* The sense in which it is now used differs from that proposed by Rabhkin et al. (1962) who used it to describe fragments of one type of kimberlite within another, i.e. merely implying two episodes of kimberlite activity. Although not originally recognized as such, autolithic kimberlite was first described by Wagner (1914) when he referred to a "pseudo-conglomerate", comprising or 5 cm rounded kimberlite lumps cemented by coarser kimberlite, on the 3250-foot level in the south-eastern part of the Kimberley Mine. More recently, it has been recognized in the Chomur pipe, Yakutia (Milashev et al. 1963) and in several Lesotho and South African kimberlites. Autolithic kimberlite is now recognized as ellipsoidal or rounded bodies of fine-grained kimberlite (autoliths) set in a matrix of either fragmental kimberlite or coarser-grained solid kimberlite, giving the appearance of a kimberlite-cemented kimberlite pseudo-conglomerate (Fig. 14); upon weathering, the coarser matrix is removed preferentially, leaving the tougher, finer-grained autoliths standing out as distinctive, rounded bodies. Many autoliths contain nuclei upon which the fine-grained kimberlite has accreted. These vary from single crystals to xenoliths of shale, dolerite or peridotite up to 15 cm in size; generally the volume of kimberlite to nucleus within an autolith varies inversely with the size of the nucleus, and on some large xenoliths the kimberlite forms only a thin veneer. Around the nuclei the crystals in the kimberlite form a concentric arrangement, this being particularly marked in the case of micaceous kimberlites where the mica platelets lie concentric or slightly tangential to the margins of the autolith (Fig. 15). Clement (1973) has drawn analogies between kimberlite autoliths and lapilli in the Schwabian tuff pipes in which cores of olivine or melilite are surrounded by glass containing melilite laths (Cloos 1941; Lorenz 1979). Both autoliths and lapilli may be explained as due to rapid exsolution of a high-pressure gas phase from the magma spraying xenoliths and wall-rock chips) into open spaces above the magma. The liquid must have been sufficiently hot and mobile to coalesce around the solid particles to form a free surface, and the combination of surface tension and spinning produced the rounded or ellipsoidal bodies of solidified magma, most with a nucleus. The fine-grained nature of these "spray lapilli" may be attributed to their forming in the exsolving high-pressure gas whose rapid expansion would cause strong adiabatic cooling. (In one autolith from the Kao pipe, the writer has observed quench calcite.) The "spray lapilli" are then cemented by lower-temperature more coarsely crystalline phases to give these autolithic rocks their characteristic texture. They also occur in a dyke in the

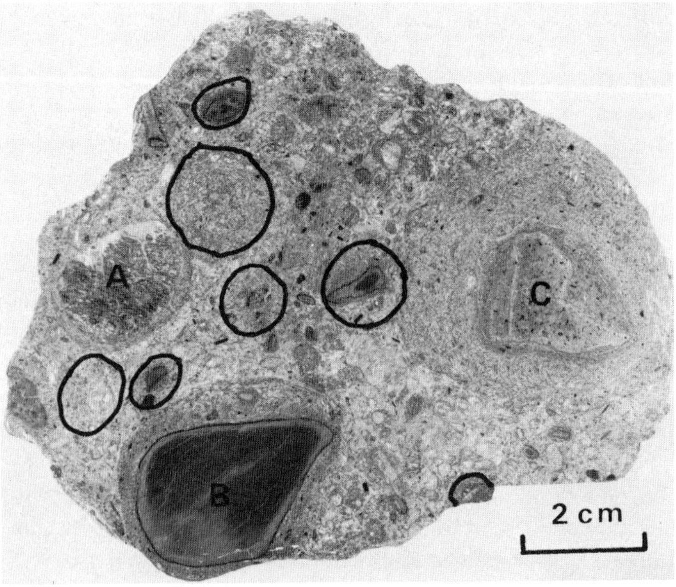

Fig. 14. Photomicrograph of nucleated autoliths in the kimberlite of the Monteleo Mine, Eastern Orange Free State. Some of the less obvious autoliths have had their margins inked in for emphasis. In the autoliths marked *A, B* and *C*, nucleii are an olivine megacryst, a shale xenolith and a dolerite xenolith, respectively. Note the relatively fine-grained nature of the kimberlite mantling the autoliths compared with the inter-autolith kimberlite

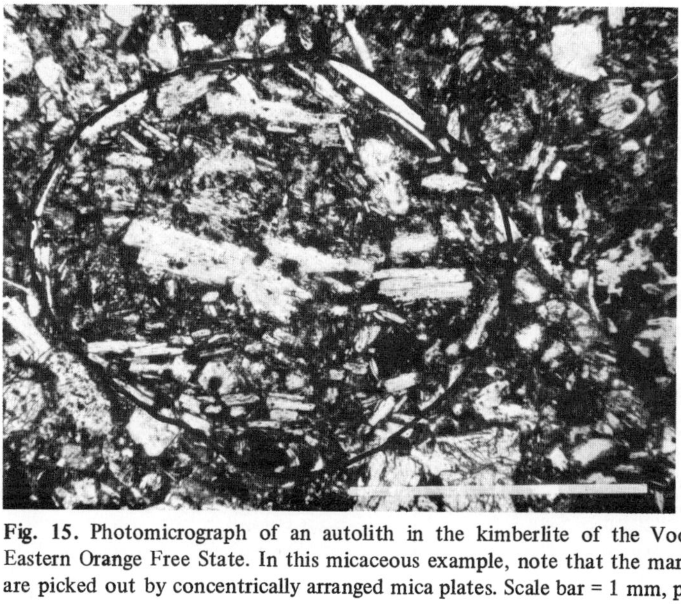

Fig. 15. Photomicrograph of an autolith in the kimberlite of the Voorspoed kimberlite, Eastern Orange Free State. In this micaceous example, note that the margins of the autolith are picked out by concentrically arranged mica plates. Scale bar = 1 mm, plane-polarized light

Theron area, S. Africa (author's observation), which indicates that they are not necessarily confirmed to diatremes. The main requirements for autolith formation appear to be (1) space, created possibly only monentarily by either explosions in a diatreme or hydraulic fracture propagation at the top of an ascending dyke; combined with (2) rapid cooling of the ejected liquid during spraying into the space provided. Absence of amygdales or vesicles in the autoliths, a common phenomenon in highly gas-charged magmas, is attributed by Lorenz (1979) to the very low viscosity of the kimberlite liquid. Although autoliths have only been recently highlighted, they have been found in several diatremes in Lesotho and S. Africa (Clement 1973, 1975; Ferguson et al. 1973b; Danchin et al. 1975), and also in the Monastery, Jagersfontein, Koffiefontein, Klipfontein, Bultfontein, Loxtonsdal, Voorspoed, Letseng (Lesotho), Hanaus II (S.W. Africa) and Sloan 2 (U.S.A.) pipes (author's own observations), and hence must be assumed to be a widespread and not uncommon phenomenon.

In addition to the clastic material described above, that can be attributed to eruptive processes, *kimberlite sediments* are found in some diatremes that have undergone little, or no, erosion since their formation. The kimberlite clasts, deposited in the crater-lakes that were the original surface expression of the diatremes, vary in size from clay to cobble grade, and are interpreted as inwash from the tuff rings surrounding the original maars. Intermixed with the volcano-clastic rocks are fragments of inwashed country-rock (Fig. 16). To date, the best-studied

Fig. 16. Kimberlite shales, sands and conglomerates forming the sediment pile in the upper part of the Orapa kimberlite pipe, Botswana; these sediments were believed to be part of the original tuff ring which has been washed back into the crater lake above the vent

Fig. 17. Thin-section, illuminated from the rear, of strongly layered kimberlite from the Lower Sill, Benfontein, showing: *A* a certain degree of cross-lamination; *B* chilled contacts picked out by dendritic calcite; *C* diapirs of calcite (clear) and opaques in the central section of the specimen. For a more detailed description the reader is referred to Dawson and Hawthorne (1973)

kimberlite sediments are those of the Mwadui diatreme, Tanzania; these are 300 m thick and provide good evidence for deltaic inwash, and transportation of coarse debris by slumping (Edwards and Howkins 1966). Other good examples of kimberlite sediments are in the Orapa pipe, Botswana (Hawthorne 1975) and in several pipes in the Bakwanga area, Zaire (Fiermans 1966).

4.3 Hypabyssal Facies Kimberlite

The second main textural type of kimberlite, *hypabyssal facies kimberlite* is confined to dykes, sills, and the deepest part of diatremes where the transition takes place between diatremes and their feeder dykes. Two main sub-types may be recognized:

a) *Massive kimberlite* that contains few or any xenoliths of country rock, being the result of comparatively gentle intrusion of kimberlite liquid (plus deep-seated xenoliths) into dykes and sills, and more rarely into cavities in diatremes that have been cleared of fragmental debris by gas explosions. The textures and structures in massive kimberlites are consistent with consolidation of kimberlite liquid. In addition to chilled margins, there are internal features in dykes such as trachytic textures and median concentration of xenoliths and megacrysts; these could be attributed to flowage differentiation or at least to interaction between crystals in a flowing liquid (see Komar 1972), although similar structures have been attributed to plastic flow (Mikheyenko 1968). The Benfontein sills, Kimberley, have provided good examples of igneous sedimentation features (cross-lamination, graded-bedding, wash-out structures), diapiric structures, and development of dendritic quench crystals of calcite and apatite (Dawson and Hawthorne 1970, 1973) (Figs. 17 and 18); concentration of high-density peridotite xenoliths, super-

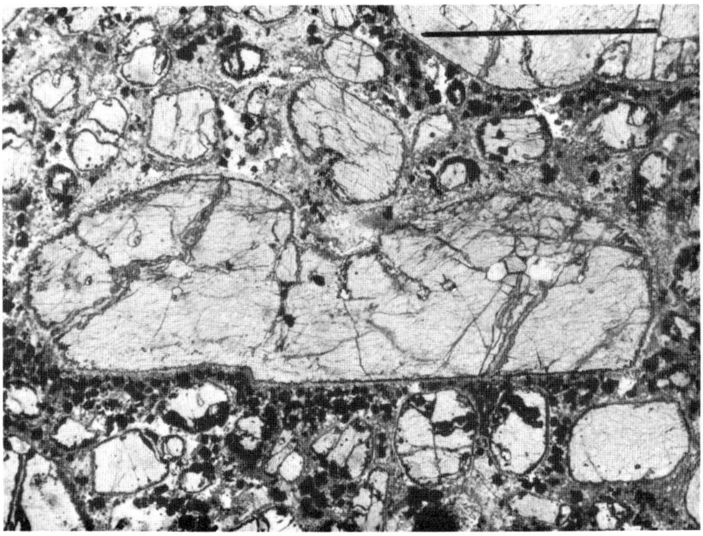

Fig. 18. Photomicrograph of sedimented olivine megacryst in kimberlite of the Middle Sill, Benfontein. Matrix perovskite and spinel has been prefenrentially concentrated on the flat, downward-facing side of the settling crystal. Scale bar = 2 mm, plane-polarized light

ficially resembling conglomerates, occurs in the lower part of kimberlite sills in Greenland (Andrews and Emeleus 1975).

All features indicate that the matrix of massive kimberlite was hot and had a low viscosity; in the case of the Benfontein and Greenland sills, the matrix has a high proportion of magmatic calcite. Another feature is that, compared with diatreme-facies kimberlites, the olivines in massive kimberlite have suffered relatively little serpentinisation.

b) *Kimberlite breccia* consists of fragments of various rock tpyes set in a matrix corresponding to massive kimberlite; a figure of $> 20\%$ by volume of fragmental material is suggested by Rabhkin et al. (1962). The breccias may be found, for example, at a contact where brittle shale xenoliths diatreme-facies material may have been percolated and cemented by massive kimberlite welling up from the sub-diatreme dyke.

It will be appreciated that where the root-zone of a diatreme grades into its feeder dyke, the relationships between kimberlite tuffs, autolithic kimberlite, kimberlite breccia and massive kimberlite may be complex and often gradational.

Chapter 5 Geochemistry of Kimberlites

In the chapter on kimberlite petrography, it was stressed that kimberlite is a complex hybrid rock, composed of varying proportions of mantle or crustal xenoliths. Moreover, mode of emplacement and degree of interaction with groundwater are very variable from intrusion to intrusion. Since all these variables, together with the partial and variable degrees of melting within the upper mantle at the onset of kimberlite generation (and subsequent modifications en route to the surface), will be expessed in the bulk compositions of different kimberlites, it is very doubtful if any bulk-rock analysis will reflect pristine kimberlite composition.

5.1 Major Elements

Bearing in mind that all these variables must be recognized, some general statements can be made about the composition of kimberlite. It is an ultrabasic rock, having a high proportion of cafemic oxides to silica and, in the micaceous kimberlites, is ultrapotassic; however for rocks of such basicity they contain unusually high amounts of K_2O, Al_2O_3, TiO_2, FeO, CaO, CO_2, P_2O_5 and H_2O, and compared with other ultrabasic rocks such as lherzolite, dunite or harzburgite (Table 6) have low Mg/Fe and very high K/Na and Fe^{3+}/Fe^{2+} ratios (Dawson 1967b). Average compositions of kimberlite such as those in Table 6, although giving a general idea as to chemistry, convey little real impression as to the wide range in the weight percentage of individual oxides that can be better appreciated by reference to Fig. 19. As a broad generalization, it can be said that SiO_2 and MgO ("olivine component") vary inversely with CaO and CO_2 ("calcite component") and H_2O; in addition increasing H_2O^+ is generally positively correlated with increasing oxidation state of the iron.

It is probably a fair comment that the major-element analyses of *most diatreme-facies kimberlites* show certain differences (often of considerable magnitude) from the probable chemistry prior to eruption and emplacement in the diatremes; this is due to (a) contamination by crustal xenoliths, (b) interaction with groundwater; and (c) loss of original volatiles (particularly CO_2), together with mobile elements that may be concentrated and transported in the gas phases. In addition, it should be noted that a false illusion of the purity of some analysed samples has been created, on the grounds of breaking out visible xenoliths from the kimberlite prior to analysis; this simplistic approach neglects the fact that

Table 6. Average of various kimberlites compared with average ultrabasic and ultramafic rocks

	1	2	3	4	5	6
SiO_2	40.6	43.4	35.2	31.1	27.6	33.2
TiO_2	0.05	0.13	2.32	2.03	1.65	1.97
Al_2O_3	0.85	2.70	4.4	4.9	3.17	4.45
Fe_2O_3	–	–	–	–	8.47[b]	10.6[b]
FeO	12.6[a]	8.34[a]	9.8[a]	10.5[a]	–	–
MnO	0.19	0.13	0.11	0.10	0.13	0.17
MgO	42.9	41.1	27.9	23.9	24.3	22.8
CaO	1.0	3.80	7.6	10.6	14.1	9.4
Na_2O	0.77	0.30	0.32	0.31	0.23	0.19
K_2O	0.04	0.06	0.98	2.1	0.79	0.79
H_2O^+	–	–	7.4	5.9	18.7	10.7
CO_2	0.04	–	3.3	7.1	–	–
P_2O_5	0.04	0.05	0.7	0.7	0.55	0.65

a Iron as total FeO; b Iron as total Fe_2O_3
1. Average ultrabasic rock (Vinogradov 1962)
2. Average ultramafic rock (Wedepohl 1975)
3. Average kimberlite (Dawson 1967b)
4. Average micaceous kimberlite (Dawson 1967b)
5. Average of 623 Siberian kimberlites (Ilupin and Lutts 1971)
6. Average of 25 Lesotho kimberlites (Gurney and Ebrahim 1973)

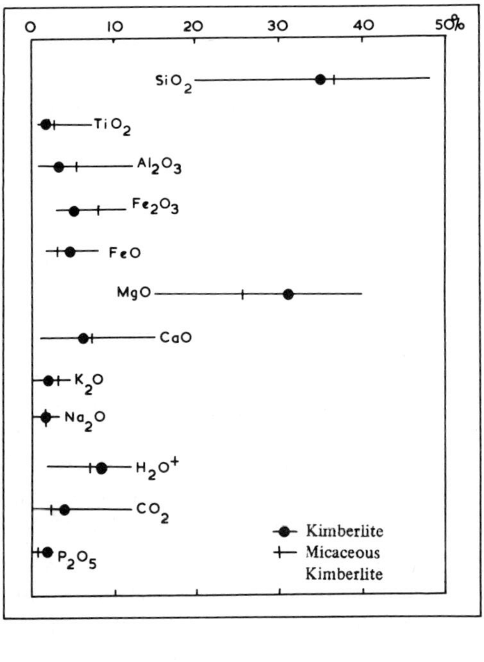

Fig. 19. Diagram showing the ranges in the oxide values for various oxides in kimberlite, together with the values for the average kimberlite and micaceous kimberlite

Table 7. Analyses of selected kimberlites

	1	2	3	4	5	6	7	8	9
SiO_2	48.5	40.7	25.3	36.12	30.73	33.2	25.19	3.41	0.52
TiO_2	1.77	0.90	0.50	1.45	1.51	2.41	1.89	12.23	0.10
Al_2O_3	4.12	5.56	1.64	4.38	1.96	3.97	2.87	7.81	0.48
Cr_2O_3	–	–	–	–	–	0.42	–	0.10	–
Fe_2O_3	8.52[a]	7.60[a]	4.77[a]	6.80	6.02	9.36[a]	3.72	19.66	1.30
FeO	–	–	–	2.68	3.90	–	6.72	12.12	0.61
MnO	0.15	0.13	0.11	0.22	0.19	0.16	0.22	0.43	0.19
MgO	23.0	23.7	10.59	22.82	29.95	23.56	29.69	11.95	8.87
CaO	4.46	3.69	26.34	8.33	10.21	9.82	13.59	17.15	44.07
Na_2O	0.57	0.29	0.03	0.29	0.25	0.48	0.01	0.26	0.41
K_2O	0.74	0.69	0.05	5.04	1.80	1.87	0.15	0.09	–
H_2O^+	–	–	–	4.89	4.94	–	1.15	2.34	0.36
H_2O^-	2.21	6.81	3.98	1.28	0.75	–	nil	0.30	0.03
CO_2	0.83	0.90	20.50	3.80	7.01	5.16	12.83	9.55	41.26
P_2O_5	0.18	0.34	0.41	1.46	1.08	1.68	2.20	2.42	1.42
LOI[b]	4.52	7.92	5.97			7.51			

a Total iron as Fe_2O_3
b Loss on ignition minus CO_2 (i.e. mainly H_2O^+
1. Average of 4 grey kimberlite samples, Premier Mine, S. Africa (Fesq et al. 1975b)
2. Average of 10 samples, Koffiefontein Mine, S. Africa (Fesq et al. 1975b)
3. Average of 10 samples, Water Fissure dyke, Bellsbank, S. Africa (Fesq et al. 1975)
4. Highly micaceous kimberlite, New Elands Mine, Boshof area, S. Africa (Dawson 1972)
5. Hypabyssal-facies kimberlite (BD1211), Jagersfontein Mine, S. Africa. Analyst: J.R. Baldwin (new data)
6. Average of 14 autoliths, Wesselton Mine, S. Africa (Danchin et al. 1975)
7. "Normal" kimberlite, Benfontein sills, S. Africa (Dawson and Hawthorne 1973)
8. Spinel-perovskite-rich cumulate, Benfontein sills (Dawson and Hawthorne 1973)
9. Carbonate-rich matrix, Benfontein sills (Dawson and Hawthorne 1973)

finely fragmented xenolithic material on a sub-microscopic scale can also create unacceptably high levels of "occult" contamination. This has been well illustrated by Fesq et al. (1975b), who show that Premier Mine grey kimberlite and Koffiefontein kimberlite (Table 7) are extremely high in SiO_2 (giving high Si/Mg ratios and reduction of other oxides) due to "occult" contamination by crustal quartzite and siltstone.

Although the effects of crustal contaminants can never be completely eliminated, they can be largely removed by analysing either (a) hypabyssal facies dykes and sills or (b) the fine-grained selvedge of nucleated autoliths. Analyses of hypabyssal-facies kimberlites (Table 7, analyses 3–5) indicate the wide range of composition of kimberlites that have neglibible crustal contamination. The high CO_2 content of the Water Fissure, Bellsbank, the high K_2O of the New Elands kimberlite, and the less extreme composition of the Jagersfontein kimberlite reflect differences in matrix calcite, mica and serpentine and, in the Jagersfontein case, megacrystal olivine; note, however, the relatively high content of retained CO_2 and P_2O_5 of these quietly-intruded kimberlites compared with that of the diatreme-facies Premier and Koffiefontein kimberlite (analyses 1 and 2). The same high calcite and apatite content is apparent from analyses of autoliths from the Wesselton Mine (analysis 6). A good example of the effects of differentiation on a single body of kimberlite (that is, moreover, virtually uncontaminated by either crustal or mantle xenoliths) is shown in the Benfontein sills (Dawson and Hawthorne 1973). Taking analysis 7 (Table 7) as the "normal" kimberlite, high-iron and high-titanium fractions (analysis 8) have developed by spinel-perovskite sedimentation, leaving complementary high calcite zones composed of groundmass carbonate (analysis 9).

Whereas by judicious choice of samples it is possible to diminish the effects of crustal xenolith contamination, the task of then eliminating the effects of mantle xenocryst input has not yet been resolved; in particular the difficulty of distinguishing between xenocrystal and phenocrystal olivine (a major constituent of most kimberlites) is formidable. At this stage, it is perhaps safest to state that the variations apparent in the chemistry of kimberlites are reflections of differing megacryst/matrix ratios and in the variations in their respective mineralogies. It appears likely that much of the MgO, FeO, TiO_2 and SiO_2 is concentrated in megacrystal olivine and ilmenite, whereas the bulk of the CaO, K_2O, H_2O and minor SiO_2, TiO_2 and FeO is concentrated in the matrix.

5.2 Trace Elements

It has been established for some time that, like the major elements, kimberlite trace elements are divisible into two major groupings (a) those in amounts usually found in ultramafic rocks; and (b) others that occur in amounts greater (in some cases considerably greater) than normally found in ultramafic rocks (Dawson 1962). Subsequent studies have confirmed this general picture and have increased the amount of data on which enhancement ratios can be calculated. Trace element data shown in Table 8 are mainly taken from the compilation of Muramatsu (1977); the concentrations of the trace elements relative to values for the average ultramafic rock (Wedepohl 1975; Wedepohl and Muramatsu 1979) are shown graphically in Fig. 20. From this it is apparent that Sc, V, Cr, Co, Ni, Cu, Zn and Cd occur in the same order of magnitude as in ultramafic rocks. Elements showing moderate enhancement (x 5 to x 10) are Cl, B, Mo, Pt, Y, Se, B, Bi and the heavier rare earths Gd-Lu. High or extreme enhancement ($>$ x 10) is found for the alkali metals (except for Na), Zr, Hf, F, Tl, Sr, Ba, Pb, Rb, Nb, Ta, U, Th, La and the light rare earths Ce-Eu; Cs is extremely enhanced.

Table 8. Trace elements in kimberlites

Element	Content in average ultra-mafic rock [a] ppm	Content in kimberlite ppm	Reference	Important host phase/s
Li	2	2–56	1	?phlogopite
Be	0.4	0–3	1	?
B	7	0–170	1, 2	?
F	97	68–7590	1	Apatite, phlogopite
S	~4000	200–8800	1	Sulphides
Cl	110	30–9800	1, 3	Apatite, phlogopite, halite
Sc	15	7–30	1	?Garnet, magnetite
V	50	21–250	1	Spinel, ilmenite
Cr	3090	550–2900	1	Cpx, garnet, spinel, ilmenite
Co	110	35–130	1	Ilmenite, olivine
Ni	1450	710–1600	1	Olivine
Cu	47	10–300	1	Phlogopite, olivine
Zn	56	15–287	1, 4	Spinel, ?phlogopite
Ga	2.5	0–30	1	?Garnet, ?phlogopite
Ge	1	0.05	1	?
Se	0.02	0.05–0.28	6	?
Rb	1.2	0–350	1, 6	Phlogopite
Sr	22	40–1900	1	Apatite, perovskite, calcite, diopside
Y	2.9	4–75	1	Apatite, perovskite
Zr	16	84–700	1	Perovskite, ilmenite, zircon
Nb	1.3	32–450	1	Perovskite, ilmenite
Mo	0.2	0.1–0.9	1	?Ilmenite
Ru	–	0.02–0.18	7	?
Rh	–	0.002–0.009	7	?
Cd	0.06	0.04–0.12	1	?Spinel
Sn	0.52	1–30	1	?
Cs	0.006	0–8.7	1, 6	Phlogopite
Ba	20	137–1970	1	Phlogopite, cpx
La	0.92	26–200	1	Apatite, perovskite, carbonates, mica (La–Lu)
Ce	1.93	45–522	1	
Pr	0.32	6.6–36	1	
Nd	1.44	25–180	1	
Sm	0.40	4.2–29.4	1	
Eu	0.16	1.1–6.5	1	
Gd	0.74	5.2–13.7	1	
Tb	0.12	0.43–2.05	1	
Ho	0.16	0.50–0.67	1	
Er	0.40	0.87–1.38	1	
Tm	0.07	0.14–0.22	1	
Yb	0.38	0.9–2.0	1	
Lu	0.06	0.09–0.22	1	
Hf	0.6	1.6–30	1	Perovskite, ilmenite
Ta	⩽ 0.1	1.4–21.3	1	Perovskite, ilmenite
Os	–	0.002–0.012	8	?

Table 8 (continued)

Element	Content in average ultra-mafic rock ppm	Content in kimberlite ppm	Reference	Important host phase/s
Ir	–	0.002–0.012	8, 10	?
Pt	0.06	0.07 –0.43	8	?
Au	0.007	0.001–0.043	1, 10	?
Hg	0.03	0.006–0.200	1	?
Tl	0.01	0.087–0.65	1	?
Pb	0.2	0.9–50	1	?Phlogopite
Bi	0.006	0.010–0.083	1, 7	?
Th	0.07	4–54	1, 9, 11	Apatite, perovskite
U	0.02	0.6–18.3	1, 9, 11	Calcite

a From Wedepohl (1975) and Wedepohl and Muramatsu (1978)
1.Muramatsu (1977); 2. Cherepanov (1967); 3. Pavlov (1969); 4. Gurney and Ahrens (1973); 5. S.F. Sobolev et al. (1975); 6. Gurney et al. (1966); 7. Dunin-Barkovskaya et al. (1974); 8. Kaminski et al. (1974); 9. Gurney and Hobbs (1973); 10. Paul et al. (1977); 11. Dawson and Milledge (unpublished data)

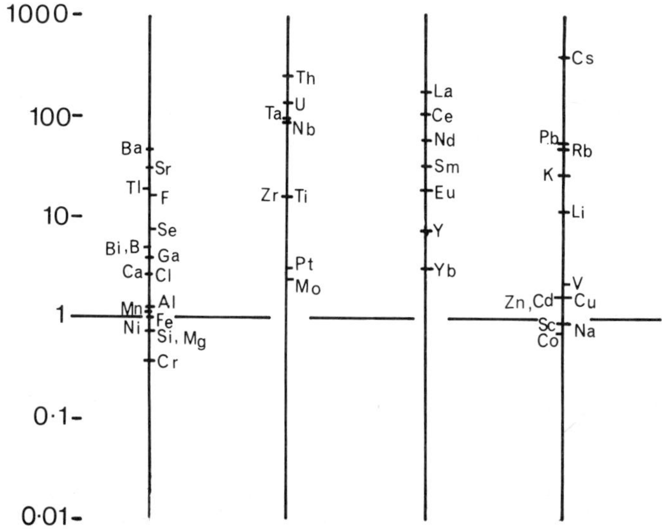

Fig. 20. Diagram showing the concentration values of various elements in kimberlite relative to the average ultramfic rock (see text). (Modified from Wedepohl and Muramatsu 1979)

It should be noted that in the ranges given in Table 8, no distinction is made between various types of kimberlite. However, it has been shown that there is good positive correlation between the following element pairs: K/Rb; K/Cs; Sr/P; La/P; Th/Sm (light REE); Cr/Ni; Ni/Co; Zr/Hf; Zr/Nb; Nb/Ta and Ti/Ta (Fesq et al. 1975b; Kable et al. 1975). Hence it is expected (and is confirmed in analyses of individual kimberlite) that elements such as Rb and Cs will be enhanced in mica-rich kimberlite; Sr, light rare earths and La will be highest in apatite-rich kimberlite; Zr, Hf, Nb, Ta will be found in perovskite- and (to a lesser extent) ilmenite-rich rocks; and high Ni, Cr, and Co will be highest in the varieties rich in olivine and xenocrystal chromite and garnet. Burkov and Podporina (1965) show enhancement of Li, Cs, Sr, B, Nb, Zr and Rb in the matrix of some Russian kimberlites, presumably due to their concentration in groundmass mica, perovskite and carbonate. Extreme enrichment may be encountered in exceptional instances when various phases occur in unusually high amounts. For example, extremely high Zr ($>$ 2000 ppm) is found in perovskite-rich cumulates in the Benfontein sills, S. Africa, and a calcite-rich layer contains high Sr ($>$ 2700 ppm), Ba (1500 ppm), La (200 ppm) and Pb (20 ppm); high U relative to Th in this layer leads to a Th/U ratio of $<$ 1, a feature only found elsewhere in carbonatites (Dawson and Hawthorne 1973).

Kimberlites have low K/Rb (88–320), K/Cs (975–3800), Ca/Sr (72), Nb/Ta (8–21) and Zr/Hf (41–49) ratios relative to ultramafic rocks (Dawson 1967b; Kable et al. 1975), indicating the relative enhancement of the second element in each ratio pair. Although differing in relative concentrations of the elements relative to ultramafic rocks, Ni/Co (15) and Cr/Ni (12) ratios for kimberlite are very similar to those for ultramafic rocks (Dawson 1967b).

During weathering and percolation of groundwater Ba, Rb and Sr tend to be readily leached (which has implications for Rb/Sr and Sr isotopic studies), whereas Zr, La, and Nb are relatively immobile (Dawson 1962). High amounts of Cl (0.9 wt%) in some Russian kimberlites result from halite being deposited from brine-rich waters percolating from nearby evaporite sequences (Pavlov and Ilupin 1973).

5.3 Rare-Earth Element Chemistry

Recent studies have confirmed early reconnaissance work (Dawson 1962) that the REE's are considerably enriched in most kimberlites relative to other ultrabasic rocks and to average chondrite. Examples of average absolute amounts are given in Table 9. The highest absolute amounts appear to be in highly micaceous dykes and sills whereas diatreme-facies kimberlite, especially ones that have been subjected to considerable wall-rock contaminatio, such as those of the Premier and Koffiefontein Mines (see p. 55), have lower absolute amounts; absolute amounts vary widely from locality to locality though in some cases different specimens

Table 9. Average values of rare-earth elements and La/Yb ratios of kimberlite from various localities

Locality	Rock type	REE (ppm)	La/Yb	Reference
Wesselton (6)	Kimberlite	540	103	1
Premier (15)	Kimberlite	83	31	2
Koffiefontein (10)	Kimberlite	148	54	2
India (5)	Kimberlite	280	107	3
Swartruggens (4)	Micaceous kimberlite	738	139	1
Bellsbank (22)	Micaceous kimberlite	807	172	2
Greenland (4)	Micaceous kimberlite	259	139	4

Numbers in parentheses indicate number of samples
1. Mitchell and Brunfelt (1975); 2. Fesq et al. (1975b); 3. Paul et al. (1975a); 4. Paul and Potts (1976)

from individual localities show wide variations. However, all are enriched in light REE's relative to heavy REE's resulting in high La/Yb ratios and in a strong slope on chondrodite-normalized distribution patterns (Fig. 21). Other rocks showing broadly similar patterns are olivine nephelinites, melilitites and carbonatites (all highly fractionated rocks), though they differ in absolute amounts and details of the distribution pattern (Mitchell and Brunfelt 1975; Frey et al. 1977). Although most kimberlites have a straight normalized pattern, some from Yakutia and South Africa (Burkov and Podporina 1966; Fesq et al. 1975) have Eu depletion anomalies. This has been attributed to the bulk-rock patterns reflecting the presence of pyrovskite (Mitchell and Brunfelt 1975) and/or high amounts of phlogopite (Fesq et al. 1975), both of which have europium-depleted patterns, or to the oxygen fugacity of the melt (Fesq et al. 1975b).

Minerals in kimberlite that contain REE are garnet, diopside and phlogopite (small amounts); and apatite and perovskite (large amounts) (Ilupin et al. 1971). The discrepancy in the total amounts of REE in calcite from the Benfontein sill compared with bulk carbonatite from Uganda (noted by Frey et al. 1977) may well be due to the presence of apatite in the latter.

5.4 Isotope Chemistry

1. Hydrogen and Oxygen Isotopes. The δ D values of the matrix serpentine and/or phlogopite, that contain most of the fixed water in kimberlite, vary from 89‰–102‰; combined with $\delta\,^{18}O$ value of – 1.08‰–12.2‰ they form a group distinct from deep-seated phlogopite megacrysts that cluster quite differently on a δ D vs $\delta\,^{18}O$ plot (Fig. 22). The conclusion is that meteoric water (presumably in the original form of groundwater) was involved in the formation of kim-

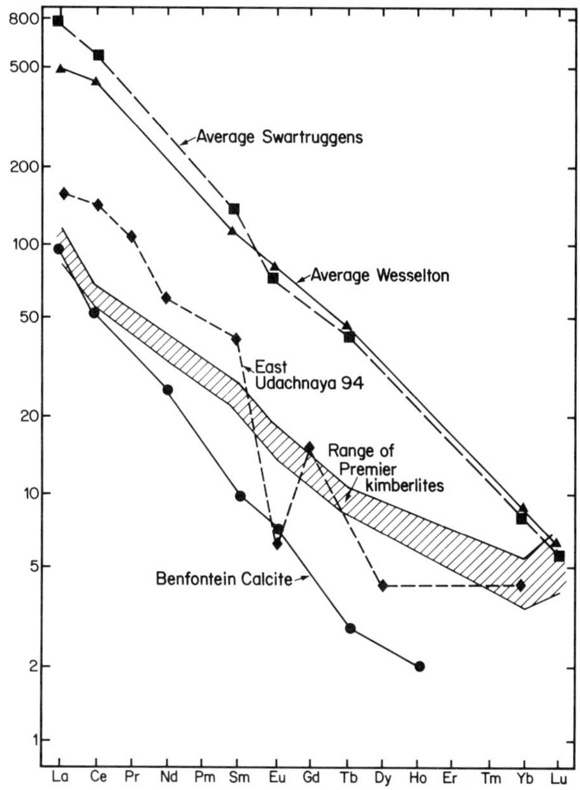

Fig. 21. Rare-earth element plots for selected kimberlites, normalized against standard chondrite. Data sources: average Swartruggens, average Wesselton and East Udachnaya 94 – Mitchell and Brunfelt (1975); Premier Range – Fesq et al. (1975b); Benfontein calcite – Frey et al. (1977)

berlite serpentine megacrysts and groundmass serpentine and phlogopite (Sheppard and Dawson 1975).

2. **Carbon Isotopes.** Most bulk-rock kimberlites have some carbonate in their matrix – mainly calcite with lesser amounts of dolomite. The carbon isotope range for bulk rocks is similar to that for carbonatites (including recrystallized varieties) and falls within the range for diamonds (Fig. 23), but reduced carbon in basic igneous rocks is relatively depleted in ^{13}C (Sheppard and Dawson 1975). On a $\delta\ ^{13}C$ vs $\delta\ ^{18}O$ plot (Fig. 24), it is apparent that (like many recrystallized carbonatites), the values for $\delta\ ^{18}O$ for most kimberlites lie in a wide field on the heavy oxygen side of the primary carbonatite field. Among nearly 200 analysed kimberlite samples, only 10 lie within the field of primary igneous carbonates.

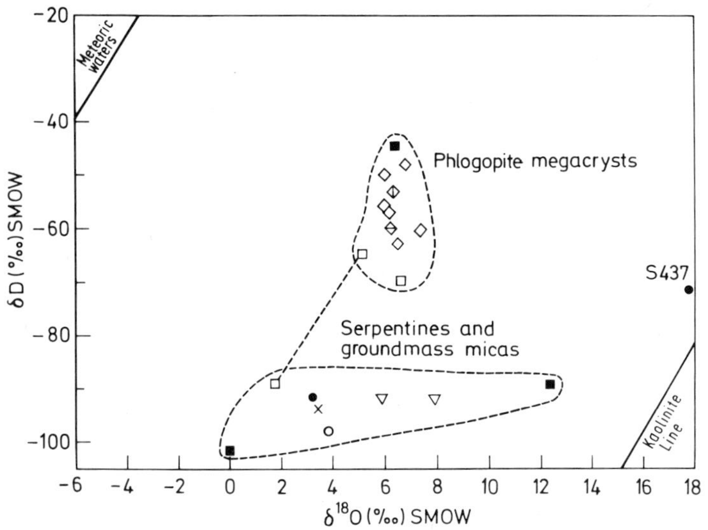

Fig. 22. δD vs $\delta^{18}O$ diagram showing the difference between phlogopite megacrysts, and groundmass micas and serpentine. (Modified from Sheppard and Dawson 1975)

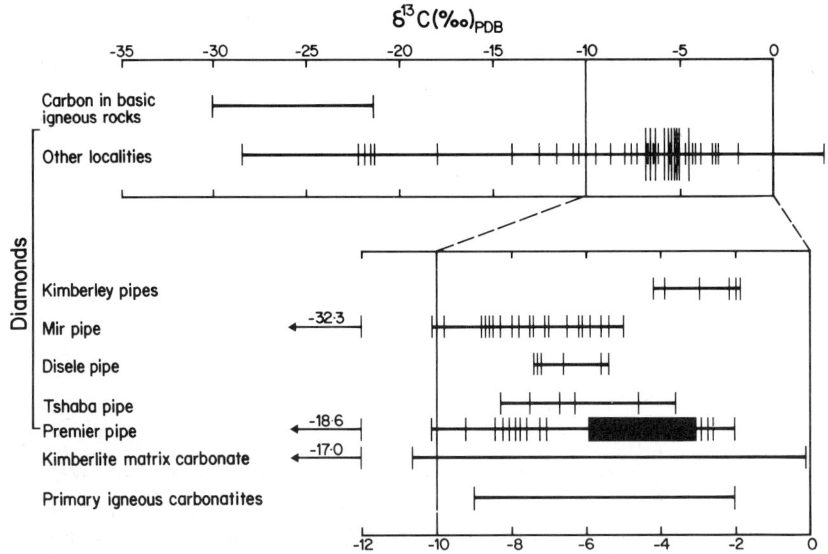

Fig. 23. Diagram showing ranges in $\delta^{13}C$ for kimberlite matrix carbonate, diamonds, carbon in basic igneous rocks and primary igneous carbonatites. The data are from a summary by Sheppard and Dawson (1975), supplemented by data from Kobelski et al. (1979) and Deines (in press); the Premier Mine "block" between $\delta^{13}C$ – 3‰ and – 6‰ contains 94 samples

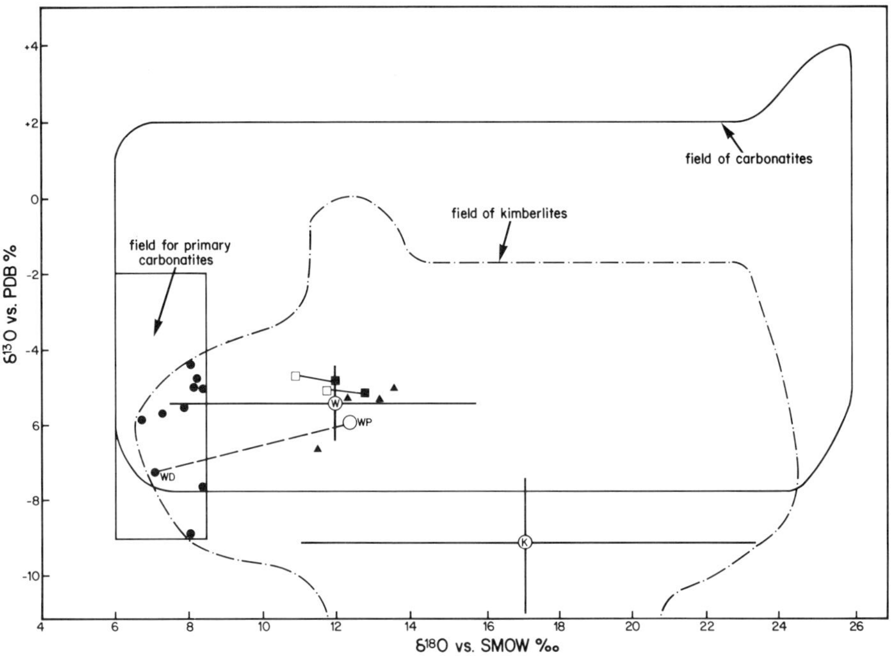

Fig. 24. Diagram of $\delta^{13}C$ vs $\delta^{18}O$ for kimberlites and carbonatites. Carbonatite data are from Deines and Gold (1973) and Suwa et al. (1975); the kimberlite data (192 samples) from Lesotho and South Africa are from Dawson and Hawthorne (1973), Deines and Gold (1973), Sheppard and Dawson (1975), and Kobelski and Gold (1977). Ornament is as follows: *solid circles* kimberlite samples (mainly from Kobelski et al. 1979) falling into the primary carbonatite field of Sheppard and Dawson (1972); *solid triangles* carbonatite dykes from the Premier Mine from Deines and Gold (1973), Suwa et al. (1975), and Kobelski et al. (1979); *open squares* and *solid squares* are dolomit-calcite (respectively) pairs from Premier Mine carbonate dykes (Sheppard and Dawson 1975); *strong vertical* and *horizontal lines* show variations in ratios (centre on means) for kimberlite samples from Wesselton *(W)* and Kao *(K)*; *WD* Wesselton dyke; *WP* Wesselton pipe kimberlites (Sheppard and Dawson 1975)

These are from the Benfontein sills (4 samples), Wesselton Mine (4 samples) and Premier Mine (2 samples). However, other kimberlites from these localities fall outside the primary igneous carbonate field and the fact that massive kimberlite from a dyke in the Wesselton Mine (WD in Fig. 24) contains "primary" carbonate whilst carbonate in fragmental kimberlite from Wesselton (WP) is enriched in ^{18}O, strongly suggests that the spread is due to interaction of kimberlites with groundwater, particularly in the case of diatreme-facies kimberlite. Another specific point of interest is that $\delta^{18}O$ values for the famous Premier Mine "carbonatite" dykes (Daly 1924) lie well outside the field of primary igneous carbonates; in these rocks dolomite is consistently enriched in ^{13}C relative to co-existing calcite.

3. Strontium Isotopes. Strontium isotope ratios in kimerlites have been reviewed by Mitchell and Crockett (1971a) and Barrett and Berg (1975). Mesozoic kimberlites from S. Africa have $^{87}Sr/^{86}Sr$ ratios of 0.7038–7104 (approximating to initial ratios, whereas ratios for Indian kimberlites (some of Precambrian age) give initial ratios of 0.7030–0.7068 (Paul 1979). Although Mitchell and Crockett (1971a) claimed no significant differences between dyke and pipe kimberlite, Berg and Allsopp (1972) and Barrett and Berg (1975) show lower ratios for fresh, hypabyssal-facies kimberlites (0.7038–0.7047), mainly for samples low in mica; higher values were found for fresh micaceous kimberlites and altered specimens. Since the groundmass carbonate in hypabyssal facies kimberlites is in the order of 0.702–0.706, high ratios may be ascribed to the high content of radiogenic strontium content of groundmass or megacrystal phlogopite that have ratios between 0.715 and 0.7780 (one from Swartruggens has 0.9594) (Allsopp and Barrsett 1975); however, groundwater percolation over a prolonged period is probably an important mechanism for raising the $^{87}Sr/^{86}Sr$ ratio of altered (mainly diatreme-facies) kimberlites.

4. Lead Isotope ratios of the fresh matrix in Cretaceous kimberlite and autoliths from S. Africa and Lesotho (initial ratios) were 206/204 17.6–20.0, 208/204 37.7–39.5 (Kramers 1977). On a $^{207}Pb/^{204}Pb$ vs. $^{206}Pb/^{204}Pb$ plot the kimberlites roughly define a slope of 0.10, corresponding to an age of 1575 m.y. The isotopic pattern is comparable to those obtained from carbonatites and ocean island volcanoes. These patterns are considered to provide a strong indication that the source region of the kimberlite beneath southern Africa is isotopically heterogeneous on a regional scale, and that individual areas within the source region existed in isolation from each other over long periods (Kramers 1977).

Chapter 6 Mineralogy of Kimberlites

6.1 Introduction

The petrography of kimberlite indicates that it is a hybrid rock consisting of fragments of rocks of upper mantle and deep crustal origin intermixed with the crystallization products of a highly volatile fluid. Some of these deep-seated rock crystals are still sufficiently fresh that they are recognizable as having been derived from a particular rock-type, whereas others may have become completely altered by interaction with the kimberlite liquid to form new mineral species. When, to this original mixture, there may be added, particularly in the case of diatreme-facies kimberlites, (a) minerals incorporated from the high-level formations forming the wall rocks of the kimberlite intrusions, and (b) later minerals formed during weathering or percolation of groundwater, it will be appreciated that the mineralogy of individual kimberlites may be extremely variable and complex.

In the compilation of the following list of kimberlite minerals, minerals of obvious, high-level exotic origin (such as, for example, minerals from basaltic or banded iron-ore wall-rock formations) have been deliberately excluded.

This section on the mineralogy of kimberlites should be read in conjunction with the chapter on the megacryst suite (Chap. 8).

6.2 Native Elements

Silver. Native silver occurs as isolated grains in the groundmass of the Somerset Island kimberlite, Canada; it contains small amounts of Fe (1%), Co (0.01%) and Ni (0.1%) (Clarke and Mitchell 1977).

Copper. Native copper, occurring as inclusions in olivine, is recorded by Haggerty (1975a) in kimberlite from the Ramatseliso intrusion, Lesotho.

Platinum Group Minerals. Wagner (1914, p. 77) reports that V. Hartog found platinum-group minerals (and possibly iridium and osmium) occurring intergrown with picroilmenite, magnetite and chromite in the kimberlite of the De Beers Mine, S. Africa. Hartog also found platinum-group metals associated with chromite in kimberlite from the Kimberley, De Beers, Bultfontein, Wesselton, Dutoitspan, Jagersfontein and Premier Mines.

Nickel-Iron is reported by Haggerty (1975a), together with metallic iron, as 5 μm inclusions in olivine from the Ramatseliso kimberlite, Lesotho. Metallic iron is also reported as very rare inclusions in the rare moissanite from the Mir and Yanvarskaya pipes, Yakutia (Marshintsev 1970).

Graphite, the low-pressure polymorph of carbon, occurs in small grains in the matrix of kimberlite from the Premier and De Beers mines (Wagner 1914), the Mir pipe (Bobrievich et al. 1959) and as intergrowths with pyrrhotite (Haggerty 1975a). As yet there are insufficient data to say whether most graphite is derived by fragmentation of the rare graphite-bearing eclogites and peridotites (see pp. 119 and 153), or whether the graphite is a true kimberlite mineral. Some in S. African kimberlites possibly results from metamorphism of Carboniferous tree-fragments incorporated into the kimberlite (Williams 1932).

Moissanite (SiC) has been discovered in the heavy concentrates from the Mir, Zarnitsa, Yanvarskaya and Aykhal pipes, Yakutia, together with a trigonal polymorph of SiC. The mineral is rare and exhibits a wide range in colour: colourless, yellow-green, blue, indogo-blue. In rare instances it is partially intergrown with rounded inclusions of metallic iron, and even less frequently with graphite (Marshintsev 1970). It should be noted that moissanite has been found in an inclusion of "porphyritic" (flaser-texture) garnet lherzolite from the Mir pipe (Bobrievich et al. 1959); hence xenoliths may be the source of the isolated moissanite grains found in kimberlite.

Diamond, the high pressure isometric polymorph of carbon, is a highly dispersed accessory mineral that occurs as crystals and cleavage fragments in some kimberlites. Even in the most diamondiferous kimberlites it is rare. Perhaps the most diamondiferous kimberlite weight for weight is the highest grade ore from the Mir pipe, Yakutia, that contains approximately one part of diamond to every one and a half million parts of kimberlite. Other "high-yield" mines are the former Kimberley Mine (1 part in 8×10^6), and the Helam Mine, Transvaal (1 part in 3×10^6, mostly very small industrial stones). Despite the fact that it is a rare accessory mineral, the economic exploitation of kimberlite for diamond was the reason for the first studies into kimberlite petrology and mineralogy.

Diamond occurs in a wide variety of isometric forms. The various growth feature, twinnings and resorption features, causing modification and rounding of original growth forms, have been the crystallographer's delight; for illustrations of the wide variations in external morphology the reader is referred to the works of Sutton (1928), Williams (1932), Tolansky (1955) and Orlov (1977). Whilst many of the more unusual diamond shapes result from twinning, it has been shown that some crystallographic forms such as rhombic-dodecahetrons and tetrahexahedrons can be derived by differential etching and modification of diamond of basic octahedral habit. Diamonds of octahedral and modified octahedral habit are the commonest type and, by analogy with synthetic diamonds, are believed to have formed at relatively high temperatures within the diamond stability field. Diamonds of cubic habit (of *relatively* low temperature origin) are comparatively uncommon, but are found in the alluvial deposits of Zaire and Brazil. In addition to diamond, there are relatively impure admixtures of diamond with graphite and various types of included material known as carbonado and boart. The hexagonal, high-density form of carbon – lonsdaleite – has not yet been found in kimberlite, though it has been found in diamond placer deposits in the U.S.S.R.

In addition to variations in crystal shape, diamonds also show a wide variety of colour; clear, yellow, brown, green, pink, blue and black. Some of these colours can be attributed to physical causes – some green colouration can be attributed to radioactive radiation, though Gnevushev and Kratsov (1960) found that green-coloured spots and flakes in the outer layers of diamond were enriched in Fe and Si. Fesq et al. (1975a) have shown that colourless and yellow diamonds have the same Fe content, and suggest that these colour differences may be due to nitrogen content; in addition, brown and green diamonds contain higher Fe and Al contents than colourless or yellow stones, and Mg was detected only in brown stones. With the blue colour of Type IIb stones probably being due to trace elements impurities (see below), it appears that diamond colour can be correlated with small amounts of impurities in the diamond; whether all these are due to actual physical distortion of the lattice by incorporation of the impurities in the diamond lattice or whether the trace elements are in the form of micro-inclusions of cogenetic minerals is a question still to be resolved. In the case of some black diamonds, colour is due to large amounts of graphite impurity.

Regardless of their colour and external morphology, diamonds can be subdivided into two fundamental types, termed Type I and Type II, whose basic diferences are listed in Table 10. They show marked differences in their transparency to ultra-violet light and in their infra-red absorption spectra. These differences were formerly believed to be due to the presence of nitrogen in Type I diamonds and its absence in Type II, but recent high-precision trace-element analyses of diamond have shown the presence of nitrogen in both types of diamond (Sellschop et al. 1974). Almost all diamonds incorporate both Type I and Type II material, but one type of material usually predominates and this determines its ultimate classification. Type I diamonds can be further subdivided into Type Ia in which

Table 10. Basic physical properties of the two main types of diamond

Physical property	Type I diamond	Type II diamond
Infra-red absorption	Show absorption between 8 and 10 μ	No absorption between 8 and 10 μ
Ultraviolet absorption	Complete beyond 3300 Å	Transparent down to 2200 Å
Photoconductivity	Weak	Strong
Birefringence	Usually strong	Weak or absent
Thermal conductivity	Very good	Extremely good
Electrical conductivity	None	Type IIb are semi-conductors
X-ray diffraction	Additional spots and streaks	Normal
Morphology	Well-formed single crystals	Poorly developed
Cleavage	Relatively uneven	Relatively perfect

the nitrogen is arranged in platelets, and Type I b in which the nitrogen is dispersed and which are paramagnetic; Type Ia diamonds comprise the majority of natural diamonds. In practice, Type I diamonds tend to have good crystal form but contain more impurities, resulting in such features as birefringence and extra spots and streaking observed during X-ray diffraction. Type IIa diamonds (which form a very small percentage of natural stones) are relatively pure, have more perfect internal structure but have poor external crystal shape; nearly all the larger diamonds (e.g. Cullinan) are Type IIa. Type IIb diamonds are an uncommon variant of Type II stones (about 0.1% of all Type II diamonds). Unlike other diamonds, they are electrical semiconductors; in addition, all natural blue diamonds are of Type IIb, the only current source of these blue diamonds being the Premier Mine. The colour of Type IIb diamonds and their semiconductivity was formerly attributed to the presence of Al in the lattice but the trace element studies of Sellschop et al. (1974) indicate that, although Al is virtually absent in Type I diamonds, it is also present in Type IIa diamonds, in some cases in amounts in excess of those in Type IIb diamonds.

Diamond also shows a variety of internal crystal defects that indicate a very complex growth history. The following is summarized from a detailed account of these feature by Frank (1969):

1. "Most diamonds, unless hardened by nitrogen precipitation, have suffered some degree of plastic deformation, that generated dislocations arrayed on (111) slip bands, made visible by birefringence, X-ray topography or electron microscopy. The implication is that they have been subjected to high shear stress at high temperatures (probably > 2000 K). Others with smaller birefringence are also full of dislocations, but the dislocations appear to have polygonized, implying a period of high temperature following deformation.

2. Most diamonds experienced numerous, often abrupt, changes in chemical environment during growth. Among the most easily visible features are growth shells of widely differing nitrogen content.

3. Within this complex phase two growths stages are reconizable and one solution stage:

a) An early stage in which the growth habit is rounded or mamillary except for limited octahedral flat facets. During this phase of growth, the impurity content differed on the octahedral facets and other surfaces growing at the same time, sectors of mamillary growth incorporating specks corresponding in composition to silica, corundum and ?garnet; these must have been present, as finally divided solids, at the time of growth. Non-incorporation of these solids in the growth sectors of octahedral habit can be explained by differences in surface energy. This stage (a) is absent in the growth history of some diamonds.

b) A stage of diamond growth with octahedral habit, with the continued occurrence of fluctuations in nitrogen uptake.

Stages (a) and (b) are generally succeeded by a stage of resorption, varying in amount from removal of a substantial amount of the earlier weight (the commonest case) to a light etch.

4. The nitrogen is normally present in very thin platelets and there are sharp boundaries between platelet-rich and platelet-free areas. It appears that nitrogen must have been incorporated into the diamond during growth at a temperature at which nitrogen diffusion was negligible, and that the nitrogen subsequently diffused to form platelets. Diffusion would result from either prolonged isothermal storage or a considerable rise in temperature after growth; the turbulent growth history of diamond suggests a tentative preference for the second explanation."

Inclusions in Diamond. Diamonds frequently contain small inclusions of various minerals. On the basis of whether they crystallized at the same time as the diamond or whether they are of much later (and of lower pressure-temperature) origin, having crystallized within fractures in the diamond, they are divided in (a) syngenetic or (b) epigenetic inclusions. According to Harris (1968) and Sobolev (1977) the syngenetic inclusions may be recognized on any, or a combination of the following criteria: (1) the mineral is believed to be a stable phase at pressures and temperatures consistent with an upper mantle origin; (2) the mineral forms discrete crystals within the diamond and may either be bounded by crystal faces typical of its normal morphology, or else form negative crystals, these being octahedra within the diamond and having precisely the same orientation as the host diamond; (3) if they are microcrystalline, they show a preferred orientation coincident with some crystallographic direction of the host diamond; (4) there is a complete absence of alteration of the inclusion, this being in strong contrast to the extensive alteration of the same species within kimberlite; (5) various grains of a given species within the same diamond have the same composition, but wide variations in the composition of the same species occur from diamond to diamond; (6) in some cases there exists a paragenetic relationship between different species within the same diamond, and these are always high-pressure parageneses.

The most abundant mineral inclusions are chromite, olivine, garnet, clinopyroxene and graphite (Harris and Gurney 1979). The order of abundance varies from intrusion to intrusion. For example, garnet is the most abundant inclusion in Finsch Mine diamonds but relatively scarce at Premier where pyroxene is commoner. Other less common inclusions are orthopyroxene, pyrrhotite and pentlandite (Sharp 1966), rutile, coesite, diamond, ilmenite, picroilmenite, magnetite and kyanite (Prinz et al. 1975). Mica, originally described as a syngenetic inclusion by Williams (1932) was later classified as epigenetic by Harris (1968); however recent finds of both biotite and phlogpite (Prinz et al. 1975; Giardini et al. 1974) establish this hydroxyl-bearing phase as truly syngenetic.

Most inclusions are the site of stresses within the diamonds, identified by birefringence or by dark-coloured microfracture infilled by graphite, pyrrhotite or pentlandite (Harris 1972); the same three minerals may also partially coat some inclusions.

In many instances, only single crystals or else a number of crystals of the same species are included within a single diamond. However, the following groups of minerals, have been recorded from single diamonds (Sobolev 1977).

1. Garnet + olivine ± enstatite ± chromite (dunite-harzburgite paragenesis).
2. Garnet + olivine + enstatite + diopside ± chromite (lherzolite paragenesis).
3. Garnet + olivine + diopside (wehrlite paragenesis).
4. Garnet + clinopyroxene + rutile (eclogite paragenesis).

It should be noted that the chemistry of the phases in these assemblages is distinctive and there is no overlap between the ultramafic assemblages and the eclogite parageneses.

The dunitic paragenesis is characterized by low-calcium, chrome pyrope (Dawson-Stephens Group 10, see p. 100), high magnesium olivine (Fo_{92-94}), high chromium spinel ($Cr_2O_3 > 60\%$) and high chromium enstatite ($Cr/Cr + Al$ = 0.20–0.40). The lherzolite paragenesis consists of Group 9-type garnet, chrome diopside, low-chromium enstatite, and olivine (Fo_{92}). The wehrlite paragenesis comprises garnet (Group 11 type), diopside and olivine (Fo_{92}). The eclogite parageneses comprise garnets of a wide range in composition (CaO 3%–14%, Group 3) together with clinopyroxenes that have wide variations in their Na_2O content (1.4%–6.9%) (and hence widely varying jadeite content), and a total absence of aluminium in four-field co-ordination.

With reagard to the single crystals, as well as the garnets noted above, inclusions of titanian pyrope (Group 1) and high almandine (Groups 4 and 5) garnets are known though low-calcium chrome pyropes (Group 10) are by far the most common. The olivines are highly forsteritic (Fo_{92-94}) with Cr_2O_3 up to 0.17 wt%. Enstatites are mainly in the range En_{93-94}, with rarer ones of Fo_{86}; the Al_2O_3 content (0.44–0.97 wt%) and Cr_2O_3 values (0.11–0.55 wt%) are generally low, though the high FeO variety noted above contain 1.32% Al_2O_3 and 0.06% Cr_2O_3. Clinopyroxenes are chromiferous diopsides or omphacites with a high jadeite content; subcalcic diopsides have not yet been reported. Clinopyroxenes rich in the ureyite ($NaCrSi_2O_6$) molecule have been reported in diamonds from the Mir pipe, the two examples containing (1) Na_2O 4.93%, Cr_2O_3 6.15; and (2) Na_2O 7.07, Cr_2O_3 11.8% (Sobolev et al. 1971). Most single chromite inclusions are magnesian chromite (MgO 10%–17%, Al_2O_3 5%–7%); however some true chromites (MgO < 1%, Al_2O_3 3%–4%) are recorded, these containing in the order of 2% ZnO (Meyer and Boyd 1972). Both ilmenite (MgO 0.16%) in Brazilian diamond, and magnesian ilmenite (MgO 11.3%) in diamond from the Mir pipe (Meyer and Svizero 1975, and Sobolev et al. 1971 respectively) are known. Mica inclusions are of titanian phlogopite (MgO 12.6%, TiO_2 10.8%) in a diamond of unknown source (Prinz et al. 1975), and biotite (MgO 7.63, TiO_2 3.63, Cr_2O_3 0.63) in a diamond from "Africa" (Giardini et al. 1974); in its titanium content the titanian phlogopite resembles phlogopite in garnet lherzolite from the Lashaine volcano, Tanzania (Dawson et al. 1970), whereas the biotite resembles kimberlite biotite from the Murfreesboro' kimberlite, Arkansas (Giardini et al. 1974).

In addition to the recognizable crystalline inclusions, some Type I diamonds contain dense white clouds of minute particles < 3 microns in diameter, and it has been suggested that these are gas or liquid inclusions (Harris 1968). Certainly gas inclusions are present in some diamonds, since in addition to nitrogen, Melton and Giardini (1974, 1975) have detected water, hydrogen, carbon dioxide, argon, ethylene, ethyl alcohol, butane and oxygen gas in diamonds from Africa, Arkansas and Brazil. The proportions of the gases vary from specimen to specimen, but water and carbon dioxide are present in all the specimens examined by these authors; large (0.6 gm) well-crystallized octahedra and hexoctahedra were consistently high in hydrogen and low in nitrogen, whereas in small (0.00025 g) flattened octahedra the reverse was true; cubes were found to contain larger amounts of H_2O and often a small amount of molecular oxygen. In addition, as a result of high-precision analyses of inclusion-free diamonds that reveal a wide variety of elements, Fesq et al. (1975a) propose that these elements are present as micro-aggregates – generally amorphous – representing droplets of the liquid from which the diamonds crystallized; those "liquid" inclusions appear to have a garnet + diopside ± olivine chemistry together with high amounts of sulphur and CO_2, and yield a primitive rare-earth element pattern.

Isotopic Composition of Diamond. Until fairly recently it was believed that the isotopic ratio of C^{12} to C^{13} varied within narrow limits ($\delta^{13}C$ – 1.9‰ to – 9.7‰) (Craig 1953; Wickmann 1956). Most of these early data were obtained on colourless diamonds, and it has subsequently been established, mainly on grey, yellow and brown diamonds from the Mir and Aykhal pipes, that most coloured diamonds fall within this range, though some from Aykhal would extend the range to $\delta^{13}C$ – 10.7‰ (Kovalski and Cherski 1972). It should be noted that the narrow range embraces not only diamonds from kimberlites in S. Africa, Zaire, U.S.S.R. and U.S.A., but also alluvial diamonds from Zaire, W. Africa, Venezuela, Brazil, Borneo and U.S.S.R. (Urals). Since the range from – 2‰ to – 10‰ for $\delta^{13}C$ is also typical for kimberlite matrix carbonates (see Fig. 23) and primary igneous carbonatites (see Sheppard and Dawson 1975), this range may tentatively be taken as the isotopic value of carbon within the upper mantle. It should be noted also that the isotopic value of diamond found in eclogite from the Mir pipe ($\delta^{13}C$ –9.0‰, Kropotova and Fedorenko 1970) also falls within this "upper mantle" range. Nonetheless, there are some exceptions that fall outside this range. Diamonds, varying in colour from grey to reddish-grey from the Ebelyakh alluvial deposit in N.E. Yakutia vary from $\delta^{13}C$ – 11.5‰ to 22.2‰, but the most disparate is a grey diamond from the Mir kimberlite pipe with $\delta^{13}C$ –32.3‰. This last value is most close to values obtained on Brazilian carbonados ($\delta^{13}C$ – 28.0‰) (Vinogradov et al. 1966). The reason for these departures is not yet understood. Kovalskii and Cherskii (1972) suggest that, in addition to upper mantle carbon (of limited isotopic range in composition), biogenic carbon and carbon from hydrocarbon sources may be the source materials for natural diamond; and Frantsesson and Kaminski (1974) go further, in proposing that carbonado is specifically of non-kimberlite origin, forming during crustal metamorphism of biogenic materials in the Precambrian. Conversely, Mitchell (1975) argues that variable diamond isotopic compositions may be due to differences in the proportions of carbon-bearing gases (CO_2, CO and CH_4) involved in gas-reduction interactions with diamond outside its stability field during ascent, as composition of the gases will vary considerably, particularly at lower temperatures. Mitchell's suggestion is supported by current experimental evidence that negligible isotopic fractionation occurs during graphite → diamond reactions at high temperatures (Kropotova et al. 1967), whereas during lower temperature formation of diamond by precipitation from hot methane gas onto diamond-seed nuclei, considerable isotopic fractionation may take place (Galimov et al. 1973).

Classification of Diamonds. Apart from the basic two-fold classification of diamond on purely physical principles mentioned above (Types I and II), the classification of diamonds has been based mainly on parameters such as shape, colour, and presence or absence of flaws and inclusions. In the accurate evaluation of gem-quality diamonds, a highly critical assessment of the combinations of these parameters is essential, and can only be built up by many years of subjective experience. Recently a more standardized classification scheme (not principally concerned with the evaluation of individual crystals) has been proposed, based on the crystal form, crystal angularity, crystal regularity, opacity, colour, inclusion content and surface features of nearly 68,000 diamonds from three kimberlite pipes (Premier, Finsch and Koffiefontein) and one alluvial source – Dreyer's Pan (Harris et al. 1975, 1979). On the basis of this classification, samples of diamonds have been uniquely assigned to one of these four investigated sources; extension of the scheme will be of use in assigning parcels of diamonds of unknown origin to distinct sources, a claim often made in the past (e.g. Wagner 1914; Sutton 1928; Williams 1932) but not hitherto subjected to rigorous statistical analysis.

6.3 Sulphides

The following sulphides have been reported from kimberlite. *Heazlewoodite* (Ni_3S_2) as the final zone in a complex series of epitaxial zones on chromite in the Somerset Island kimberlite, Canada (Clarke and Mitchell 1975). *Pentlandite, pyrhottite* and *chalcopyrite* occur in South African kimberlite (Haggerty 1975a); and pyrhottite, sphalerite, wurtzite, galena, chalcopyrite, cinnabar, marcasite and molybdenite in the Yakutian kimberlites (Bobrievich et al. 1959, 1964; Milashev et al. 1963). In addition, pyrite is very common in many South African kimberlites where it is assigned to a secondary origin, being always associated with secondary minerals such as serpentine (Wagner 1914). Pyrrhotite is a common inclusion in diamond (Sharp 1966), and hence it appears that the sulphides must form over a very wide pressure-temperature range. Whereas the sulphur in the pyrrhotite diamond inclusions must be of upper-mantle origin, the source of this element in the lower temperature sulphides is less certain. In the case of some of the sulphides in Yakutian kimberlite, Ilupin (1962) suggests that, whereas the metallic ions were an integral part of the kimberlite, the sulphur was derived from Cambrian sulphate evaporite horizons penetrated by the kimberlite intrusions. In South Africa, pyrite is quite common and large nodules are exceptionally abundant in some kimberlites that intrude pyritiferous Karroo shales, as at Roberts Victor and the Bellsbank intrusions; in these instances a genetic relationship might at first sight appear to be valid. However, in a preliminary study by Dr. A. Sasaki, Geological Survey of Japan (on specimens collected by J.B. Dawson), two specimens of pyrites from the Saltpeterpan dyke (60 km S of Kimberley) yielded $\delta^{34}S$ values of 1.9% and 2.5%, whilst three from the Main Dyke at Bellsbank gave values of 1.2%, 7.2% and 16.1%; the lower values are consistent with a primary igneous origin (Thode et al. 1961) but those of 7.2% and 16.1% have affinities with sulphides of biogenic origin. Another sample of possible mantle origin is a pyrite nodule from the Premier Mine (Tsai et al. 1979b).

6.4 Halides

Halite has been found in the kimberlite in the Mir pipe (Bobrievich et al. 1964). However, it should be remembered that some of the Yakutian kimberlite, including Mir, have penetrated Cambrian evaporite horizons en route to the surface. It is possible that low-temperature phases such as halite and some sulphates may be incorporated and go into solution at the time of emplacement of the kimberlite, and are redistributed and reprecipitated during cooling or later weathering of the bulk rocks. The possibility of a high-level origin of sulphate sulphur has been mentioned earlier. Another hypothesis, proposed by Pavlov and Ilupin (1973), is that the halite is precipitated by brines percolating through the kimberlites subsequent to their emplacement.

6.5 Oxides and Hydroxides

Periclase has been tentatively indentified as 2–3 μ inclusions in monticellite from the "Monticellite" kimberlite, Yakutia and it is also claimed to be present in the groundmass in the kimberlite of the Originalnoye, Sluchaynoye and Viktoriya pipes (Nikishov and Nikishova 1966); on the basis of this, plus the report of minute, isotopic crystals around the periphery of olivine crystals in the Novinka, Komsomolskaya and Zimnayaya kimberlites (reported by A.D. Kharkiv), Nikishov and Nikishova (1966) suggest that periclase may be a more common mineral than recognized hitherto in the kimberlite groundmass. However, in view of the extreme tendency of periclase to hydration, the suggestion of its persistence in the highly hydrous kimberlite matrix must be treated with circumspection.

Spinel Group Phases. Only in recent years, with the use of the electron microprobe, have intensive studies been made on the small opaque phases generally occurring in grains <1 mm in size in the groundmass of kimberlite. It is convenient to discuss first magnetite, which is mainly of secondary origin, and secondly the more chemically complex spinels of primary origin.

Magnetite in the form of dust-like aggregates is very common in the serpentine pseudomorphs after olivine. Other larger crystals may be found in druses or cavities, indicating a low-temperature or hydrothermal origin; they contain virtually no magnesium, chromium, titanium chromium or nickel. Aluminous magnetite crystals (Al_2O_3 1 wt%) up to 1 cm long occur in immiscible carbonate globules in the Benfontein sills, S. Africa (Dawson, unpublished data).

Spinels. In addition to the secondary spinels occurring in association with perovskite in reaction mantles on picroilmenite, the following types of spinel have been recognized in the groundmass of kimberlite (Haggerty 1975a): (1) xenocrystal spinel; (2) primary groundmass spinels; and (3) secondary spinels mantling the xenocrystic and primary groundmass spinels.

The xenocrystal spinels fall compositionally within the quaternary system $FeCr_2O_4 - MgCr_2O_4 - MgAl_2O_4 - FeAl_2O_4$. They can be matched with spinels occurring as primary phases in a wide variety of peridotitic xenoliths (Smith and Dawson 1975) and as secondary phases in the "kelyphite" rims on garnets in garnet lherzolites. It is reasonable to assume that these spinels were shed into the kimberlite during the fragmentation of peridotite xenoliths.

By contrast, some primary groundmass spinels and the secondary mantles on xenocrystic spinels are more usually in the system $Fe_2TiO_4 - FeAl_2O_4 - FeCr_2O_3$ or within the system $Fe_3O_4 - FeAl_2O_4 - FeCr_2O_4$; there is considerable overlap between these two ternary systems by the additional presence of $MgFe_2O_4$ and Mg_2TiO_4. Generalizations and simplifications are therefore difficult, and it is difficult to lay down firm criteria for what is primary and what is secondary spinel on both chemical and textural criteria. However the overall trend that emerges is one of enrichment in titanium and iron, especially Fe^{3+}, with increasing degrees of crystallization or with increasing solid-state reaction.

The composition of primary spinels reported and summarized by Haggerty (1975a) are not unique to kimberlite, but the zonal trends and the composition of spinels intermediate between the cores and the mantles are unique. Most compositions are mainly restricted to the join $MgCr_2O_4 - FeCr_2O_4$ with limited solid solution of these end-members with not more than 25 mol% $FeAl_2O_4$, not more than 10 mol% Fe_3O_4 and rarely containing amounts of TiO_2 greater than 5 wt%. In this last respect, they do resemble many peridotite spinels.

It is only in the mantle of some spinels from Lipelaneng, Nquechane and Kao, Lesotho (reported by Haggerty 1975a) and in the cores of the spinels of the Benfontein sills (Dawson and Hawthorne 1973) that the high TiO_2 and Fe_2O_3 are found. These are distinctive of kimberlite, as are the compositions of secondary spinel assemblages mantling picroilmenite in the Liqhobong kimberlite, Lesotho

(Haggerty 1973). In addition, Clarke and Mitchell (1975) report magnetite cores overgrown by ulvospinel and ilmenite, and magnesian chromite overgrown by high TiO_2/high Fe_2O_3 spinel and perovskite from the Somerset Island kimberlite, Canada.

In the overall chemistry of kimberlite spinels, the theoretically expected divalent ionic substitutions (Fe^{2+}–Mg–Mn) and trivalent ionic substitutions (Fe^{3+}–Al–Cr) are satisfied, and in an analysis of oxide correlation parameters Haggerty (1975b) found the following correlations between the following oxide pairs: (a) positive linear correlation: Mn vs Ti, Fe^{2+} vs Ti, Fe^{3+} vs Ti, Fe^{3+} vs Fe^{2+}, Al vs Mg, Fe^{2+} vs Mn; (b) negative linear correlation: Cr vs Ti, Fe^{3+} vs Cr, Fe^{2+} vs Mg, Fe^{3+} vs Mg; (c) hyperbolic correlation: Ti vs Al, Mn vs Al, Fe^{2+} vs Al, Fe^{3+} vs Al.

However, it must be stressed that the implications of variations in the chemistry and texture of kimberlite spinels are not yet fully understood, and many more textural and chemical data are needed.

Corundum as both sapphire and ruby has been found in the concentrates of the Frank Smith, Jagersfontein and Koffiefontein Mines (Harger 1911) and in the Aykhal pipe, Yakutia (Marshintsev 1970b); it appears probable that it is derived from xenoliths of corundum-eclogite or of crystalline crustal rocks.

Magnesian Ilmenite. Magnesian ilmenite is one of the most characteristic minerals of kimberlite, and may be regarded as a more reliable kimberlite tracer mineral than pyrope garnet. Its relatively low density (varying from 4.1 to 4.6 depending on its MgO content), bright lustre, hackly fracture and occurrence in well-rounded grains, make it distinctive; and its tough, non-brittle nature permits it to survive transportation for considerable distance from its kimberlite source following erosion.

Within individual intrusions the ilmenite varies widely in amount. For example, in S. Africa the Premier, Monastery and Frank Smith kimberlites contain the relatively high amounts of 2.6, 0.9 and 0.5 vol% respectively (Frick 1973a) whilst most kimberlites contain considerably less; and in the U.S.S.R. relatively high amounts are found in the Iskorka and Morkoka pipes, Daldyn-Alakit area (3.4 and 3.9 wt% respectively) whilst at the other end of the spectrum it has not been found in the Aldan kimberlites (Bobrievich et al. 1964).

The following paragenetic types of ilmenite have been recognized, the first five by Haggerty (1975a).

1. Large discrete rounded crystals or composite nodules.
2. Primary groundmass ilmenite.
3. Subsolidus exsolution ilmenite in titanomagnetite.
4. As a member of spinel-ilmenite-perovskite mantles on xenocrystal ilmenite.
5. Forming intergrowths with rutile or with rutile + armalcolite.
6. Intergrown with diamond (Sobolev 1977), and as inclusions in diamond (Meyer and Svizero 1975).

7. Forming inclusions within olivine (Dawson and Hawthorne 1973; Boyd and Nixon 1973a), and zircon (Kresten 1973c).

These parageneses are additional to the granular or lamellar intergrowths of ilmenite with olivine, enstatite, diopside, low-Cr pyrope, zircon or phlogopite; and additional to the ilmenite occurring in peridotites and eclogites, all of which are discussed elsewhere.

In the following review it has been assumed that the data pertain to the larger, more-easily investigated discrete nodules of ilmenite, unless specifically stated otherwise.

1. The *discrete nodules of ilmenite* are typically rounded or ellipsoidal and may be up to 10 cm in size. The surface is smooth, but may contain deep invaginations which, together with interior tubules, suggest alteration and removal of former inclusions (Fig. 25); inclusions of olivine (often serpentinised), garnet (usually fresh) and clinopyroxene (usually altered) are quite common. The following types of nodules have been recorded, the description being a composite from several authors (Frick 1973a; Haggerty 1973, 1975a; Mitchell 1973a).

a) Inclusions consisting of single crystals that exhibit undulose extinction, kink-banding and concentric fractures.
b) Polycrystalline inclusions in which large grains with undulose extinction and kink-banding form the strained cores, with a peripheral rim of smaller, deformation-free crystals.
c) Polycrystalline inclusions consisting of fine-grained (0.1–0.05 mm) strongly anisotopic grains, with granoblastic polygonal texture.
d) Polycrystalline inclusions of weakly anisotopic, coarse (5 mm) grains with granoblastic texture, curved grain-boundaries and equal-angle triple-points; these inclusions may be cut by zones of fine-grained polygonal crystals (Mitchell 1973a).

Taken over-all, types (a) to (d),,with their textural overlapping, have the appearance of a gradational deformation/recrystallization/annealing sequence.

Exsolution phenomena are not uncommon. Danchin and d'Orey (1972) have described exsolution of chrome spinel along (0001) planes in ilmenite from the Premier Mine, and magnetite lamellae are known in ilmenites from Zarnitsa (Yakutia), Koidu (Sierra Lenone) and the Monastery Mine (S. Africa) (Smirnov 1959; Grantham and Allen 1960; and Mitchell 1973a respectively). In addition, Haggerty (1975a) records crystallographically oriented exsolution lamellae of chromite, titanian chromite and aluminous titano-magnetite in S. African ilmenite, these being the result of subsolidus reduction.

The ilmenites have a wide compositional range, the total ranges in the main components of the examples listed in Table 11 being TiO_2 41.40–55.90, Fe_2O_3 5.75–2768, FeO 15.00–30.28 and MgO 4.86–23.1. The ranges are found not only within a total province, as for example, in Yakutia (Table 11), but also within an individual pipe such as Monastery (Frisch 1970). For example, in Yakutia there are significant variations in the MgO content; ilmenite from the No. 2 pipe

Fig. 25. Ilmenite megacrysts from the Thaba Putsoa kimberlite, Lesotho. Note the rounded nature of the nodules, external corrosion and internal tubules

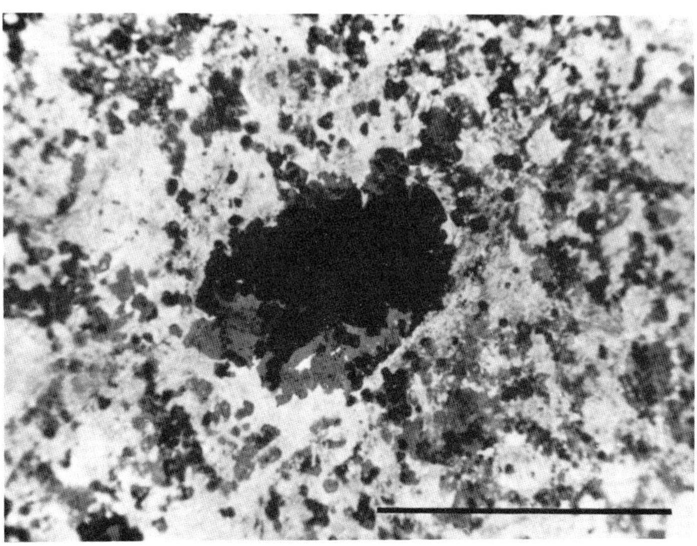

Fig. 26. Photomicrograph showing opaque ilmenite rimmed by semi-translucent perovskite, in kimberlite from the Monastery Mine. Scale bar = 2 mm, plane-polarized light

Table 11. Average compositions (A) and compositional ranges (B) of ilmenites from selected kimberlite intrusions

Locality	Yakutia (14 intrusions)		Mir, Yakutia		Zarnita, Yakutia		Sytikanskaya, Yakutia		No. 2 pipe, Ulukit, Yakutia		Monastery, S. Africa	
n	41 [a]		8 [b]		8 [c]		3		2		3	
Reference	Frantsesson (1970)		Frantsesson (1970)		Frantsesson (1970)		Frantsesson (1970)		Frantsesson (1970)		Frisch (1970)	
	A	B	A	B	A	B	A	B	A	B	A	B
TiO_2	45.58	41.40–55.09	45.83	41.40–48.40	48.40	47.10–49.52	49.17	47.02–50.75	56.10	55.09–57.12	50.5	49.3 –52.5
Al_2O_3	–	–	–	–	–	–	–	–	–	–	0.28	0.08– 0.54
Cr_2O_3	0.44	0.04– 1.74	–	–	–	–	1.48	1.33– 1.74	0.51	0.49– 0.55	0.60	0.21– 0.82
Fe_2O_3	15.80	5.75–22.68	20.27	16.72–27.68	15.65	10.0 –18.64	14.78	13.69–15.92	7.14	5.75– 8.53	10.9	7.51–13.0
FeO	25.75	22.24–29.46	26.47	25.20–28.80	27.13	25.56–30.24	24.67	23.36–26.23	22.80	22.24–23.36	28.4	27.3 –30.6
MnO	0.17	0.00– 0.35	0.15	0.10– 0.20	0.15	0.10– 0.18	0.13	0.08– 0.22	0.12	0.12– 0.13	0.08	0.06– 0.10
MgO	9.04	4.86–14.13	6.42	4.86– 7.92	8.45	5.60– 9.88	9.78	8.60–10.93	13.98	13.83–14.13	9.47	7.63–11.1
NiO	–	–	–	–	–	–	–	–	–	–	–	–

Locality	Bultfontein, S.Africa		Wesselton, S.Africa		Pofadder, S. Africa		Premier Mine, S.Africa		Artur de Paiva, Angola		Ison Creck, U.S.A.	
n	7		16		10		100		100		4	
Reference	Mitchell (1973)		Mitchell (1973)		Frick (1973a)		Gurney et al. (1973)		Boyd and Danchin (1974)		Mitchell (1973b)	
	A	B	A	B	A	B	A	B	A	B	A	B
TiO_2	51.44	49.48–51.75	52.23	43.91–55.90	52.12	46.12–55.65	–	–	46.01	38.42–48.17	53.38	53.21–54.35
Al_2O_3	0.29	0.18– 0.38	0.35	0.03– 0.81	–	–	–	–	0.66	0.29– 0.84	0.54	0.35– 0.77
Cr_2O_3	0.54	0.40– 0.92	1.12	0.23– 2.37	1.32	0.85– 1.67	1.20	0.64– 3.37	0.75	0.44– 1.01	0.83	0.06– 2.06
Fe_2O_3	8.98	6.61–11.68	9.02	5.02–17.66	11.36	5.86–20.95	–	–	16.12	13.88–30.48	7.63	6.72– 8.17
FeO	27.65	26.40–28.81	22.40	18.24–27.83	22.01	15.70–23.84	32.4 [d]	26.6 –40.1 [d]	28.86	23.19–31.04	22.47	22.16–22.75
MnO	–	–	–	–	0.47	0.17– 1.99	–	–	0.36	0.16– 0.53	–	–
MgO	10.42	8.72–12.44	13.64	6.43–19.26	10.80	7.67–13.16	16.3	9.4–23.1	6.94	3.90– 9.61	14.31	13.97–15.05
NiO	0.23	0.20– 0.27	0.27	0.22– 0.36	0.15	0.06– 0.22	–	–	0.05	–	–	–

a For MnO, n = 37; for Cr_2O_3, n = 25. b For MnO, n = 6. c For MnO, n = 6. d Total iron as FeO of the Ulukhit group have 14 wt% MgO whereas at the opposite end of the spectrum ilmenites from Mir contain only 4.8–7.9 wt% MgO (see Table 11); and in S. Africa the data in Table 11 suggest a similar MgO variation, and Boyd and Nixon (1973) suggest a high MgO content in ilmenite from Frank Smith Mine compared with those from the Monastery Mine and N. Lesotho. In addition, Mitchell (1977) has shown that ilmenite suites from the Kao, Sekhameng and Frank Smith pipes are distinct not only in their major element values, but also in their trace element contents; for example, the Frank Smith ilmenites contain less Zr, but more Cu, than ilmenites from the two other pipes. Despite these inter-pipe variations, the high MgO, intermediate Fe_2O_3 content, and the presence of Cr_2O_3 serve to give these kimberlite ilmenites a highly distinctive composition. The only other ilmenites of similar composition are reported from alnöite at Malaita, Solomon Islands (Allen and Deans 1965); ilmenite nodules occurring as high-pressure megacrysts in the melanephelinite breccia at Kakanui, New Zealand, contain lower MgO – 3.4% (Reay and Wood 1974).

The main trace elements in kimberlite ilmenites are Co (139–366 ppm), Sc (19–27 ppm), Zr (385–1950 ppm), Hf (29–40 ppm), Nb (477–2574 ppm), Ta (73–218 ppm), Ni (173–2000 ppm), Cu (1–173 ppm) and Zn (133–219 ppm) (Gurney et al. 1973; Mitchell et al. 1973b; Nixon and Kresten 1973; Fesq et al. 1975b; Mitchell 1977). In particular the kimberlite ilmenites are enriched in Zr, Hf, Cr, Nb, Ni, Co and Sc, but depleted in Mn, Cu and Zn, compared with magnesium-poor ilmenites in basic igneous rocks, though Nb and Ta values are similar to those in ilmenites from granitic or syenitic rocks. The relatively constant Zr/Hf ratios (29–40) are similar to those for ilmenites from basic igneous rocks, and are not similar to the later stages of differentiation when Zr/Hf ratios may rise to 64 (Brooks 1970). In addition, Fesq et al. (1975) report small amounts of Na (0.222% ± 0.005%), La 7.1 ppm, $\sigma \pm 6.1$), Eu (0.33 ppm $\sigma \pm 0.23$) and Lu (0.25 ppm $\sigma \pm 0.04$) in South African kimberlite ilmenites. Small amounts of lead (0.04 wt% PbO) are recorded in magnesian ilmenite (MgO 11.7, Cr_2O_3 4.14 wt%) from kimberlite at Bakwanga, Zaire (Fieremans 1967); although this is the only report of lead in kimberlite ilmenites, it should be appreciated that this element is not usually sought during analysis. In this context, it is relevant that lead-bearing ilmenite, termed "senaite", occurs in diamondiferous gravels in Brazil.

Although compositional variations or zoning within individual ilmenites are not common, Ilupin et al. (1974) have shown increased Mg, Cr, and Mn in the rims, relative to the cores, of ilmenites from the Udachnaya, Dalnaya, NIIGA, Geofizicheskaya and Yasnaya kimberlites, Yakutia; Ni, V and Zr are variable, and in some instances show relative depletion in the rims. In addition, Ilupin et al. (1974) describe lower amounts of Zr and V in microcrystalline (recrystallized) areas compared with coarsely crystalline areas of the same crystal, whereas Ni and Cr may be the same or greater in the microcrystalline areas.

Correlation matrices run by Mitchell et al. (1973b), Haggerty (1975b) and Mitchell (1977) show consistent correlations between various element pairs, as follows:

a) Linear or hyperbolic positive correlation: Ti – Fe^{2+}, Ti – Mg, Ti – Mn, Mg – Mn, Mg – Ni, Fe^{3+} – Al, total Fe – Sc, Zr – Hf, Zr – Nb, Nb – Ta, Cr – Ni and Cr – Cu.
b) Linear or hyperbolic negative correlation: Ti – Fe^{3+}, Ti – Al, Ti – Nb, Ti – Zr, Fe^{2+} – Fe^{3+}, Fe^{2+} – Al, Fe^{2+} – Cr, Fe^{2+} – Ni, Mg – Fe^{3+}, Mg – Nb, Mg – Zr, Mn – Fe^{3+}, Mn – Al, Mn – Cr, Ni – Zr, Ni – Nb, Cr – Sc.

Consistent negative correlation exists between the groups Mg, Ni vs Nb, Zr; and between the groups Ti, Fe^{2+}, Mn vs Fe^{3+}, Al. Most correlations for major and minor elements may be explained by the substitutions (a) divalent ionic: Fe^{2+} – Mg – Mn; (b) trivalent ionic: Fe^{3+} – Al – Cr; and (c) coupled: $2(R^{3+})$ – $(R^{2+} + R^{4+})$. However, in addition to the predictable linear and hyperbolic element correlations, parabolic correlations (not as yet understood) exist between the following element pairs: Ti – Cr, Mg – Cr, Al – Cr, Fe^{3+} – Cr, Mg – Fe^{2+}, and Mg – Al. It may be noted that Cr, when high, occurs in high-Mg ilmenites (such as the Ison Creek ilmenite) though, as noted above, no overall Mg – Cr positive correlation exists.

2. *Primary groundmass ilmenites* occur in angular grains 0.02–0.05 mm in size. In autoliths from Lesotho and Wesselton Mine, Danchin et al. (1975) have found that the ilmenites in the kimberlite mantle of the autoliths are considerably more magnesian (10–15 wt%) than ilmenite (?xenocrystal) that forms the nuclei of the autoliths (7.6 wt% MgO). This observation correlates well with the increased MgO content of later ilmenite in overgrowth zones around original ilmenite (see below). In addition to the isolated grains mentioned above, Mitchell (1973) has noted highly magnesian ilmenite (~ 19 wt% MgO) replacing rutile in the groundmass of the Wesselton Mine kimberlite; whilst it appears to be a groundmass reaction, it should be noted that ilmenite partially replaces rutile in some eclogite xenoliths (Frick 1973a) and fragmented eclogite could contribute ilmenite/rutile intergrowths to the kimberlite matrix. Nonetheless, the bimodal distribution of MgO content of later ilmenite in overgrowth zones around original ilmenite (see below). In addition to the isolated grains mentioned above, Mitchell (1973) has

3. *Ilmenites occurring in titanomagnetites* are attributed to sub-solidus oxidation (Haggerty 1975a); they are characterized by low MgO (1 wt%) and Cr_2O_3 (< 0.5 wt%).

4. *Secondary ilmenite* forms part of a zoned reaction mantle on xenocrystal ilmenite from the Liqhobong kimberlite, Lesotho (Haggerty 1973). The zoned mantle is (from core to rim) xenocrystal ilmenite → magnesioferrite → ulvöspinel → secondary picroilmenite → titanian-ferrian pleonaste → perovskite. The xenocrystal ilmenite is inversely zoned with 11% MgO at its core and up to 16% MgO at its rim; Fe_2O_3 is uniformly low (0.5%) from core to rim. The secondary picroilmenite, sandwiched between the two spinel zones, contains higher MgO content (up to 20%) and approximately 2% Fe_2O_3. This secondary ilmenite (plus the spinels and perovskite) is interpreted as resulting from variable and changing redox conditions of a volatile gas phase in association with metasomatic transfer.

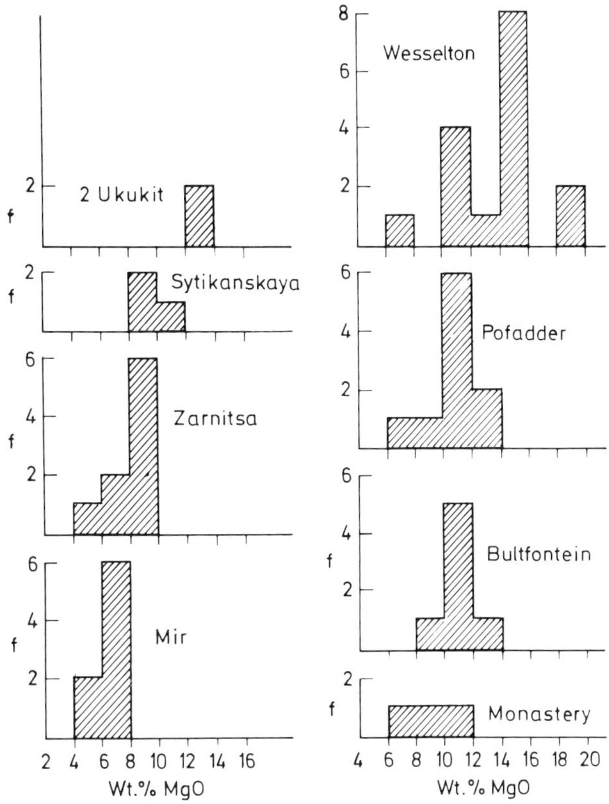

Fig. 27. Histograms of MgO in ilmenites from selected Russian and South African localities. (Data sources as in Table 11)

5. Picroilmenite also occurs *in ilmenite/rutile intergrowths* (Haggerty 1975a); although bearing some resemblance to ilmenite exsolved from TiO_2, the intergrowths are more likely to result from thermal decomposition of armalcolite ($Fe\,Mg\,Ti_2O_5 \rightarrow Fe\,Mg\,TiO_3 + TiO_2$).

6. *Ilmenites intergrown with diamond* from Yakutia have a relatively high TiO_2 content (55.0–57.8 wt%), MgO 9.6%–11.3%, and insignificant Fe_2O_3, thereby differing from most normal kimberlite ilmenites (Sobolev 1977). By contrast, an ilmenite inclusion in a Brazilian diamond is nearly pure iron ilmenite, containing only 0.16 wt% MgO (Meyer and Svizero 1975).

7. *Ilmenite* occurring as *inclusions in olivine* from the Benfontein sill is high in MgO (13.8 wt%) but not unusually high Cr_2O_3 (1.5 wt%) (Dawson and Hawthorne 1973). By contrast, an ilmenite inclusion in olivine from the Ison Creek kimberlite, Kentucky, though highly magnesian (15.74 wt% MgO), is much richer

in Cr_2O_3 (5.04 wt%) (Boyd and Nixon 1973a). It should be noted that all the Ison Creek ilmenites are highly magnesian (see Table 11) as is ilmenite in clinopyroxene-ilmenites lamellar intergrowths from this locality (Gurney et al. 1973)[2]. The *picroilmenite* included *in zircon* from the Kao Mine, Lesotho, is not unusual, containing 8.77 Wt% MgO and 0.57 wt% Cr_2O_3 (Kresten 1973c).

Alteration. In addition to features such as gross fracturing and recementation by calcite (as for example, in the Monastery Mine), there is abundant evidence that ilmenite nodules are not in equilibrium with their kimberlite host. In addition to the spinel-secondary-oerovskite mantles described above, perovskite rims are very common (Fig. 26).

The perovskite itself may be replaced by sphene (sometimes of dendritic habit) and very rarely the ilmenite may be directly replaced by sphene (Frick 1973a). Furthermore, titanomagnetite-ilmenite intergrowths are reported to be replaced by sphene + schorlomite (Haggerty 1975a). The rounded and pitted nature of the ilmenite nodules was mentioned earlier; in fine detail, the surfaces are intricately sculptured and exhibit a wide variety of micro-relief (Sergeyeva 1971).

Perovskite (Ca TiO_3) is one of the characteristic accessory minerals in kimberlite and may exceptionally be a major constituent, as in perovskite/spinel cumulate layers in the Benfontein sills (Dawson and Hawthorne 1973). It occurs (1) as discrete crystals in the groundmass; (2) as reaction rims around magnesian ilmenite (Fig. 20); or (3) as a member of a series of epitaxial zones overgrowing other oxides or silicates, as for example in the following zone sequences (from core to rim):

olivine – perovskite
olivine – perovskite – spinel (Fig. 28)
olivine – perovskite + serpentine – rutile
ilmenite – magnesian ulvospinel – perovskite
ilmenite – serpentine + rutile – perovskite + rutile
chromite – spinel – perovskite
chromite – carbonate – perovskite – rutile – (± heazlewoodite)

(see Dawson and Hawthorne 1973; Haggerty 1973, 1975a; Clarke and Mitchell 1975; Reid et al. 1975a). Compositionally there is no significant difference between discrete, reaction-rim and zone perovskites, the composition being essentially Ca TiO_3; there is limited substitution by FeO (up to 2.4%) and MgO (up to 0.22%), and MnO and Al_2O_3 are present in very small amounts. There is a limited degree of substitution of the couple Na^+ Nb^{5+} for Ca^{2+} Ti^{4+}, the Na_2O content being up to 0.88% and Nb_2O_5 to 1.45% (Grantham and Allen 1960; Mitchell

2 In Gurney et al. (1973) the clinopyroxene-ilmenite intergrowth KEN 1 is described as being from Riley County, Kentucky, U.S.A. The Riley County kimberlites are in Kansas, and Dr. J.J. Gurney (personal communication) states that the specimen is in fact from the Ison Creek kimberlite, Elliot County, Kentucky

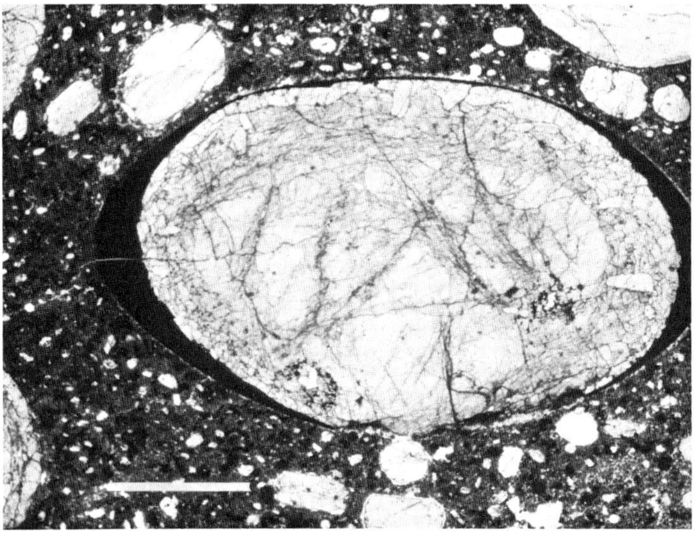

Fig. 28. Photomicrograph of an opaque rim consisting of an inner zone of perovskite and an outer zone of spinel, coating olivine in the kimberlite flow at Igwisi, Tanzania. Scale bar = 2 mm, plane-polarized light

1972; Haggerty 1973; Reid et al. 1975); although perovskite is the principal niobium-bearing mineral in kimberlite, it contains relatively little when compared with carbonatite perovskites. Perovskite is also the principal rare-earth-bearing mineral in kimberlite; the rare-earths, substituting for Ca, comprise up to 3.6 wt%, and are dominated by the light rare-earths, particularly La, Ce and Nd (Nikishov and Gamyanina 1970; Ilupin et al. 1971; Reid et al. 1975). A typical rare-earth distribution in perovskite is from the Poiskovaya pipe Yakutia (Ilupin et al. 1971), and is as follows: La 23.3, Ce 41.7, Pr 7.6, Nd 19.5, Sm 2.8, Gd 7.3, Dy 1.7, Y 5.3 (in wt% of the total rare-earth oxides).

Pyrochlore is rare, having been reported in carbonate-rich kimberlite explosion breccias in the Kuonam area, Yakutia (Frantsesson 1969); (see also "monazite", p. 90).

Iron hydroxides occur in practically all kimberlite intrusions, specially in the upper horizons of breccia-filled diatremes; they are the main colouring matter in "yellow-ground".

Quartz. Secondary quartz occurs in cracks and cavities in many kimberlites, together with secondary calcite and pyrite. Primary quartz occurs with magnetite and ankeritic siderite in immiscible globules in the kimberlite of the Benfontein sills, S. Africa (Dawson and Hawthorne 1973).

Chalcedony occurs in internal silicification zones in some kimberlites, and at the contacts with the county rocks where, in some cases, extensive silicification may occur, e.g. the so-called McKeowns breccia at the contact of the Mwadui Mine, Tanzania (Edwards and Howkins 1966).

TiO_2 Polymorphs. *Rutile* is the commonest TiO_2 polymorph, occurring in the groundmass of many kimberlites, but only rarely as individual, discrete grains. It is generally mantled by ilmenite and, more rarely, contains exsolution lamellae of ilmenite (Haggerty 1975a). Rutile is an accessory mineral in kimberlite xenoliths (being quite common in eclogite but rare in peridotite) and it is probable that fragmented xenoliths are the source of most of the rutiles, prior to their mantling with kimberlite ilmenite. In addition, Haggerty (1975a) believes the rare rutile intergrown with ilmenite to result from thermal decomposition of armalcolite. Some rutile, which may be tentatively ascribed as precipitating from the kimberlite liquid itself, forms narrow rims (with nickeliferous pyrite) on perovskites in the Peuyuk kimberlite, Somerset Island, Canada (Mitchell and Clarke 1975b). *Anatase,* together with haematite, results from ilmenite decomposition in deeply weathered kimberlite, and *brookite* has been positively identified only in concentrates (locality unspecified) (Haggerty 1973).

Rare Titanium-Rich Minerals. *Armacolite,* a pseudobrookite-like mineral with the general formula $Fe^{2+}_{0.5}\ Mg_{0.5}\ Ti_2O_5$, was first discovered in lunar samples brought back by the Apollo 11 mission. Armalcolite has now been reported in minute, discrete crystals 25–50 μm in dimension in association with rutile and magnesium-rich ilmenite in the kimberlite of the Dutoitspan Mine (Haggerty 1975a). The only other terrestrial occurrence of armalcolite is in a potassic lamprophyre from Smoky Butte, Montana (Velde 1975) which (although differing from kimberlite in being more siliceous and containing sanidine) like kimberlite does contain titanian phlogopite, titanian diopside and potassic richterite; unlike the lunar armalcolites, the Smoky Butte armalcolite contains ferric iron, and it may be noted that Haggerty (1975a) states that the presence of ferric iron cannot be unambiguously excluded from the Dutoitspan armalcolite.

Three further rare titanium-rich minerals, as yet unnamed, have been found in various kimberlites. A complex barium- and vanadium-bearing titanate (TiO_2 53.2, MgO 3.66, BaO 4.07, FeO 11.3, Cr_2O_3 18.0, V_2O_3 8.46 wt%) mantled by perovskite + ilmenite, and enclosed in phlogpite, has been identified in kimberlite from the De Beers Mine (Haggerty 1975a). A second mineral, forming mantles on primary picro-ilmenite in the Kao Mine kimberlite, falls within the intermediate region of a solid-solution series between perovskite and haematite: it ranges in composition from 50 to 75 mol% Ca TiO_3 (Haggerty 1975a). The third mineral, a hydrated manganese-rich titano-silicate occurs as 50–100 μm globular grains, together with barytes and calcite, in intergranular serpophite "pools" in a dyke kimberlite cutting the main diatreme-facies kimberlite of the Dutoitspan Mine; its composition is SiO_2, 16.4, TiO_2 29.8, Al_2O_3 0.14, FeO 15.6, MnO 12.0, MgO 17.4, total 91.8 (Fig. 29) (J.B. Dawson, unpublished data).

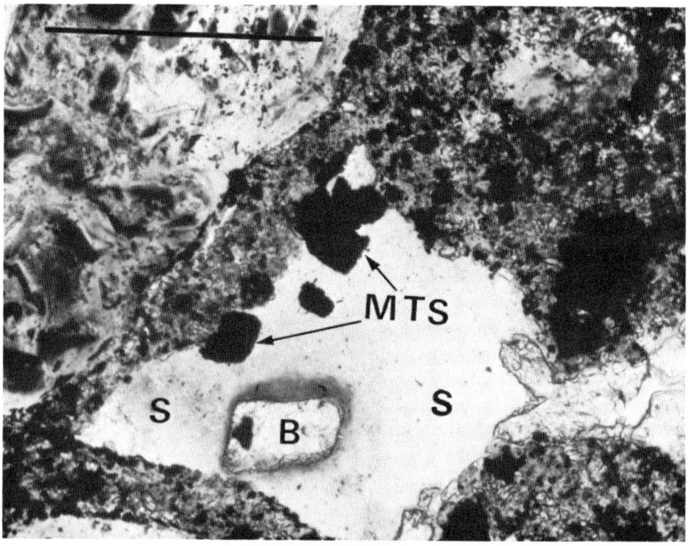

Fig. 29. Photomicrograph of an interstitial "pool" in kimberlite from the Dutoitspan kimberlite, South Africa, showing serpophite *(S)*, barytes *(B)* and an unnamed manganese-rich titanosilicate *(MTS)*. Scale bar = 1 mm, plane-polarized light

Baddeleyite has been found in two modes of occurrence in kimberlite (1) As fine-grained coatings (together with γ ZrO_2) on zircon crystals (Nixon et al. 1963; Kresten 1973a); and (2) as well-developed small (up to 0.5 mm) single crystals in the kimberlite of the Aykhal pipe, Yakutia (Marshintsev 1970b) and in rounded grains in concentrates from the Bakwanga pipes, Zaire (Fieremans 1966). Semi-quantities spectrographic analyses of Aykhal baddeleyite indicate the presence of very small amounts of Ti, Nb, Hf, Sn, Fe, Al, Mg, Mn, Cr, Ce and Sc.

Brucite Group Minerals. Despite the fact that kimberlite is a rock rich in magnesia and water, thereby containing potential brucite (especially in specimens low in silica), brucite has only recently been reported for the first time in kimberlites from the E. Udachnaya and Novinka pipes, Yakutia (Malkov 1974); it was identified by both X-ray and thermogravimetric methods. Malkov (1974) also proposes, on the basis of an excess of MgO over SiO_2 in terms of serpentine stoichiometry, that cryptocrystalline brucite is a component of a number of other Yakutian kimberlites. This suggestion appears highly plausible, particularly in view of the difficultiesof optical differentiation between brucite and the serpentine-group minerals, and it could transpire that brucite is more common in the kimberlite matrix than suspected hitherto.

A mineral of the brucite group, named amakinite, has been reported from the E. Udachnaya pipe; its general formula is $Fe(OH)_2$, with limited substitution of Mg and Mn for Fe^{2+} (Kozlov and Levshov 1962).

6.6 Carbonates

The following carbonates have been reported from kimberlite: calcite, magnesian calcite, siderite, dolomite, aragonite, strontianite and shortite; of these only calcite is an important mineral.

Calcite is present in most kimberlites, varying from a very minor to an important constituent. Prior to 1960, it was generally believed that calcite was either a late-stage deuteric phase or the product of weathering. Whilst some undoubtedly has these origins, there is a growing body of evidence that at least some calcite in certain intrusions is a primary phase. One of the best examples of this is in the layered Benfontein sills (Dawson and Hawthorne 1973) where much of the transporting liquid and intercumulus material was high-temperature calcite; textures of the calcite vary from poikilitic to columnar, sphenoid and dendritic – the last being due to liquid quenching (Figs. 17 and 30) and trachytic texture is also present (Figs. 31 and 32). Calcite may also replace serpentinised olivine. The calcite in a rock consisting purely of kimberlite calcite (the result of differentiation) in the Benfontein sill complex contains appreciable amounts of strontium, yttrium, barium, lanthanium, lead, uranium and thorium; these, plus a Th/U ratio of < 1 indicate affinities with carbonatite calcites (Dawson and Hawthorne 1973). Isotope studies on primary kimberlite calcites further emphasize their affinities with carbonatite calcite; the $^{87}Sr/^{86}Sr$ ratio of calcite from Bachelor Lake,

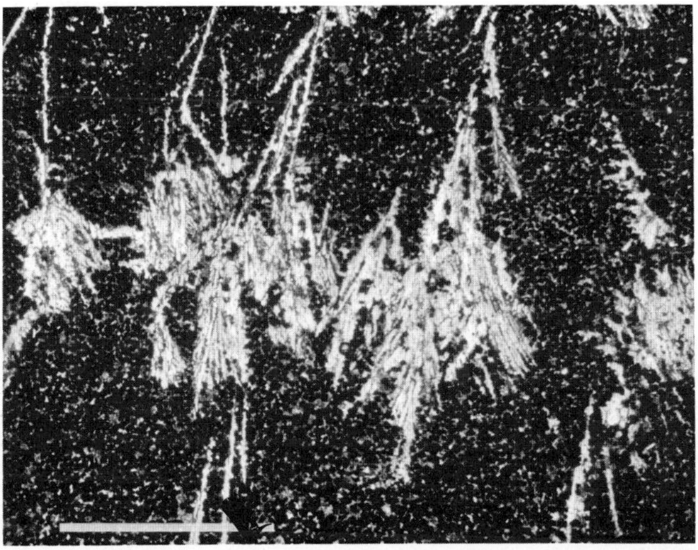

Fig. 30. Photomicrograph showing dendrites of rapidly quenched calcites in a matrix rich in spinel and perovskite in the Lower Sill, Benfontein, South Africa. Scale bar = 1 mm, plane-polarized light

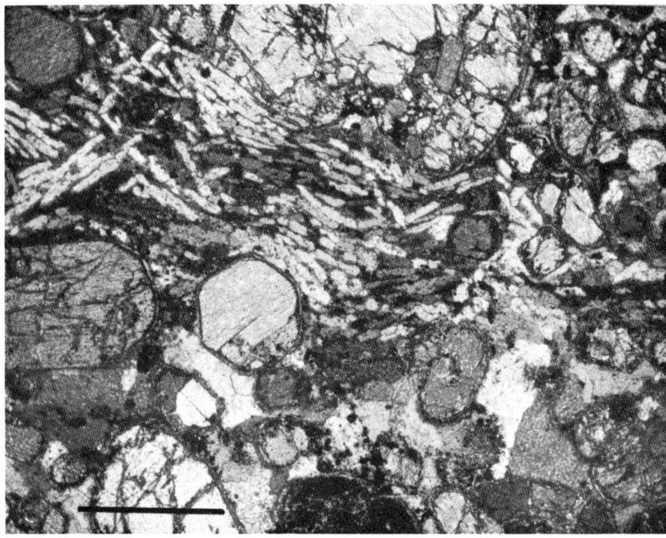

Fig. 31. Photomicrograph of trachytic texture in kimberlite of the Middle Sill, Benfontein, picket out by lath-shaped crystals of calcite; the other crystals are mainly olivine. Scale bar = 2 mm, crossed polarizers

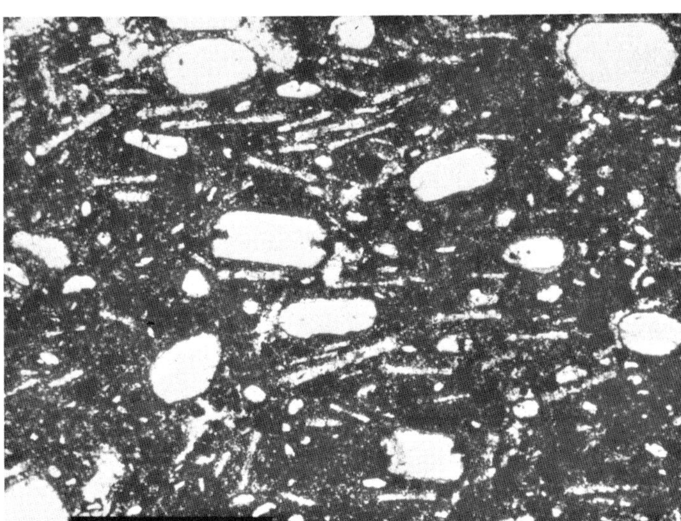

Fig. 32. Photomicrograph of trachytic texture picked out by lath-shaped crystals of calcite in the Igwisi Hills kimberlite flow, Tanzania. Large rounded crystals are of olivine. Scale bar = 2 mm, plane-polarized light

Quebec and Riley County, Kansas, vary from 0.7028–0.7046 for primary calcites; 0.7051–0.7067 for calcite replacing earlier phases; and 0.7079–0.7093 for post-emplacement veins (Brookins 1967; Brookins and Watson 1969). The strontium content of the primary calcite is high, varying from 2200–9200 ppm; in the Bala pipe, Kansas, the strontium content of primary calcite increases with increasing depth (Brookins 1967).

The carbon and oxygen isotopes for kimberlite calcites give a more complex picture. $\delta^{13}C$ vs $\delta^{18}O$ for kimberlite bulk rocks (reflecting generally that of the dominant carbonate-calcite) show a wide spread (see Fig. 13). Most lie within the $\delta^{13}C$ range for carbonatite and diamond but some lie outside this range. For $\delta^{18}O$, there is a wide spread (6‰–24‰) with only a very few in the primary carbonatite range. In the few cases where coexisting calcite/dolomite pairs have been analysed, the pairs have virtually the same $\delta^{13}C$ values, but the dolomites have higher $\delta^{18}O$ values, in the order of 2‰ (Sheppard and Dawson 1975).

As noted above, other carbonates are relatively rare, although *dolomite* often occurs replacing earlier magnesian phases. *Aragonite,* in the form of fibres, encrustations and radial stellate aggregates, is known in the Udachnaya, Malyutka, Yakutskaya and Marshrutnaya pipes in Yakutia; analytical data on a specimen from the Malyutkaya intrusion contains small amounts of MgO (2.47 wt%), SrO (0.03%) and BaO (0.12%) (Bobrievich et al. 1959, 1964). Stellate aragonite needles also occur in vesicles in the Recent kimberlite flow at the Igwisi Hills, Tanzania (J.B. Dawson, unpublished data). *Strontianite* occurs in the upper parts of the Zarnitsa and Udachnaya pipes, where it fills cracks and cavities in the kimberlite; in both cases they form overgrowths on celestite, from which they are apparently derived (by replacement?); the strontianite contains 14%–17% calcite molecule (Bobrievich et al. 1959). *Shortite* has been identified in the groundmass of a micaceous kimberlite dyke in the Upper Canada gold mine, Ontario (Watkinson and Chao 1973); it has been suggested that the presence of shortite in this particular rock may be due to the subsolid breakdown of nyrereite (Cooper and Gittins 1974).

Differences in carbon and oxygen ratios of carbonates within single intrusions or even hand-specimens have been noted by Sheppard and Dawson (1975). Differences in carbonate species within a single intrusion are well illustrated in the kimberlite sill-complex at Benfontein, S. Africa. In one unit, carbonate concentrated by differentiation in one particular layer has migrated to form rooted and detached diapirs in the overlying layer. The carbonate in the root-layer is magnesian calcite ($Ca_{99}Mg_1$) and it co-exists with a spinel of complex composition (approx. spinel 15%, ulvospinel 42%, titanospinel 13%, magnesioferrite 26% molecular); the detached diapirs are zoned with an outer rim of dolomite ($Ca_{55}Mg_{44}Fe_1$), and an inner core of ankeritic siderite ($Ca_{13}Mg_{11}Fe_{76}$) that may be intergrown with aluminous magnetite and/or quartz (Dawson and Hawthorne 1973; Dawson, unpublished data). These differences between the bulk-rock and diapiric assemblages are atrributed to liquid immiscibility, but must also reflect differences in crystallization temperatures and fO_2. In carbonatitic kimberlite dykes in the Saguenay River valley, Quebec, Gittins et al. (1975) recognize both phenocrystal and groundmass calcite; the phenocryst calcite contains higher iron and manganese, but lower strontium, than the groundmass calcite.

Strontian calcite and ferroan magnesite are included in pyrope garnets from Arizona kimberlite (McGetchin and Besancon 1973), and calcite inclusions occur in olivines in the kimberlite flow at Igwisi, Tanzania (Reid et al. 1975a). For these carbonates, at least, an upper-mantle origin is beyond dispute.

6.7 Sulphates

Sulphates occur in low-temperature phases in veins, and cavities in the kimberlite matrix. *Barytes* occurs in several of the Kimberley Mines and in the Newlands Mine, S. Africa (Wagner 1914) and in several of the Yakutian pipes, including Udachnaya, Sitikanskaya, Zagadochnaya and Dolgazdannaya (Bobrievich et al. 1964). One analysed barytes from the Zagadochnaya pipe contains CaO 2.18 wt% and SrO 1.64 wt%. In all these cases, both S. African and Russian, the barytes occurs in the upper horizons of the pipes. Nonetheless Dawson (unpublished data) has found barytes in serpophite pools in a hypabyssal-facies dyke on the 1300-foot level in the Dutoitspan Mine. *Gypsum,* like barytes, is found in the near-surface zones of weathered kimberlites, and has been found in the Mir, Udachnaya and Yanvarskaya pipe, Yakutia (Bobrievich et al. 1964). *Anhydrite* is recorded in the Mir pipe by the same authors, who also report *epsomite* in the Obnazhennaya and Udachnaya pipes. *Celstine* occurs in the upper parts of the Udachnaya and Zarnitsa pipes, varying in colour from white to azure or rose; azure celestine is fairly pure in composition (1% CaO, 0.38% BaO) whereas rose and white celestine contain less BaO but higher CaO (up to 4 wt%) (Bobrievich et al. 1964).

6.8 Phosphates

Apatite and monazite are the only phosphates so far recorded from kimberlite. Apatite is quite common, occurring in many kimberlites; it exhibits a wide range of morphology. Most are stumpy, well-rounded prismatic crystals but Wagner (1914, Fig. 2) illustrates idiomorphic skeletal crystals, and skeletal crystals are also reported from the kimberlite flow at Igwisi, Tanzania (Reid et al. 1975a). Acicular apatite, in rosette or sheaf-like aggregates, interpreted as quench products, are recorded from the Benfontein sills kimberlite (Dawson and Hawthorne 1973), the Kao diatreme (Clement 1975) and dykes in the Saguenay valley, Quebec (Gittins et al. 1975); Wagner (1914, p. 70) reports that "acicular crystals . . . may sometines be seen to be moulded with their base upon the surface of an olivine phenocryst".

Few analyses of kimberlite apatite are available; however analyses of two from intrusions in the Lena area, Yakutia, show them to be fluor-hydroxyl-apatites containing appreciable amounts of rare-earth oxides (~ 0.5wt%) and smaller amounts of SrO (0.23 wt%) (Khomyakov and Frantsesson 1971); in these respects they resemble carbonatite apatites. The rare-earths are dominated by Ce and Nd (see below).

Monazite is rare, being reported in kimberlite explosion breccias in Kuonam area, Yakutia; these breccias have a high calcite content and, as they contain baddelyite, columbite, monazite, pyrochlore and fenitised rock fragments, are regarded as transitional to carbonatites (Frantsesson 1969). The composition of the monazites is dominated by the light rare earths, though the element distribution (see below) differs from that in kimberlite apatites; the examples are taken from Ilupin et al. (1971):

	TR_2O_3 wt%	La	Ce	Pr	Nd	Sm	Eu	Gd	Tb	Dy	Er	Yb	Y
Apatite, Marichka (Lena)	1.23	* 18	41	6	22	4	1	3	1	1	1	1	10
Monazite, Kuonam	–	43	47	2	7	1	–	–	–	–	–	–	–

* Weight percentage of total rare-earth oxides.

6.9 Silicates

6.9.1 Nesosilicates

Olivine Group

1. *Magnesian olivine* together with its alteration products is the commonest constituent of kimberlite, with the exception of some micaceous kimberlites where phlogopite is more abundant. It commonly occurs as large rounded megacrysts, usually in the size range 1–10 mm though rare crystals up to 15 cm are recorded (Boyd 1974); a smaller, less-obvious generation of euhedral crystals occurs in the groundmass. The olivine may be fresh, when it has a lime-green colour and vitreous lustre, or partially or wholly replaced by serpentine-group minerals with attendant exsolved magnetite that give the mineral a dark brown or black colour. Even the freshest grains are separated from the matrix by a thin rim of serpentine-group minerals which in some instances is overgrown by hydrophologpite (Dawson 1962). Also in the Benfontein sills and the Igwisi kimberlite flow some well-rounded and recrystallized megacryst are overgrown by thin layers of perovskite or perovskite + spinel (Fig. 28) (Dawson and Hawthorne 1973; Reid et al. 1975a).

Morphologically, two types of olivine may be distinguished – (a) anhedral, rounded grains, spanning a wide grain-size spectrum; and (b) euhedral or subhedral crystals less than 2 mm in size. In the case of the larger anhedral grains, some of green colour can actually be seen in the immediate vicinity of fragmenting peridotite xenoliths, so derivation from this source seems to be beyond doubt in these instances. The composition of these (Fo_{89-94}; NiO 0.3–0.5 wt%; and Ca,Cr, Ti and Al all < 0.1 wt%) is in the same range as most peridotitic olivines. A similar origin may be attributed to isolated megacrysts that show strain features such as kinking, undulose extinction or marginal recrystallization or are intergrown with (or contain inclusions of) enstatite, pyrope, diopside, chromite or phlogopite. Both Mitchell (1973b) and Boyd (1974) have estimated that the breakup of an inordinately large number of ultramafic xenoliths would have been necessary to contribute all the large megacrysts in kimberlite, and both authors suggest that many of the rounded forsteritic megacrysts are phenocrysts that have undergone attrition. Other large megacrysts (up to 15 cm) are dark brown in colour and range in composition from Fo_{80}–Fo_{88}, and NiO decreases with increasing FeO content (Boyd and Nixon 1973a; Mitchell 1973b; Boyd 1974; Emeleus and Andrews 1975); one example from the Monastery Mine (cited by Boyd 1974) is intergrown with high-titanium pyrope.

Some of the smaller euhedral crystals may attain almost perfect morphology. They may contain small inclusions of ilmenite, rutile or spinel (Fig. 33), usually concentrated in the periphery, that may exhibit perfect zonal arrangement parallel to the crystal faces (Wagner 1914; Dawson and Hawthorne 1973). Marginal zoning

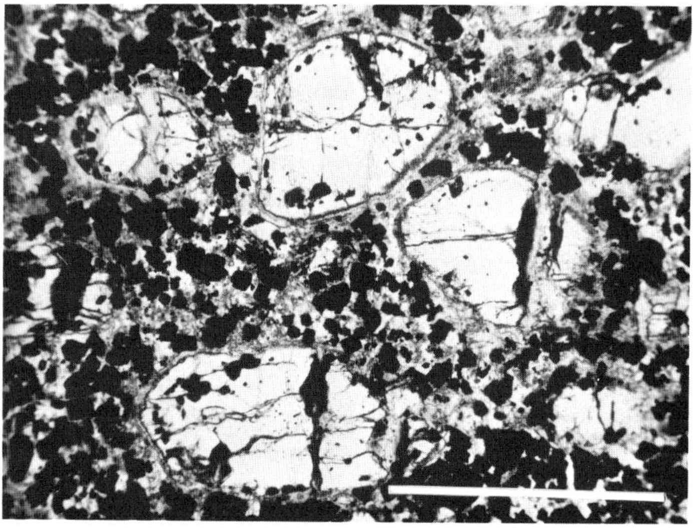

Fig. 33. Photomicrograph showing groundmass olivines in the matrix of the Benfontein kimberlite sill; note inclusions of groundmass opaques paralleling the crystal margins. The matrix is mainly spinel, perovskite and calcite. Scale bar = 1 mm, plane-polarized light

is recorded in euhedral matrix olivines in kimberlite from Wesselton, Benfontein, Ison Creek (Kentucky) and S.W. Greenland when the compositional changes are confined to the extreme margins. The zoning is in most cases towards more iron-rich rims but there are examples of more magnesian rims; the major-element zoning is, however, very limited, usually causing less than 1% Fo variation between core and rim, though there are rare crystas showing 4% variation. Minor-element variations are more apparent, with NiO decreasing sharply from core to rim often with a complementary rise in CaO; nontheless in some crystals NiO depletion in the rims may coincide with increased MgO content (Dawson and Hawthorne 1973; Emeleus and Andrews 1975). Accumulations of euhedral olivines and small composite nodules of euhedral olivine occur in the Benfontein sills (Dawson and Hawthorne 1973). Furthermore, it should be appreciated that many of the smaller anhedral olivine crystals may possibly have been euhedral originally but became rounded during transport and intrusion; in this context it is pertinent to note that the thin, compositionally different margins of the euhedral crystals would be obliterated by only a very small degree of peripheral destruction.

Whereas most of the small euhedral olivines noted above are in the compositional range Fo_{89}–Fo_{94}, Boyd (1974) records more iron-rich (Fo_{86}) examples in the kimberlite of the Monastery Mine where megacrysts of the same composition are present; zoning is not reported.

In addition to inclusions of silicate minerals, other more unusual inclusions have been recorded. Ilmenite has been noted in olivines from Benfontein (Dawson and Hawthorne 1973) and from Isonville, Kentucky (Boyd and Nixon 1973a), the latter being unusually high in Cr_2O_3 (5.04 wt%). Roedder (1965) records liquid CO_2 inclusions from kimberlite olivines, and Haggerty (1975a) reports Ni-Fe alloy inclusions in olivine from the Ramatseliso kimberlite, Lesotho. Olivines from the Udachnaya pipe, Yakutia, contain inclusions of chrome-spinel, knorringitic pyrope-almandine, ortho- und clino-pyroxene, pyrrhotite, pentlandite and chalcopyrite (Lazko and Afanasev 1974), and calcite, pyrope, pyroxene, chromite and phlogopite are included in olivine megacrysts in the kimberlitic lava flow at Igwisi, Tanzania (Reid et al. 1975a). Small spheroidal inclusions of kimberlite, and sulphide droplets occur within a large olivine megacryst in kimberlite from the Monastery Mine (Haggerty and Boyd 1975). Phlogopite in the host kimberlite contains less TiO_2 (3.6–6.3 wt%) than mica in included kimberlite (0.9–4.8 wt%) and ilmenites in the included kimberlite are richter in MgO than those in the host kimberlite; spinels in the host kimberlite are Mg-Cr-titanomagnetites compared with Mg-titanomagnetites in the inclusions. The sulphide inclusions are pyrrhotite-pentlandite-chalcopyrite. With reference to chromite crystals observed in some olivines, Burns (1975) suggests that these may be due to Cr^{2+} ions originally held within the olivine lattice under extreme pressures, and the chromite now present in the olivine may be due to pressure-release oxidation.

2. *Monticellite* was formerly regarded as a rare mineral in kimberlite, but recent studies of the groundmass of some kimberlites suggest that it is more

common than previously suspected. It was tentatively identified, on optical grounds, in several Lesotho kimberlites (Dawson 1962) and it forms 80% of the groundmass of the "Monticellite" kimberlite body, Yakutia, where it also forms reaction rims on forsterite megacrysts; at this locality the monticellite contains 20%–30% of the ferrimonticellite molecule (Nikishov and Nikishova 1966). Monticellite rims on forsterite are also reported from the Gwena kimberlite, Zaire, where its presence is attributed to reaction between forsterite and the calcic groundmass of the kimberlite (Verhoogen 1938). It also occurs as a primary groundmass phase in an olivine-phyric "peridotite" that forms a small intrusive on the flank of the Gross Brukkaros kimberlite/carbonatite volcano, S.W. Africa (Janse 1971), and in the kimberlitic diatremes of the Bearpaw Mountains, Montana (Hearn 1968). Monticellite (confirmed by probe analysis) comprises 50% of the groundmass of the kimberlite of the younger intrusive stages in the De Beers Mine, and it has also been identified optically in the groundmass of kimberlites from Wesselton, Dutoitspan, Bultfontein, Jagersfontein and Matsoku (Lesotho) (Clement et al., in press).

3. *Titan-olivine,* reported from the Ruslovaya pipe, Yakutia (Voskresenskaya et al. 1965) has been re-investigated and proved to be titan-hydroclinohumite (Shchelkova and Brovkin 1969).

Garnet

Garnet is one of the commoner minerals in kimberlite and the wine-red-purple pyrope is regarded as one of the more diagnostic tracer minerals used in prospecting for kimberlite. Whithin any kimberlite intrusion the amount of garnet varies widely, as does the size of the individual grains, and their colour. The size of wine-red-purple garnets are in the range 1–3 mm, wheras red-brown garnets may be up to the size of a large grapefruit. Besides red, brown, yellow and purple garnets, rare green garnets are known from some intrusions. The garnets occur as rounded, isolated megacrysts within the kimberlite matrix from which they are separated by a rind of "kelyphite" that has developed at the expense of the garnet (Fig. 34); in some cases the kelyphite – a turbid, microcrystalline aggregate of spinel, phlogopite, chlorite, amphibole and ?plagioclase – may completely replace the garnet. During this alteration the garnet is distinctively etched (Fig. 35).

On the basis of variation in the CaO, MgO, FeO, TiO_2 and Cr_2O_3 contents, Dawson and Stephens (1975, 1976) have recognized twelve chemically coherent groups of garnet in kimberlites, in the ultramafic and mafic xenoliths, and occurring as inclusions in diamonds. The means of distinguishing oxides for the various groups are shown in Table 12. A brief description of the garnets comprising the various groups is as follows: *Group 1* garnets are red-brown titanian pyropes. Those occurring in deformed (flaser-textured or "sheared") lherzolites are relatively small, whereas isolated megacrysts from kimberlite may be up to 5 cm. Megacrysts of this composition are known to form nodular intergrowths with magnesian ilmenite and smaller crystals occur as inclusions within magnesian ilmenite nodules.

Fig. 34. Photomicrograph of garnet with a reaction rim of "kelyphite" in the kimberlite of the Frank Smith Mine, South Africa; the kelyphite comprises a micro-crystalline, optically irresolvable aggregate of phyllosilicates and spinels. Scale bar = 2 mm, plane-polarized light

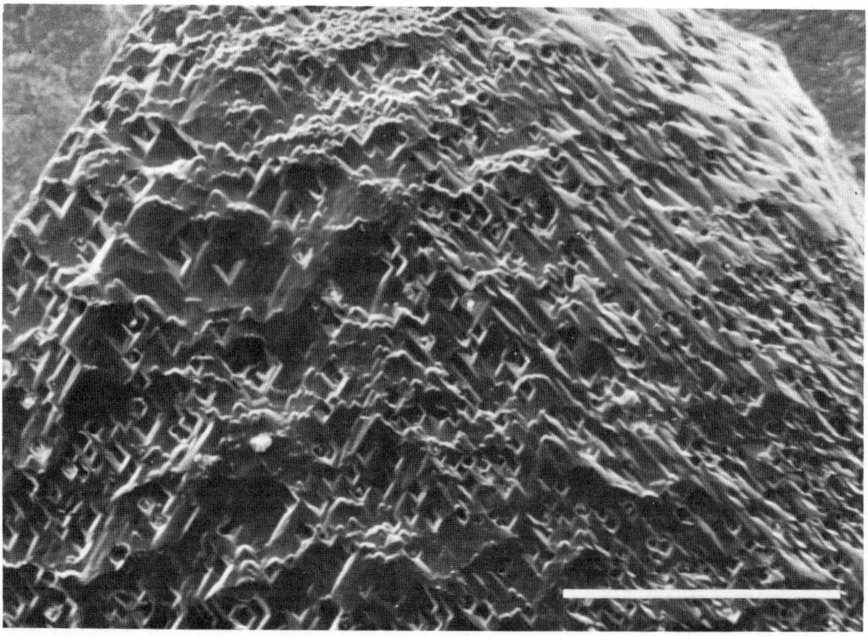

Fig. 35. Scanning electron microphotograph of etch pits on the surface of garnet from which the kelyphite has been removed; from the Excelsior Mine, South Africa. Scale bar = 500 μ

Table 12. Means of distinguishing oxides and occurrence of various types of garnets in groups of garnets recognized by Dawson and Stephens (1975)

Group	Name	TiO_2	Cr_2O_3	FeO	MgO	CaO	Occurrence [a]
1	Titanian pyrope	0.58	1.34	9.32	20.00	4.82	K, GL, GOW, D
2	High-titanium pyrope	1.09	0.91	9.84	20.30	4.52	K
3	Calcic pyrope-almandine	0.31	0.30	16.49	13.35	6.51	K, GL, GOW, EC, D
4	Titanian, calcic, magnesian almandine	0.90	0.08	17.88	9.87	9.41	K, EC, D
5	Magnesian almandine	0.05	0.03	28.33	7.83	2.44	K, EC, D
6	Pyrope-grossular almandine	0.24	0.27	10.77	10.38	14.87	GP, EC, GR
7	Ferro-magnesian uvarovite-grossular	0.29	11.52	5.25	8.61	21.60	K, GS
8	Ferro-magnesian grossular	0.25	0.04	6.91	4.69	24.77	GR
9	Chrome-pyrope	0.17	3.47	8.01	20.01	5.17	K, GL, GOW, GH, EC, D
10	Low-calcium chrome-pyrope	0.04	7.73	6.11	23.16	2.13	K, GS, D
11	Uvarovite-pyrope	0.51	9.55	7.54	15.89	10.27	K, GL, GWH, D
12	Knorringitic uvarovite-pyrope	0.18	15.94	7.47	15.40	9.51	K, GS

a Abbreviations: K = kimberlite, GL = garnet lherzolite, GH = garnet harzburgite, GD = garnet dunite, GW = garnet wehrlite, GS = garnet serpentinite, GOW = garnet-olivine websterite, GP = garnet pyroxenite, GWH = garnet wehrlite, GR = grospydite, EC = eclogite, D = Diamond inclusion

Group 2 garnets are megacrysts of high titanium pyrope from kimberlite. They contain more TiO_2 than Group 1 garnets but there is a gradation between them. Garnets of this composition are of widespread occurrence. Some garnets of this group may be intergrown with magnesian ilmenite. *Group 3* garnets are calcic pyrope-almandines of brown colour. The group is dominated by garnets from both common and diamondiferous eclogites, though there are cases from kimberlite and deformed peridotites. *Group 4* comprises brown garnets rich in FeO (17.9 wt%) and TiO_2 (0.90 wt%); they occur as isolated crystals in kimberlite and in eclogites. *Group 5* garnets are magnesian almandines (28.3 wt% FeO); the group contains both kimberlite and eclogite garnets from the kimberlites of the Colorado Plateau. *Group 6* are moderate-high lime garnets, being dominated by cases from peraluminous eclogites (containing kyanite and/or corundum); no kimberlite garnets occur in this group. *Group 7* garnets are rare green ferro-magnesian grossular-uvarovites. Most reported cases in the group occur in Russian kimberlite, but they have also been reported from the Newlands Mine, S. Africa (Lawless 1974; Clarke and Carswell 1977); one case is known from a xenolith of garnet-bearing serpentinite (altered dunite) from Russia. *Group 8* garnets are all from grospydites; they are yellow-brown in colour and are ferro-magnesian grossular (CaO 24.8%). This type of garnet has not, as yet, been recorded from kimberlite. *Group 9* consists of wine-red chrome-pyrope garnets containing negligible TiO_2 but moderate amounts of Cr_2O_3. They are of widespread occurrence in kimberlite, and in the common type of garnet lherzolite xenoliths from many localities. *Group 10* consists of lilac- or violet-coloured low-calcium chrome pyropes; compared with Group 9 garnets they contain more MgO, and Cr_2O_3 and exceptionally low CaO (2.1 wt%). The group is dominated by garnets from some kimberlites (e.g. Finsch Mine) and these types of garnet are also common as inclusions in diamond. Only two cases from xenoliths are known to occur in this group. *Group 11* garnets are lilac or green unvarovite-pyropes; although containing moderate amounts of MgO (15.6%), they also contain high amounts of CaO (10.3%), Cr_2O_3 (9.5%) and TiO_2 (0.5%). These types of garnet occur in kimberlites, and in peridotitic xenoliths (some with flaser or sheared texture) that have apparently equilibrated at high temperature. *Group 12* garnets are knorringitic uvarovite-pyropes that are similar compositionally to Group 11 garnets but by comparison TiO_2 is low (0.18%) and Cr_2O_3 is very high (15.9%). The group includes the garnet on which the garnet end-member knorringite ($Mg_3Cr_2Si_3O_{12}$) was established (Nixon and Hornung 1968), and is dominated by greenish-violet or purple garnet from kimberlite; only one garnet from a xenolith (a diamondiferous garnet serpentinite from Yakutia) is known in the group.

Of these twelve groups, nine contain garnets occurring both as individual crystals in kimberlite and also in the various types of xenolith; two of the groups contain only garnets from xenoliths and one consists purely of kimberlite garnets. It could be argued that the kimberlite garnets in the nine kimberlite + xenolith groups have been derived from fragmented xenoliths containing garnets of the appropriate composition; however in three of these groups there is some justification for caution in this point. In the case of low-calcium chrome-pyropes (Group 10), occurrences of isolated crystals in kimberlite are reasonably common, whereas xenoliths containing garnet of this composition are very rare; also the possibility that these rare xenoliths are cumulates cannot be discounted. The same also applies to knorringitic pyrope-uvarovites (Dawson-Stephens Group 12). The third "ambiguous" group (Dawson-Stephens Group 1) contains titanian pyropes from deformed garnet lherzolites from Lesotho kimberlite intrusions, together with large garnet megacrysts from kimberlite; the size of the megacrysts (1–3 cm) indicates that they are unlikely to have been derived from fragmented xenoliths. In their large size, they resemble the high-titanium pyropes comprising Dawson and Stephen's Group 2 which consists only of kimberlite garnets. The composition of the high-titanium pyropes resembles that of pyrope megacrysts occurring in basanitic breccias in E. Scotland and in mela-nephelinite breccia at Kakanui, New Zealand, and they are tentatively believed to be high-pressure phenocrysts.

Garnet of similar composition to Group 2 garnets has been synthesized at 1500°C, 30 kb, "dry", from pyrolite-less-40%-olivine composition; and titanian and high-titanium garnets (compositionally similar to those in Groups 1 and 2) coexist with magnesian ilmenite in experiments with the same material but at lower temperatures (950°–1100°C) and in charges containing 0.3% H_2O (Green and Sobolev 1975).

In summary, large titanian pyrope megacrysts (of Groups 1 and 2) are possibly high-pressure phenocrysts associated with the kimberlite magmatism. Two other garnet types (those comprising Groups 10 and 12) are ambiguous in that, although their main occurrence reported to date is in kimberlite, rare ultramafic nodules containing garnets of this composition are known; whether these nodules are truly xenolithic or are cognate cumulates is not known, though the fact that Group 10 garnets are of reasonably common occurrence as inclusion in diamond indicates some link with the kimberlite event.

Most kimberlite garnets analysed to date are apparently of uniform composition; or, to be more accurate, compositional variations have not been reported. However, Sobolev et al. (1972) report a zoned garnet from the Mir kimberlite that has a violet iron-rich chrome-pyrope core and a grey-green uvarovite-pyrope rim that is enriched in calcium and titanium relative to the core; inclusions of titanium-rich magnesian chrome-spinels are scarce in the core but abundant in the rim.

A variety of co-genetic inclusions is recorded in kimberlite garnets – olivine, pyroxene, chromite, magnesian ilmenite, and phlogopite. Rarer inclusions are of diamond, titano-clinohumite, amphibole and carbonate. The clinohumute and amphibole inclusions occur in pyropes in the Moses Rock and Cane Valley kimberlites, Arizona, respectively (McGetchin and Silver 1970; McGetchin et al. 1973). Carbonates in pyrope from the Garnet Ridge and Red Mesa kimberlites, Arizona, vary from strontian calcite to ferroan magnesite (McGetchin and Besancon 1973).

Rare low-temperature hydrogarnet is recorded in metamorphosed limestone xenoliths from the "Chomur" pipe, Yakutia (Bobrievich et al. 1964), and schorlomite mantles rutile/ilmenite intergrwoths in the Monastery Mine kimberlite (Haggerty 1975a).

Trace elements. As yet there are very few data available on the trace-element content of garnets in kimberlite. Data by Nixon et al. (1963) and Gurney et al. (1973) show all types of garnet contain small but variable amounts of Ni, Sc, Co, V, Zr, Na and rare earths; bearing in mind the *caveat* of very limited data, there is some suggestion that the large discrete nodules of titanium-rich pyrope contain higher Zr and rare-earths than purple grains of chrome pyrope (? derived from garnet lherzolite).

Zircon usually occurs in rounded grains up to 1 cm in length, though rarer crystals up to 2.5 cm and rare euhedral crystals are known. The zircon crystals fluoresce in UV light and, like diamond, in X-rays; in fact, zircons are nowadays recovered together with diamond during the X-ray sorting of diamond concentrates. The colour varies from white to yellow or honey brown; unlike zircon from other sources kimberlite zircons show well-developed cleavage and are often covered with a white coating.

Chemically they are very pure, the ZrO_2 content being 65 ± 1 wt% (Schutte 1966; Lupin and Kozlov 1970; Kresten 1973c; Kresten et al. 1975); the other principal oxide is HfO_2 which varies from virtually zero to > 3 wt%. This gives rise to widely varying Zr/Hf ratios (17–81), though Kresten et al. (1975) have shown that for any given diatreme the range of Zr/Hf ratios in zircon appears to be of limited range and also characteristic of the intrusion; as examples they cite Kao (Lesotho) 17–24; Mothae (Lesotho) 31–40; Nzega (Tanzania) 45–57. However, zircons from some dyke intrusions show wider ranges. Minor elements detected in gravimetric analyses of zircons (that may include impurities) are Al_2O_3, Fe_2O_3, BeO, MgO, CaO, U, Th and rare earth oxides. Schutte (1966) reports that kimberlite zircons contain more MgO but less Y_2O_3 and U_3O_8 than zircons from S. African granitic rocks. Uranium values of 7.28 ppm and Th contents of 1.8–7.2 give low Th/U ratios of 0.23–0.70, which are in the same range as zircons from granites (Ahrens et al. 1967). Total rare-earths are very low, < 0.003 wt%, and are dominated by the heavy rare-earth elements (Ilupin et al. 1971).

The white coating on the zircons, mentioned above, has been identified as both monoclinic (γ) and tetragonal (β) ZrO_2 (Kresten 1973a). Thus it is an admixture of both γZrO_2 and baddeleyite, not pure baddeleyite as reported by Nixon et al. (1963), and could have formed, in part, at temperatures as low as 500–800°C, in contrast to the temperatures of > 1150°C required to form baddeleyite alone.

Kimberlite zircons are generally very pure but inclusions of chromiferous diopside, phlogopite and ilmenite are recorded (Kharkiv et al. 1972; Kresten 1973c).

Kyanite has been found in kimberlite concentrates in S. Africa and Russia; it is believed to be derived from fragmented kyanite eclogites.

Humite Group

Titano-clinohumite containing in the order of 5 wt% TiO_2 occurs as isolated grains, in intergrwoths with olivine (Fo_{90-93}) and as inclusions in pyrope garnet in the kimberlites of the Moses Rock dyke, Utah (McGetchin et al. 1970); one pyrope grain contains inclusions of olivine, ? mica, titanoclinohumite, and picro-ilmenite. Titanoclinohumite is also recorded as isolated grains in the Buell Park kimberlite, Arzizona (Aoki et al. 1976), and in the Ruskovaya pipe, Yakutia (Shchelkova and Brovkin 1969).

Datolite is reported in a "meimechite" sample from the Obnazhennaya pipe, Yakutia (Cherepanov 1967).

Spurrite-Afwillite Group

Thaumasite is recorded as partially or wholly replacing xenoliths of limestone in the Novinka and Komsomolskaya pipes, Yakutia (Bobrievich et al. 1964, p. 146). *Afwillite* occurs with calcite and bultfonteinite infilling cavities in dolerite and shale inclusions in the Dutoitspan Mine kimberlite, S. Africa (Parry and Wright 1925). *Bultfonteinite* has the same mode of occurrence as afwillite; it has been reported from the Bultfontein, Dutoitspan and Jagersfontein kimberlite pipes (Parry et al. 1932). In the original paper on afwillite (Parry and Wright 1925), the associated bultfonteinite was incorrectly identified as natrolite.

Titanite (sphene) occurs rarely in kimberlite since its place is taken by its silica-undersaturated equivalent, perovskite. It has been recorded in the Ukukitskaya and Chomur pipes, Yakutia (Milashev et al. 1963), and has also been reported as an alteration product of ilmenite.

6.9.2 Sorosilicates

Melilite group. Although melilite occurs in the olivine melilitites that are often found in the same areas as kimberlite, there is, as yet, no authenticated record of a melilite-group mineral in kimberlite. Wagner (1914) states that, in some reported occurrences, apatite may have been misidentified as melilite. In many instances the former presence of melilite has been proposed on the basis of the presence of tabular calcite laths, alleged to be pseudomorphs after melilite (see discussion by Dawson and Hawthorne 1973). Experimenta work on the system akermanite $-CO_2$ indicates that under high partial pressures of CO_2, akermanite would be unstable, calcite + diopside being preferred (Yoder 1975).

Epidote group minerals are rare in kimberlite. Epidote occurs in one of the Zaire pipes (Fieremans 1967), and clinozoisite in the New Elands pipe, S. Africa (Dawson 1968).

Prehnite is stated as occurring as a deuteric phase in S. African kimberlites (Williams 1932), but no details of localities are given.

6.9.3 Inosilicates

Pyroxene is one of the less common megacryst minerals in kimberlite, though emerald-green chromiferous diopside is regarded as an important tracer mineral in kimberlite prospecting. The amount, size and colour of pyroxene crystals and cleavage fragments vary widely between individual intrusions. Some dark, bottle-green diopsides and dark-brown orthopyroxenes may be up to 15 cm in size, whereas most of the conspicuous chromiferous diopsides are not greater than 2–3 mm. Crystals are generally rounded and separated from the host kimberlite by a microcrystalline reaction rim, and many are partially replaced along cleavage planes by hydrous phases (amphibole, serpentine-group phases, phlogopite) and carbonate.

In the course of statistical comparison between the chemistry of pyroxenes in kimberlite and associated xenoliths, with a view to testing the assumption that some kimberlite pyroxenes derive from fragmented xenoliths, Stephens and Dawson (1977) distinguished five chemically distinct orthopyroxene and ten distinct clinopyroxene groups from the TiO_2, Al_2O_3, Cr_2O_3, FeO, MgO, CaO and Na_2O contents. The means of oxides in these groups is given in Table 13. There is some colour difference between the various types. For example the enstatites (Groups 1–3) are green-yellow, whereas the titanian enstatite (Group 4) and the high-Ti bronzite (Group 5) are brown, the last particularly so. The clinopyroxenes are various shades of green, most of them non-distinctive; however the sub-calcic diopsides (Group 1) are grey-green, the low-Cr diopsides (Group 4) are a dark bottle-green, and the chrome-diopsides (Group 5) are emerald green.

On the basis of chemical composition, pyroxenes from both kimberlite and xenoliths are common to all five orthopyroxene groups and to six of the ten clinopyroxene groups; hence, on composition alone, it could be argued that the kimberlite pyroxenes are xenolith-derived. However, when size is considered as a basis for comparison, it is apparent chemical criteria must be treated with caution when used in isolation from other parameters. For example, brown orthopyroxene megacrysts from the Letseng kimberlite, Lesotho, fall compositionally into orthopyroxene Groups 3, 4 and 5. However, their size (up to 4.5 cm) is much larger than those of comparable composition occurring in xenoliths. Alternatively their composition gradation could be interpreted as due to their being phenocrysts from a fractionating magma whose systematic change is reflected in changing Mg/Mg+Fe ratios and TiO_2 and Cr_2O_3 contents of the precipitating pyroxenes. This interpretation has been postulated by Gurney et al. (1979) for orthopyroxene megacrysts from the Monastery Mine which show consistent gradational compositions. Of particular interest are megacrysts of high-Ti bronzite (Group 5) that occur in irregular intergrwoths with ilmenite; these have not been recorded as yet from xenoliths. Similarly amongst the clinopyroxenes, there are large megacrysts (the so-called discrete nodules – Nixon and Boyd 1973b) of sub-calcic diopside (Group 1) and low-Cr diopsides (Group 4) megacrysts (frequently intergrown with lamellar ilmenite) that, on the basis of size, could not derive from any xenoliths known at present (see Chap. 7). Whereas the sub-calcic diopside megacrysts have been interpreted as high-pressure phenocrysts, the source of the pyroxene intergrown with ilmenite is obscure (Gurney et al. 1979). However, they form a not uncommon paragenesis in kimberlite, having been found in twenty kimberlites (five in S. Africa, four in Lesotho, two in both Zimbabwe and S.W. Africa, three each in U.S.A and U.S.S.R., and one in Angola). On experimentally-calibrated Mg and Fe^{2+} partitioning between pyroxene and intergrown ilmenite, Bishop (1977) estimates that the orthopyroxene/ilmenite intergrowths equilibrate at temperatures of 1100°–1200°C (at 50 kb), and the low-Cr diopside/ilmenite intergrowths mainly equilibrate in the range 1300°–1400°C; combined with their size, the temperatures of the intergrowths would suggest that they are not derived by breakup of normal xenoliths, though Rawlinson and Dawson (1979) suggest that they are derived from fragmented upper-mantle pegmatites.

In addition to the megacrysts, diopside that is an undoubted primary phase crystallizing from kimberlite is found as minute (< 1 mm) crystals in the matrix of several micaceous S. African kimberlites (Dawson et al. 1977); they have distinct pyroxene morphology (unlike megacrysts) (Fig. 36) and sometimes form overgrowths on xenocrystal clinopyroxene that was used as a nucleation site. Compared with xenolith-derived diopsides, the groundmass diopsides contain relatively low Cr, Al and Na, and have very little enstatite in solid solution (Table 13).

On the basis of present data it appears that cooling kimberlite magma may crystallize high-pressure phenocrysts of enstatite/bronzite, subcalcic diopside, and groundmass diopside. The relationship to kimberlite of the high-ureyite diopsides (Group 7), that are intergrown with diamond, still requires clarification, as does that of the large apple-green Group 2 diopside megacrysts (up to 5 cm) that occur in the Bultfontein, Wesselton and Kamfersdam kimberlites.

Within some kimberlites there are orthopyroxene megacrysts that are veined with chrome-pyrope and/or chrome-diopside. In some intrusions there also occur quite large (up to 5 cm) apparently randomly oriented aggregates of orthopyroxene, the "aggregates" being cemented by chrome-pyrope and/or chrome-diopside; Boyd (1974) has termed aggregates of this type from the Frank Smith Mine "exsolved pyroxenites", but it is the authors opinion that these "aggregates" are, in fact, deformed crystals derived from lherzolite xenoliths of inequigranular texture; such peridotites contain large crystals of orthopyroxene that have been broken up by numerous micro-faults (often with rotated segments being cemented by exsolved garnet and diopside) and which are set in a finer-grained matrix of olivine, diopside and garnet; lherzolites of this texture are admittedly rare but are known from the Monastery Mine (Dawson and Smith 1973), Wesselton and Bultfontein (Dawson et al. 1980).

Amphiboles. Amphiboles are rare in kimberlite. A pale-green actinolite is recorded in the Jagersfontein and Dutoitspan Mines (Wagner 1914) and tremolite (? of retrograde origin) occurs in the "metakimberlites" of the Côte d'Ivoire and

Table 13. Means of oxides in groups of pyroxene megacrysts in kimberlite and in associated xenoliths (Stephens and Dawson 1977)

	TiO_2	Al_2O_3	Cr_2O_3	FeO	MgO	CaO	Na_2O	Source [a]
Orthopyroxenes								
Group								
1 Enstatite	0.02	0.82	0.29	4.52	36.2	0.41	0.08	K, P, D
2 Cr-Al enstatite	0.02	2.83	0.65	4.90	35.2	0.60	0.03	K, P
3 Na-Ca enstatite	0.14	1.45	0.32	5.31	33.8	1.42	0.43	K, P
4 Ti-enstatite	0.14	0.93	0.25	5.98	34.5	0.71	0.17	K, P
5 High-Ti bronzite	0.19	1.06	0.10	8.64	32.1	1.07	0.22	K, P, D
Clinopyroxenes								
Group								
1 Sub-calcic diopside	0.31	2.51	0.43	5.17	20.7	13.8	1.58	K, P, D
2 Diopside	0.26	2.69	0.71	4.16	16.9	18.4	1.78	K, P, E, D
3 Ti-Cr diopside	0.80	3.86	1.02	2.61	16.0	19.5	1.94	K, P, D
4 Low-Cr diopside	0.50	3.19	0.09	5.86	16.9	17.5	1.85	K, E, D
5 Chrome diopside	0.09	2.50	1.45	2.02	16.8	20.7	1.68	K, P, D
6 Ureyitic diopside	0.27	3.14	2.99	2.37	15.2	17.9	3.11	K, P, D
7 High ureyite diopside	0.19	3.14	11.8	1.68	19.3	10.6	7.07	D
8 Jadeitic diopside	0.44	7.61	0.10	6.10	11.5	14.5	4.50	P, E, D
9 Omphacite	0.27	11.34	0.15	3.29	10.3	14.6	5.09	E, D
10 Diopsidic jadeite	0.22	16.87	0.02	2.42	6.36	10.2	7.64	E
Typical groundmass diopside [b]	0.70	0.36	0.12	3.45	17.0	23.9	0.32	

a Source K = from kimberlite; P = from peridotite or pyroxenite (inc. websterite); E = from eclogite; D = occurring as inclusion in, or intergrown with, diamond

b From Dawson et al. (1977, Table 2, anal. 2)

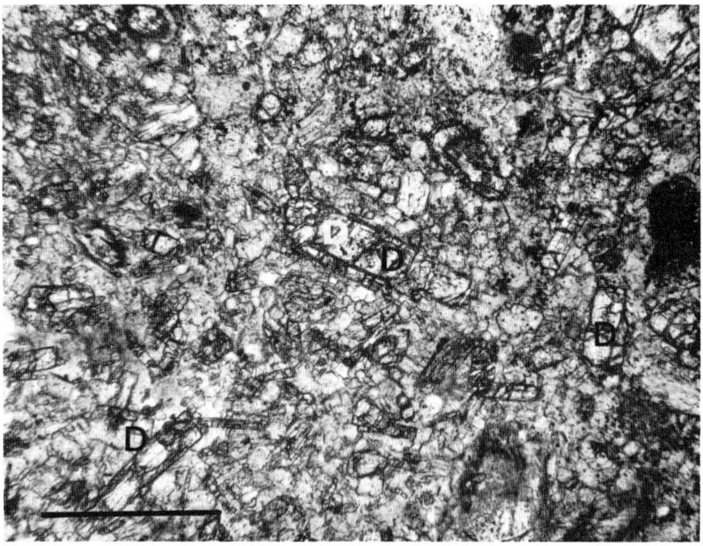

Fig. 36. Photomicrograph of groundmass diopsides *(D)* in the mica-rich matrix of kimberlite from the Roberts Victor Mine, South Africa. Scale bar = 0.5 mm, plane-polarized light

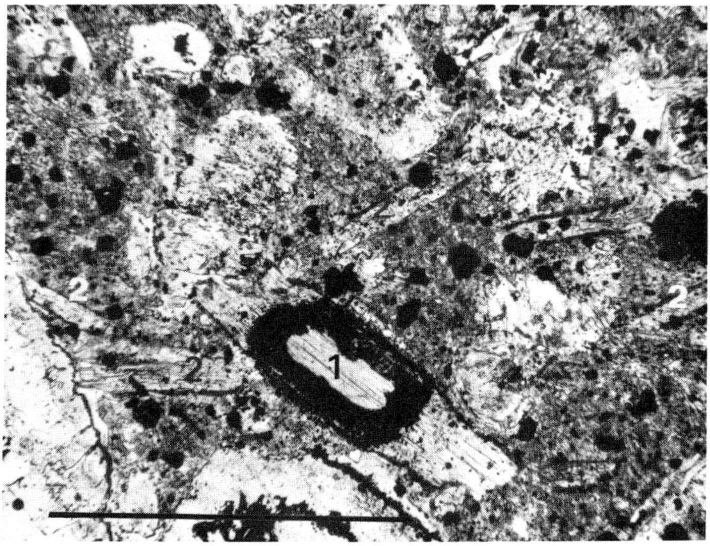

Fig. 37. Type I *(1)* and Type II *(2)* micas in matrix of the Hololo kimberlite, Lesotho. The type I mica has been severely altered by reaction with the kimberlite groundmass and has a reaction rind containing a large amount of dark magnetite; the altered ends of the crystal are now clear white as opposed to the brown core of the crystal. Note that both Type I and Type II micas contain inclusions of groundmass opaques. Scale bar = 1 mm, plane-polarized light

Gabon (Bardet 1973); sodic tremolite occurs in the Buell Park kimberlite, Arizona (Aoki et al. 1972). Of particular interest is the occurrence of potassic richterite in the dyke kimberlite at Pniel, Barkly West, S. Africa (Erlank 1973); in its high alkali ($K_2O + Na_2O = 8.0$ wt%) low Al_2O_3 and Cr_2O_3, and high TiO_2 contents this groundmass amphibole resembles the potassic richterites found in the mica-amphibole-rutile-ilmenite-diopside (MARID-suite) nodules (see Chap. 7.5.1.4).

6.9.4 Phyllosilicates

Phyllosilicates occur either as high-temperature phases (biotite and phlogopite), as pseudomorphs after pre-existing silicates, or as late-stage minerals crystallizing within the kimberlite groundmass. Compared with the other mineral groups in kimberlite, the phyllosilicates have been relatively neglected, as is apparent from the paucity of references to them in the kimberlite literature.

Apophyllite, together with zeolites, occurs within cracks and fissures in dolerite xenoliths in the Bultfontein Mine, where its presence is attributed to the action of kimberlite fluids on the xenoliths. It also occurs scattered throughout the kimberlite in the Koppiesfontein dyke, S. Africa (Wagner 1914).

Talc has been reported in the kimberlite of the Dvoynaya, Nemagnitonye and Pozdiyaya intrusions in Yakutia, where it was identified by optical means (Milashev et al. 1963) and in the Moses Rock and Sloan kimberlites, U.S.A., where X-ray techniques confirmed its presence (McGetchin and Silver 1970; McCallum and Eggler 1971). Due to its optical similarity to serpentine-group minerals, talc has probably been overlooked in many kimberlites.

Phlogopite varies widely in its presence in kimberlite; in some olivine- or serpentine-rich kimberlites it may be absent, whilst in others – the micaceous kimberlites – it may form up to 50 vol% of the rock, giving it a bronzy, lustrous sheen. The mica occurs either as megacrysts or as a constituent of the groundmass. The megacrysts are described more fully in Chapter 8.

By contrast with the megacrysts, the groundmass micas often contain inclusions of magnetite and perovskite; they are often euhedral and may be zoned, sometimes with the rims of bright red-brown mica which has anomalous pleochroism (anomite). In detail, however, there is considerable variation in the groundmass micas. both within an individual intrusion and from intrusion to intrusion. On the basis of the *cores* of groundmass micas, Smith et al. (1978) recognize two distinct types of groundmass micas that are less than 1 mm across (Fig. 37). Very rare Type I micas are relatively iron-rich with *mg* (= Mg/(Mg + Fe) 0.45–0.65, TiO_2 3–6 wt%; Al_2O_3 14–16 wt%; no Fe^{3+} required in tetrahedral sites; low NiO (approximately 0.02 wt%); and relatively high *na* ($Na_2O/Na_2O + K_2O$) 0.02–0.03. The much more abundant Type II micas are more variable in composition but relative to Type I micas are more magnesian (*mg* 0.08–0.93); lower in TiO_2 (0.7–4.0 wt%) and Al_2O_3 (6.8–14.2 wt%); have substantial Fe^{3+} in tetrahedral sites; and have relatively low *na*. The rims of both types of mica have compositions indicative of mica-"serpentine" mixtures resulting from reaction with a highly aqueous fluid. Inter-grain variations in compositions of Type II micas may result from establishment of local reservoirs on a millimetre scale, consequent upon mechanical mixing and competition of other phases for minor elements such as chromite for Cr or serpentine for Ni. Both Type I and Type II micas may occur within the same kimberlite.

Of the authors who have investigated the zoning on groundmass phlogopites, Rimsaite (1971) reports that groundmass mica in kimberlite from the Upper Canda Mine, Ontario, has dark-brown titanian biotite cores (3.7 wt% TiO_2), mantled by a zone of less titanian (2.0% TiO_2) phlogopite that is itself overgrown by pale rims of titanium- and iron-poor phlogopite that resembles the phlogpite intergrown with calcite in the matrix of the rock. Similarly Gittins et al. (1975) record in the kimberlite dykes of the Sagueney Valley, Quebec, phlogopites with brown titaniferous biotite cores, pale-brown less-titaniferous phlogopite mantles, and green rims that are more phlogopitic and titanium-poor. In Greenland kimberlites, Emeleus and Andrews (1975) record phlogopites with light-coloured cores zoning towards anomite rims (Al-poor tetra-ferriphlogopite). They resemble phlogopites from the Roberts Victor kimberlite that also may be zoned from brown cores to rims either of titaniferous phlogopite or bright-orange anomite (Dawson, unpublished data). The decrease in the Fe/Mg ratio and decrease in TiO_2 in most rims is the reverse of what might be expected in an igneous process; it could, however, be correlated with increasing fO_2 resulting in the precipitation of, and preferential partitioning of iron and titanium into, titaniferous magnetite. In this context, it is pertinent that zoned spinel in the Upper Canada Mine kimberlite also zones towards a rim rich in titanium and ferric iron (Rimsaite 1971).

Regarding minor elements, the Type I groundmass micas contain 240–630 (average 470) ppm Rb_2O and 1030–4300 (average 2670) ppm BaO; the abundant Type II groundmass micas contain 90–820 (average 420) ppm Rb_2O and 470–5800 (average 2620) ppm BaO (Smith et al. 1979). These give average K/Rb ratios of 209 and 330, and K/Ba ratios of 42 and 74 for the Type I and Type II micas respectively, and the negative correlation between K/Rb and K/Ba may possibly result from mica having higher K/Rb and lower K/Ba than coexisting liquids. The Type I groundmass micas have K/Rb and K/Ba ratios very similar to those in micas from some carbonatites (Smith et al. 1979). Other trace-element data are given by Rimsaite (1971) who reports the following values for groundmass phlogopite from the Upper Canada Mine: Li_2O 0.01, ZnO 0.05, BaO 0.3, Rb_2O 0.09, SrO 0.04, F 0.3, Cl 0.08 (all values in wt%).

Biotite. In addition to the biotite cores in groundmass phlogpites reported above, biotite containing 17%–21% FeO is recorded in the Moses Rock and Cane Valley kimberlites, Utah (McGetchin et al. 1973), and the Murfresboro', Arkansas, kimberlite (Giardini et al. (1974).

Hydrophlogopite is found fringing phlogopite megacrysts where it may be regarded as an intermediate stage in a vermiculitization process. This may be illustrated by analyses of "phlogopite" reported from Yakutia kimberlites in which the K_2O content ranges from 1.9 to 9.0 wt%, and decreasing K_2O correlates with increasing water content (Milashev et al. 1963). In addition it has been identified as reaction fringes on serpentine pseudomorphs and in the matrix of some Lesotho kimberlites (Dawson 1962).

Montmorillonite, together with *nontronite, halloysite, illite, kaolinite* and *sepiolite,* is reported in the highest levels of highly *weathered* (lateritised) kimberlites in Yakutia, Sierra Leone and South Africa (see summaries by Fairbairn and Robertson 1966; Ruotsala 1975). They form at the expense of hydromicas and serpentine-group minerals that replace earlier micas, orthopyroxene and olivine. These phases are typical of weathering, which may be imposed on the late stage ("saponite stage") deuteric effects (see below).

Stevensite, vermiculite and *saponite* are recorded in kimberlites from Russia (Bobrievich et al. 1964), Sierra Leone (Fairbairn and Robertson 1966) and Lesotho (Kresten 1973b). They are a distinct mineral suite, typical of late-stage deuteric alteration of kimberlite, and form at the expense of earlier chlorite- and serpentine-group phases; in some instances, the vermiculite may exhibit the external appearance of the original mica (Dawson 1962). The diagrammatic representation of the association of these and other phases during the alteration of kimberlite is shown in Fig. 38.

Chlorite is a common late-stage mineral in many kimberlites. It forms at the expense of phlogopite and may be seen replacing it along cleavage planes; in many intrusions this replacement appears to be a near-surface effect (Bobrievich et al. 1959). In most instances the chlorites are of normal composition, but Mc Getchin

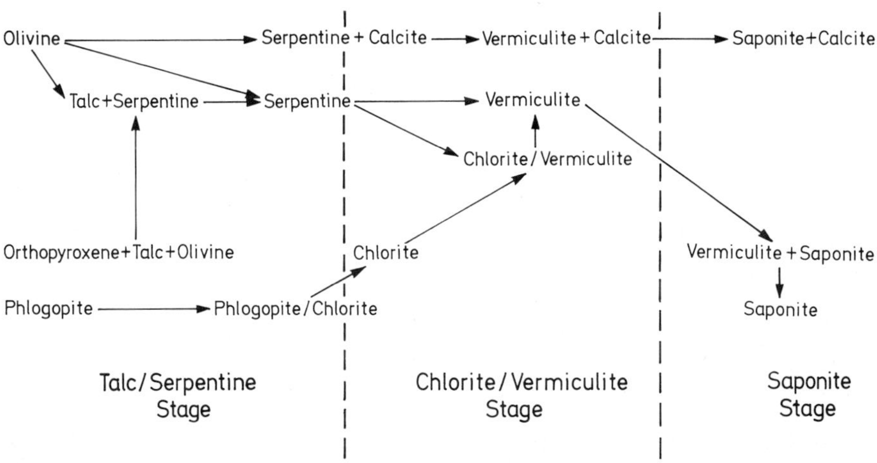

Fig. 38. Schematic diagram showing the alteration of olivine, pyroxene and mica. (Adapted from Kresten 1973b)

et al. (1973) have reported chrome-bearing chlorites (Cr_2O_3 2.4–3.9 wt%) from the Cane Valley kimberlite, Utah; the more chromiferous chlorites are pink in colour.

Serpentine-group minerals are common as pseudomorphs after olivine, where it is usually accompanied by exsolved magnetite or haematite. Little systematic work has been carried out on the identification of serpentine-group species within most kimberlites, and most are referred to only as "serpentine". However antigorite has been identified by Gittins et al. (1975) in Canadian kimberlites, and lizardite and antigorite are found in the Randolph County, Kansas, kimberlites where chrysotile is rare (Brookins and McDermott 1970). By contrast, in addition to single- and double-layered lizardite-chrysotile and antigorite (the latter pseudomorphing enstatite), fibrous chrysotile is a common vein-filling in the Sloan diatreme, Colorado (McCallum and Eggler 1971).

Serpophite [structureless, isotropic (originally gel-like) hydrated magnesium-ironsilicate] occurs in the interstices of the serpentine network in olivine pseudomorphs, and also as irregular nodules and lenticles. The nodules described by Bobrievich et al. (1959) from Yakutia kimberlites, are compact and porcellaneous with conchoidal fractures and matt or waxy lustre; segregations of fine-grained magnetite occur often in the periphery of the nodules what may be up to 15 cm in length. Three analyses of Yakutian serpophite show water contents of 14 to 17 wt%. Isotropic serpophite also occurs in late-stage "pools" or interstitial patches in many South African kimberlites (Fig. 29). In other South African kimberlites, serpophite segregations are structurally inhomogeneous, varying from areas that are optically isotropic to areas that show very low birefringences. Circular aggregates may be fringed by euhedral calcite and magnetite; some patches in kimberlite from a late-stage dyke in the Wesselton Mine include small crystals of barytes and of an opaque hydrated magnesium-iron titano-silicate containing 13%–14% MnO (J.B. Dawson, unpublished data). Although calcite appears to be coexisting with the serpophite in some cases, Clement (1973, 1975) puts forward a convincing case that in many cases, earlier interstitial or globular calcite has been replaced by later serpophite. In the case of Yakutian kimberlite serpophites (Bobrievich et al. 1959) and serpophitic patches in kimberlite from the Jagersfontein, Dutoitspan and Monastery Mines (J.B. Dawson, unpublished data), the serpophites are highly magnesian with MgO/FeO ratios of between 20:1 and 26:1; however, in Greenland kimberlite, Emeleus and Andrews (1975) record groundmass serpentine that is exeptionally rich in iron (up to 11.6 wt% FeO).

Serpentine megacrysts (pseudomorphing olivine) from the Monastery Mine and the Kao pipe have high negative δ D values and relatively low $\delta^{18}O$ (Sheppard and Dawson 1975). In addition, values of $\delta^{18}O$ and δ D for whole-rock kimberlites, in which the main hydroxyl-bearing phases are serpentine, micas and hydromicas, plot in a distinctive field on a δ D v. $\delta^{18}O$ diagram, together with the serpentine megacrysts (Fig. 22). This field is distinct from the field of phologpite megacrysts.

6.9.5 Tektosilicates

Nepheline Group. Nepheline occurs in olivine nephelinites, melilitites and some of the "picrite porphyries" that occur in the same areas as the kimberlites in S. Africa and Yakutia, but none is so far recorded for kimberlite. It appears likely that the high pH_2O during kimberlite crystallization would preclude its formation.

Danburite occurs in the Arbabyt pipe, Yakutia (Cherepanov 1967).

Zeolite Group. Minerals of low birefringence and moderate relief, probably zeolites, occur in the groundmass of some kimberlites, and also infill cavities around fractures in basaltic or doleritic xenoliths. Few species have been identified, though Williams (1932) reports chabazite and natrolite in S. African kimberlites. It should be noted, however, that some older occurrences of "natrolite" have subsequently been identified as bultfonteinite.

6.10 Summary

The mineralogy of kimberlite is hybrid and complex. Many megascopic crystals have been derived by fragmentation of peridotites, pyroxenites and eclogites; other larger crystals are of broadly comparable composition to those derived by fragmentation of peridotites but in their chemistry, particularly in their high TiO_2 and low Cr_2O_3, they are distinct from xenolith phases. In addition these larger crystals, the megacryst suite, are often associated with ilmenite and are usually of greater size than minerals derived by fragmentation of xenoliths.

The xenocrysts and phenocrysts are a high-temperature assemblage and are set within a matrix comprising minerals, which crystallized at considerably lower temperatures. Within the matrix, typical minerals are phlogopite, carbonate, perovskite, apatite, a very characteristic spinel and in some cases diopside and monticellite. These phases have crystallized following the emplacement of the kimberlite; in some instances, as in the case of diopside microlites, crystallization was rapid, but in others crystallization was protracted, giving rise to complex zoning. In addition to the oxides such as CaO, CO_2, iron oxides and K_2O, which are so obviously concentrated within the kimberlite matrix, there is also concentration of Mn, Ba, Ti, S, P and the rare-earth elements.

There has been considerable interaction between the xenocrysts and phenocrysts and late-stage fluids to provide a new suite of low-temperature minerals; for example, garnets may be replaced by chlorite-amphibole assemblages and ilmenite may be partially or wholly replaced by perovskite and sphene. Olivine in particular is replaced by serpentine and following the serpentisation may be replaced by various combinations of calcite, dolomite, talc, phlogopite and perovskite; in short, many of the xenocrysts, phenocrysts and also xenoliths are replaced by minerals which are typical of the kimberlite matrix (Fig. 39). During this process

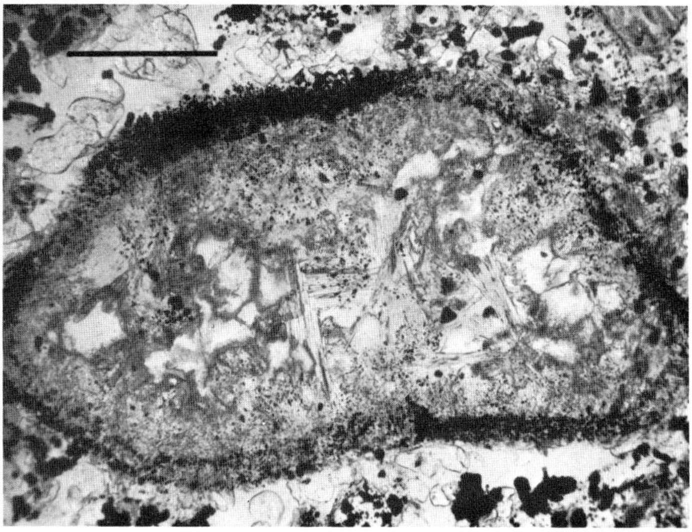

Fig. 39. Photomicrograph of a micro-xenolith of shale being metasomatised and replaced by a combination of calcite, phlogopite and perovskite; from the Monastery Mine, South Africa. Scale bar = 0.5 mm, plane-polarized light

many of the high-temperature minerals are reduced in size having been chemically corroded; an example of this is diamond which, during its "degradation" from high-temperature sharp-edged octehedra to the rounded forms which are developed by etching, may lose up to 45% of its weight by corrosion.

In particular, the oxide phases present abundant evidence of a complex cooling history and changes in redox conditions; following cooling with subsolidus oxidation or reduction together with subsolidus exsolution of specific phases, interaction between the spinels and groundmass material resulted in the formation of alteration and reaction mantles particularly on picro-ilmenite.

The matrix carbonate is of extreme interest; when seen in its pristine state (i.e. not having been subjected to recrystallization or redistribution by groundwater) its morphology, trace-element content, and isotopic characteristics are similar to carbonate in carbonatites; or, to be more correct, both these sets of high-temperature carbonates, presumably originating within the upper mantle, show the same characteristics whether they are derived by fractionation and immiscibility from nephelinitic magmas or else are the end stage of fractionation of kimberlite magma.

Post-consolidation changes have been effected by percolating groundwater causing redistribution of phases such as calcite and introducing low-temperature phases such as halite, ?pyrite and sulphates.

Chapter 7 Xenoliths in Kimberlite

Within kimberlite intrusions there are included fragments of a wide variety of rock types. Minerals derived from their fragmentation are incorporated into the host kimberlite, thereby giving it an extremely complex and hybrid mineralogy.

Xenoliths from five differing sources may be recognized:

1. Fragments of the immediate wall-rock of the intrusion.
2. Fragments derived from earlier formations, existing at the time of kimberlite emplacement, which have subsequently been removed by erosion.
3. Blocks from buried formations.
4. Granulites derived from deep-seated metamorphic terrains.
5. Xenoliths of rock-types believed to derive from the earth's upper mantle.

7.1 Wall-Rock Fragments

By contrast with the xenoliths from greater depths, these are frequently angular in shape. They may be of considerable size and, in some diatremes, may be concentrated in the peripheral parts of the intrusion.

7.2 Fragments from Earlier, Eroded Formations

One of the principal features of interest in these xenoliths is that they are often the only firm evidence of the earlier extent of formations that are now no longer in existence or whose outcrop area is at present much smaller than at the time of their deposition. Examples are the fragments of Karroo basalt in the Kimberley diatremes, some 350 km to the west of the nearest present-day outcrops (Hawthorne 1975); and the xenoliths of Lower Palaeozoic sediments in the diatremes intruding the Precambrian granitic basement of the Frontier Range of the Colorado-Wyoming area, U.S.A. (McCallum et al. 1975). Three features are worthy of note:

a) These xenoliths, in contrast to those from the next three groups, have slumped into the diatremes and are often at considerable depths below their original stratigraphic position. For example, Hawthorne (1975) estimates that the Karroo basalt xenoliths in the Kimberley area have descended at leat 1500 m, and Hearn (1968) records blocks of sediment that have descended 1300 m in the diatremes of northern Montana.

b) Some of the descending xenoliths are of immense size and are referred to by the miners as "floating reefs". A good example is the enormous block of Precambrian quartzite, 800 m long that almost bisects the Premier Mine diatreme (Fig. 9).

c) When their stratigraphic age can be determined, these included blocks can give a maximum age to intrusions.

7.3 Blocks Derived from Recognizable Underlying Formations

These are mainly blocks derived from formations in the basement terrains traversed by the kimberlites in their ascent. These formations, often blanketed by later strata in the immediate vicinity of the kimberlite intrusions, usually crop out some distance away, or may be exposed as wall-rock in some of the deeper-mined diatremes. Examples of this type of xenolith are the Archaean granite and amphibolite blocks in the Kimberley diatremes (Williams 1932); Precambrian banded ironstone in the kimberlites in northern Lesotho (Stockley 1948); a wide variety of metamorphic rock types derived from the crystalline Siberian Platform in the kimberlites of northern Yakutia (Bobrievich et al. 1959); and blocks of amphibolite and granite in the kimberlite diatremes of the Colorado Plateau (McGetchin and Silver 1970).

7.4 Granulites

In some kimberlite pipes are blocks of granulites that are comparable with the granulites exposed in some basement complexes. Compared with peridotite and eclogite xenoliths, they have been relatively neglected, but have recently received more attention in view of the potential information they can provide about the lower crust. They are reported in the Dolgozdhannaya, Zarnitsa and Udachnaya kimberlites of the U.S.S.R (Bobrievich et al. 1959); in the Stockdale kimberlite pipe, Kansas (Brookins and Meyer 1974); in the Gondolo and M'Bo pipes of Zaire (Verhoogen 1938); in pipes in S.W. Africa (Janse 1975); in dykes in the Victoria West area, and the Monastery Mine, S. Africa (Rickwood et al. 1969; Dawson, unpublished data); and in the Liqhobong, Letseng, Kao and Matsoku pipes, Lesotho (Cox et al. 1973; Nixon et al. 1963; Dawson 1977; Rogers 1977; Griffin et al. 1979). It has been suggested that the granulites in the Lesotho and most of the S. African pipes may be derived from the 1100–1200 m.y. Namaquaride fold belt bordering the Kaapvaal craton (Dawson 1977; Griffin et al. 1979), but granulites are also known from the Klipfontein, Newlands and Orapa pipes within the craton (Dawson, unpublished data; Shee and Gurney 1979).

7.4.1 Petrography

The ovoid nodules are generally not greater than 15 cm in diameter and normally have a rough or pitted surface. In some, banding is a conspicuous feature. The grain size is generally in the range 1–2 mm, grains are equigranular and triple-grain boundaries of approximately 120° are common; however, kyanite, where present, occurs in small clusters of radiating needles.

In the Lesotho and South African garnet granulites, the most common assemblage is garnet-plagioclase-clinopyroxene ± smaller amounts of orthopyroxene. Accessory minerals include rutile (a fairly constant accessory), scapolite, amphibole, biotite, apatite, quartz, and occasionally kyanite and sphene. Rare accessories are sapphirine and sillimanite, reported from the Stockdale xenoliths (Meyer and Brookins 1976).

There is a wide range in the abundance of plagioclase + scapolite vs garnet + pyroxene from nodule to nodule; many contain 30–40 modal% dark minerals, but ultramafic varieties comprising only garnet, clinopyroxene and orthopyroxene are transitional to rocks of eclogitic appearance consisting purely of garnet and clinopyroxene. The absence of modal plagioclase in these garnet "pyroxenites" and "eclogites" is controlled mainly by bulk composition; only rocks $>35\%$ normative (An + Ab) and $<45\%$ normative (Ol + Px) contain modal plagioclase.

Minute fluid inclusions consisting of CO_2 (either liquid or liquid + gas) exist in the plagioclases in some granulite xenoliths from Matsoku, Lesotho (Bilal 1978), and resemble those found in granulite xenoliths from the carbonatite/ankaramite volcano Lashaine, Tanzania (Bilal 1978), and in the granulite terrains of S. Norway (Touret 1974); it should be noted that the fluids are pure CO_2 with no water or hydrocarbon present.

Most granulite xenoliths exhibit some sign of secondary alteration. The scapolite, plagioclase and mica alter most readily, and garnet is the most resistant. The commonest type of alteration is calcitization; in granulites from the Monastery Mine, a complete gradation can be traced, from lightly calcitized xenoliths to nodules of coarse rhombohedral calcite containing only relict garnets (Dawson, unpublished data); the carbon and oxygen isotopes of this carbonate ($\delta^{18}O$ = 13.31‰; $\delta^{13}C$ = 6.7‰) are similar to those in other kimberlite groundmass carbonates (Sheppard and Dawson 1975).

7.4.2 Phase Chemistry

The *garnets* which are generally unzoned (though there are a few exceptions) tend to have a fairly constant grossular content, between 15 and 25 mol%; however, they have variable pyrope/almandine ratios, varying from approximately 55% pyrope to 15% pyrope, with inverse almandine content. The *clionopyroxenes,* which may be slightly zoned, tend to have high molecular proportions of jadeite (10%–33%) and low molecular proportions of tschermakite. Alumina in the orthopyroxenes (Mg/Fe^{2+} ratio $\simeq$ 0.80) is between 1.4 and 2.6 wt%. The composition of the plagioclases ranges widely between An 5%–77%, and the jadeite/(jadeite + tschermakite) ratio in the co-existing clinopyroxene is roughly proportional to the albite/anorthite ratio in the plagioclase. *Scapolites* are of mizzonitic composition and are a sulphur-rich variety (SO_3 = 3.5–6.8 wt%); *amphiboles* are edenitic and pargasitic hornblende; and the mica is dominantly pholopitic (Griffin et al. 1979; Jackson and Harte 1977).

7.4.3 Bulk Chemistry

Bulk rock analyses of granulite xenoliths from Lesotho kimberlites show them to range from basic to basic/intermediate composition, approximating in composition to alkali or high alumina basalts (Table 14). Their alkalic nature is due to

Table 14. Analyses of granulite xenoliths in Lesotho kimberlite

	1	2	3	4
SiO_2	45.90	47.05	49.87	51.24
TiO_2	1.11	1.01	0.39	0.54
Al_2O_3	13.47	17.88	18.60	16.69
Fe_2O_3	n.d.	1.47	–	10.58 [b]
FeO	11.84 [a]	7.71	7.25	–
MnO	0.22	0.16	0.14	0.18
MgO	14.70	7.52	8.25	8.60
CaO	11.35	9.59	8.91	8.47
Na_2O	0.71	3.85	3.69	2.86
K_2O	0.09	0.93	1.06	1.50
P_2O_5	0.10	0.19	0.01	0.05
H_2O^+	n.d.	2.36	1.06	n.d.
H_2O^-	0.0	0.24	0.11	n.d.
CO_2	n.d.	n.d.	0.21	n.d.
L.OI	1.66	–	–	–
	100.32	99.96	99.56	100.70
Ba ppm		1,800	400	1,700
Rb ppm		12	18	23
Sr ppm		860	1,738	625
Zr ppm		24	307	21
Y ppm		7	1	4
K/Rb		650	498	541

a Total iron as FeO. b Total iron as Fe_2O_3

1. Granulite, LBM54, Matsoku (Cox et al. 1973) (Total includes 0.67 S^2)
2. Granulite, PHN2852, Matsoku (Rogers 1977)
3. Granulite, BD1351, Matsoku (Dawson 1977)
4. Granulite, PHN2588, Liqhobong (Rogers 1977)

a high Na_2O content but at the same time they are not noticeably depleted in K_2O as is common in some deep-seated metamorphic terrains (Lambert and Heier 1968). The granulites are generally olivine-normative, and sometimes nepheline-normative.

They are relatively low in Rb, giving rise to high K/Rb ratios and low Rb/Sr ratios, indicative of strong Rb depletion (Dawson 1977).

Rare-earth element patterns in specimens not contaminated by host kimberlite show moderate slopes, but with a very pronounced positive europium anomaly, due to the high amount of modal plagioclase (Rogers 1977).

7.4.4 Conditions of Formation

The continuity in the variations in bulk chemistry, and in the proportions of gt + cpx + opx + plag within the suite suggest that the rocks of the garnet granulite – garnet "pyroxenite" – "eclogite" suite may have formed by metamorphism of related rocks under the same PT regime.

Using a temperature based on iron/magnesium distribution between garnet and clinopyroxene which yield temperatures in the order of 839°C, together with the method of pressure determination of Schmid and Wood (1976), Jackson and Harte (1977) calculate pressures of 21.5 kb for a kyanite-bearing garnet granulite. (There is a certain degree of uncertainty depending upon the entropy and enthalpy values adopted which may drop the pressures to 17 kb). These authors believe, on these bases, that the Matsoku granulite suite may have equilibrated at depths below the present-day Moho in Southern Africa. The crust/mantle stratigraphic cross-section thus appears to be similar to that envisaged by Brookins and Meyer (1974) for Kansas, U.S.A., where granulite facies xenoliths are believed to be derived from rocks straddling the Moho. However, after a detailed analysis of the current methods of geobarometry and geothermometry used in estimates of the equilibration conditions of these granulites, Carswell and Griffin (in press) using temperatures derived from Råheim and Green (1974), together with the geobarometer of Wood (1974), with an additional calculation taking into account the oxidation state of the iron used in the temperature calculations, believe that the mean PT values for most individual nodules are consistent with derivation from the lower crust, i.e. at temperatures between 550° and 700°C and 5–13 kb, corresponding to depths of some 18–39 km. For the most part this places them above the Moho which has been determined for this part of South Africa at a depth of approximately 37 km (Hales and Sacks 1959). Their overall chemistry, especially their high K/Rb ratios suggest that they may be the basic residuum from rocks of intermediate composition that have undergone partial anatexis (Dawson 1977).

7.5 Mantle-Derived Xenoliths

7.5.1 Introduction

Many kimberlite inclusions contain blocks of rock-types, the density and mineralogy of which suggest formation under higher pressure and temperature conditions than those normally pertaining in the crust. By inference they are believed to originate in the upper mantle or the deepest parts of the crust. Because they represent the widest variety of rock-types from the upper mantle and, in some cases, provide information as to the depth of origin and to events connected with formation of the host kimberlite, these xenoliths have received a great deal of attention from petrographers, experimental petrologists and geophysicists, and are treated here more extensively than the preceding four xenolith groups.

The deep-seated xenoliths can be divided into five broad groups:

1. The peridotite-pyroxenite suite.
2. Eclogites and grospydites.
3. Metasomatized peridotites, rich in amphibole and/or mica.
4. Glimmerites and MARID-suite rocks.
5. Miscellaneous xenoliths.

Representatives of these five groups vary considerably in volume from intrusion to intrusion. Together they usually comprise not greater than 2% of the bulk rock (Davidson 1967) but in some exceptional intrusions, such as Matsoku, Lesotho, they form 20%–30% of the rock (Dawson 1968). Most commonly, xenoliths of peridotite predominate, though there are some rare intrusions where pyroxenites are relatively common (e.g. Matsoku, Lesotho), and others where eclogite comprises a high proportion or the whole mantle xenolith suite [e.g. Roberts Victor, Bobbejaan, Rietfontein (S. Africa); Orapa (Botswana), Garnet Ridge and Moses Rock (U.S.A.) and Zagadochnaya (U.S.S.R.)].

7.5.1.1 Peridotite-Pyroxenite Suite

1. Nomenclature. The rocks of the peridotite-pyroxenite suite are lherzolites, harzburgites, dunites, websterite and pyroxenites. The nomenclature is based upon the modal proportions of olivine, orthopyroxene and clinopyroxene, according to the International Union of Geological Sciences suggested classification ("Streckeisen classification", Fig. 40); however an exeption is made in the case of the definition for "lherzolite", for which > 5% modal clinopyroxene is required in the I.U.G.S. scheme. Till very recently most workers have regarded the presence of only two or three grains of clinopyroxene in a hand-specimen to suffice for the rock to be called lherzolite. For example, the modes of 75 out of 91 garnet "lherzolites" from South Africa reported by Dawson et al. (1980) contain less than 5% clinopyroxene (Fig. 41), and according to the I.U.G.S. scheme, should strictly be

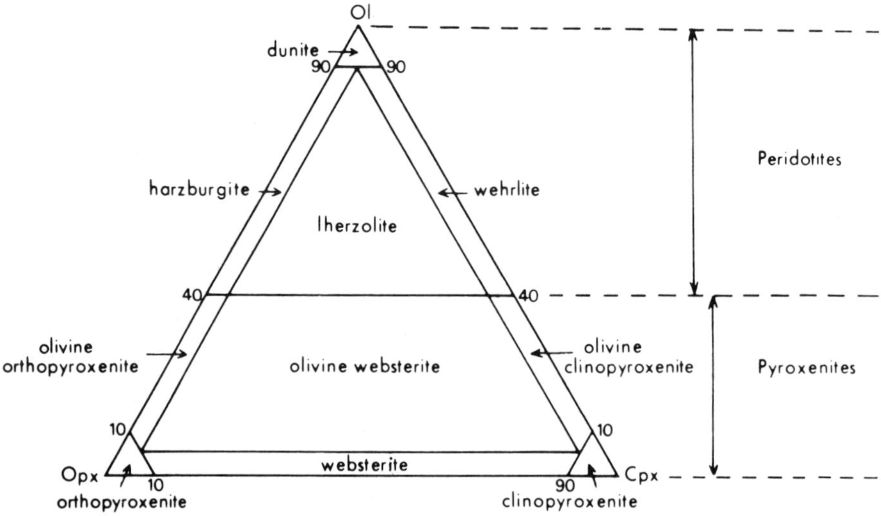

Fig. 40. Classification of peridotites and pyroxenites. (Recommended by the International Union of Geological Sciences) . Based on volume percentage

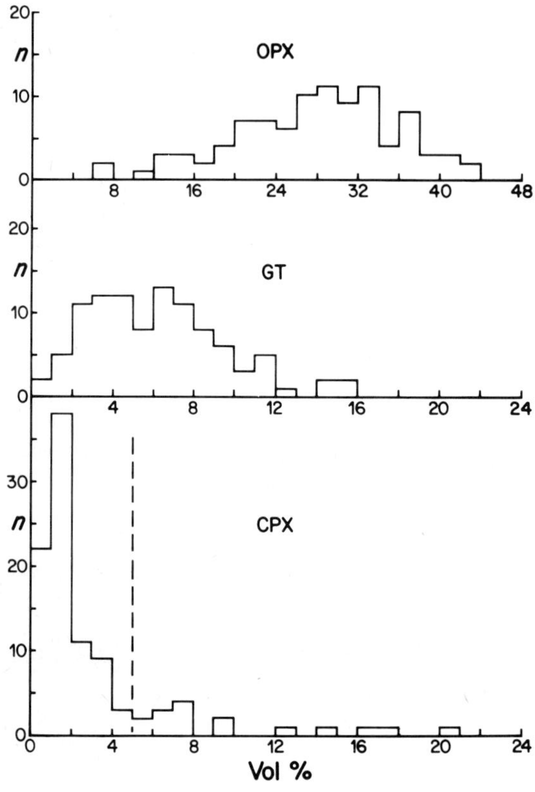

Fig. 41. Histogram of garnet and pyroxene modal percentages in garnet "lherzolites" (Dawson et al. 1980)

termed garnet-clinopyroxene harzburgites. The term lherzolite is used here in the rather looser sense to enable ultilization of data on "lherzolites" described in the literature, with the caveat that they may in fact contain less than 5% clinopyroxene. In addition it should be apprecited that the term websterite covers the terms pyroxenite or olivine pyroxenite used by some earlier workers. Also the term dunite covers the rock termed olivinite by Russian petrographers, their distinction being based on magnetite as the accessory spinel, as opposed to chromite in dunite.

Petrography. The xenoliths are ovoid or discoidal in shape (Fig. 42); they are generally not greater than 30 cm in size, though rarely may be over 1 m in length. The surface may be extremely smooth and polished, particularly in the case of peridotite blocks with a serpentinised skin, or it may be pitted due to differential weathering. Many are rounded, and the rounding and smoothing of the blocks was formerly attributed to the sand-blasting effects during fluidization intrusion (Dawson 1962), but the presence of rounded xenoliths in the non-fluidized hypabyssal dykes and sills indicates that, whereas fluidization may effect the final polishing, the initial rounding must be due to some other process. There is increasing evidence that many xenoliths are cut by fractures that are often at a high angle to any observed banding. It is possible that a type of upper-mantle jointing may have been imposed upon the rocks in the vicinity of ascending kimberlite, and that this joint-pattern contributed to the initial equidimensional aspect of the entrained xenoliths, with magmatic corrosion and fluidization completing the rounding process. However, fluidization is not a necessary factor (see above) since extremely rounded microxenoliths occur in the kimberlite lava flow at Igwisi, Tanzania (Reid et al. 1975a). There are also some blocks that have both rounded and planar faces; these planar faces are invariably coated with mica or amphibole, and the interference is that these particular blocks were originally rounded but subsequently fractured along planar veinlets filled by these minerals.

Many xenoliths are *apparently* structureless but banding is observed in specimens from the Obnazhennaya pipe, where garnet lherzolite grades into "eclogite" (?garnet-pyroxene segregations), and dunite grades into chromite harzburgite (Sobolev and Sobolev 1964); in xenoliths from the Matsoku pipe (Harte et al. 1975), and in numerous xenoliths from the Bultfontein Mine (J.B. Dawson, unpublished). It is the author's experience that banding is common in large specimens, suggesting that its absence in smaller specimens is an effect of scale.

Mineralogically, the rocks consist of varying combinations of magnesian olivine, magnesian orthopyroxene, green chromiferous diopside, red to purple magnesian garnet, magnesian spinel, phlogopite and pargasitic amphibole (very rare), together with the alteration products of these minerals. Rare accessory minerals are diamond, graphite, moissanite, various sulphides, rutile, ilmenite and zircon. Diamond has been found in 16 garnet-peridotites from the Aykhal and Udachnaya pipes, Yakutia (Pokhilenko et al. 1977), in mica garnet lherzolite from Mothae, Lesotho (Dawson and Smith 1975b) and from altered garnet lherzolite from a

Fig. 42. Xenolith of garnet lherzolite in kimberlite, Liquobong South-east Pipe, Lesotho

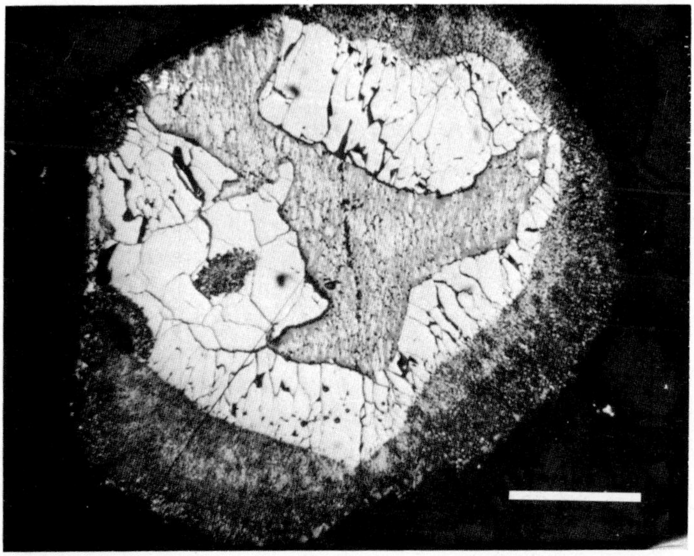

Fig. 43. Photomicrograph of a sulphide oxide globule, in chromite lherzolite from De Beers Mine, South Africa. A mottled core of white pentlandite and grey magnetite is surrounded by massive pentlandite; the finer-grained rim with a poor polish is an alteration rind. Scale bar = 0.2 mm; reflected light

Wyoming diatreme, U.S.A. (McCallum and Eggler 1976). Graphite occurs in garnet lherzolite from Jagersfontein (Wagner 1911), in lherzolite from the Jagersfontein and Premier Mines (Mathias et al. 1970), in mica garnet lherzolite and garnet-spinel lherzolite from Thaba Putsoa; in garnet lherzolite from Mothae; in garnet-spinel harzburgite from Mothae; and lherzolite from Kao (all Lesotho) (Boyd and Nixon 1975). Moissanite is very rare, having been recorded in only one specimen of garnet lherzolite from the Mir pipe (Bobrievich et al. 1959). Ilmenite is also rare, having been found in garnet lherzolite, Thaba Putsoa (Dawson 1960), in diamondiferous and diamond-free garnet lherzolite from Udachanya (Green and Sobolev 1975; Pokhilenko et al. 1977) and in garnet lherzolites from the Thaba Putsoa and Matsoku pipes, Lesotho (Boyd and Nixon 1975). Primary metasomatic ilmenite is found in veins cutting garnet lherzolite blocks from Matsoku (Lesotho) (Harte and Gurney 1975b) and in polymict peridotites from De Beers and Bultfontein (Lawless et al. 1979).

Various sulphides have been found as globules in peridotites (Fig. 43) (reviewed by Bishop et al. 1975). They include pentlandite, mackinawite, chalcopyrite, pyrrhotite, and magnetite often in complex aggregates and in variable textural relationships to each other. The most common assemblage is a core of pyrrhotite containing pentlandite inclusions, surrounded by a rim of chalcopyrite. Like the composite sulphide blebs in some eclogite xenoliths these may be attributable to sulphide-silicate immiscibility, followed by crystallization and subsolidus re-equilibration. Irregular texture involving magnetite, mackinawite, millerite and chalcocite result in part from low-temperature reactions down to below 100°C.

Differing combinations of the major minerals give rise to different rock types, their broad nomenclature having been outlined earlier. In a survey of 295 xenoliths from South African kimberlite, Mathias et al. (1970) found the four most abundant assemblages to be (1) olivine-opx-cpx-garnet (garnet lherzolite) 43%; (2) olivine-opx-garnet (garnet harzburgite) 18%; (3) olivine-opx (harzburgite) 16%; and (4) olivine-opx-cpx (lherzolite) 14%. In the case of the garnet-bearing rocks, there is a complete modal gradation from garnet dunite through garnet harzburgite and lherzolite to garnet websterite, garnet pyroxenite and garnet orthopyroxenite. Whereas the olivine-rich garnet lherzolites are the commonest ultramafic xenoliths, their olivine-poor variants are much rarer; these have now been found in the Dutoitspan Mine (Nixon et al. 1963), the Matsoku pipe, Lesotho (Mathias and Rickwood 1969; Cox et al. 1973), the Mir pipe (Frantsesson 1970), and Klipfontein, South Africa (Dawson, unpublished). Modes of garnet lherzolites are shown in Table 15, together with modes of garnet websterite and garnet pyroxenite for comparison. This modal gradation is real, because the constituent minerals do have the same chemical composition as those in the garnet lherzolites (see p. 134 et seq.). There are, however, some garnet websterites such as those from the Matsoku pipe, Lesotho (Gurney et al. 1975; Carswell and Dawson 1978), Bultfontein (Dawson, unpublished) and the Mir pipe, Yakutia (Frantsesson and Lutts 1970; Sobolev 1970) that are not part of this gradational series, being composed

Table 15. Modes of garnet lherzolites and their variants

	1		2		3	4	5
	Mean	Range	Mean	Range			
Olivine	65	(55–83)	66	(50–75)	20	5	–
Orthopyroxene	26	(7–36)	25	(12–34)	25	63	58
Clinopyroxene	4	(0.2–15)	3	(1–17)	35	11	37
Garnet (+kelyphite)	6	(1–6)	6	(1–15)	20	21	5
Others [a]	1	(0.2–7)					

a Mainly phlogopite and accessory chromite
1. 10 garnet lzerzolites, Lesotho (Dawson 1960)
2. 23 garnet lherzolites, S. Africa (Mathias et al. 1970)
3. Garnet pyroxenite, Dutoitspan (Nixon et al. 1963)
4. Garnet-olivine websterite, Bultfontein (Bishop et al. 1978)
5. Garnet websterite, Klipfontein (Bishop et al. 1978)

of minerals that are more iron-rich than those in the garnet-lherzolites or their olivine-poor variants. In this context, some of the garnet websterites in the suite of specimens from the Matsoku pipe which gives the best example of a complete *modal* gradation from peridotite to pyroxenite (Mathias and Rickwood 1969), contain more iron-rich phases than those in the peridotites, and Harte et al. (1975) record layering, possibly attributable to cumulitic origin. Thus, these particular websterites (unlike those in which the texture and phase chemistry are like those in the granular peridotites) appear to be part of a quite different and distinctive rock-suite. This will be emphasized in the discussion of the chemistry.

Lherzolites containing Al-rich spinel are rare and have to date only been reported from the Green Knobs diatreme, U.S.A. (Smith and Levy 1976).

3. **Textures** in rocks of the peridotite suite are very variable. They may be divided into two main groups – those resulting from deformation, recrystallization and annealing of originally coarse, equigranular-textured rocks, and those indicative of sub-solidus interaction between phases.

The first group of textures has been reviewed by Harte (1977); the essential features leading to classification of textures of this type are summarized in Table 16.

The phases in ultramafic xenoliths react to increasing strain at different rates; olivine reacts the most readily, orthopyroxene less readily, followed by clinopyroxene, garnet or spinel react only to considerable strain. The olivine and pyroxenes respond mainly by recrystallization, but garnet or spinel respond by fragmentation (disruption); some orthopyroxene porphyroclasts may be cut by microfaults, presumably as a result of the strain rate exceeding the ability of the grain to relieve the strain by recrystallization. The classification shown in Table 16 is based

Table 16. Summary of textural classification of rocks in peridotite-pyroxenite suite. (After Harte 1977)

Rock type	Porphyroclasts	Grain size	Grain boundaries	Subtypes
Coarse	Absent	Average grain size greater than 2.0 mm	Variable. Often largely straight or smoothly curving but may be less regular	(1) Equant (2) Tabular
Porphyroclastic	Present. More than 10% of olivine occurs as porphyroclasts	Two major populations, one represented by porphyroclasts (1–2 mm), the other by finer-grained matrix (<0.5 mm)	Irregular in porphyroclasts, straight in neoblasts. Some neoblasts equant, others tabular	(1) Neither disrupted nor laminated (2) Disrupted (3) Fluidal (4) Laminated and disrupted
Mosaic-porphyroclastic	Present. Less than 10% of olivine occurs as porphyroclasts	Usually two populations, one represented by porphyroclasts, the other by finer-grained matrix. One fine-grained population in porphyroclast-free rocks	Irregular in porphyroclasts, straight in neoblasts. Some neoblast equant, others tabular	(1) Neither disrupted nor laminated (2) Fluidal (3) Laminated and disrupted
Granuloblastic	Absent or rare for all mineral species	Small size range for each mineral species (excepting rare porphyroclasts). Grain size generally less than 2.0 mm	Straight or smoothly curving, bounding grains of polygonal tabular shape	(1) Equant (2) Tabular

Fig. 44. Photomicrograph showing coarse texture in garnet lherzolite from the Bultfontein Mine. Note straight and curved grain-boundaries, often intersecting at 120°. Scale bar = 2 mm, plane-polarized light

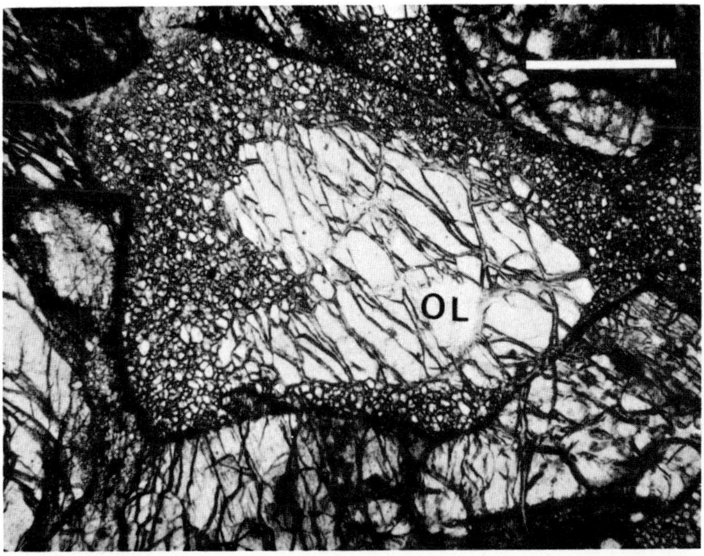

Fig. 45. Photomicrograph showing porpyroclastic texture in lherzolite from the Bultfontein Mine. Porphyroclast of olivine *(OL)* is surrounded by a zone of strain-free small olivine neoblasts. The other grains are orthopyroxene. Scale bar = 1 mm; crossed polarizers

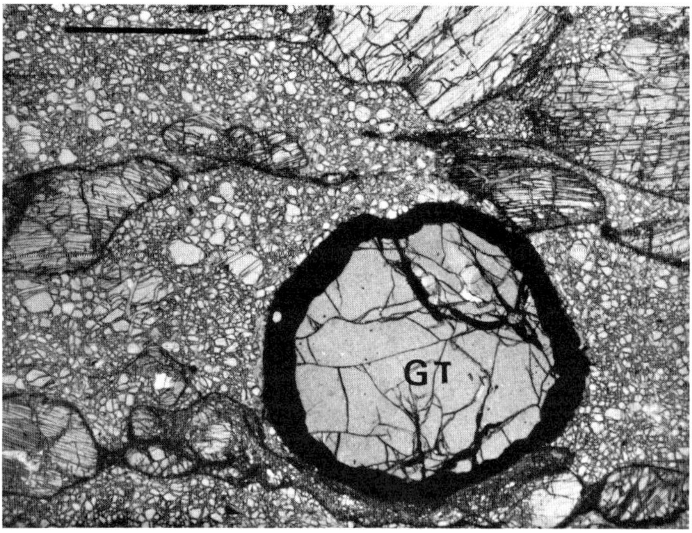

Fig. 46. Photomicrograph showing mosaic porpyroclastic texture in garnet lherzolite from the Jagersfontein kimberlite. Olivine has recrystallized to a fine-grained neoblast matrix and porphyroclasts of orthopyroxene are developing tails of orthopyroxene neoblasts; garnet *(GT)* is unaffected. Scale bar = 1 mm; plane-polarized light

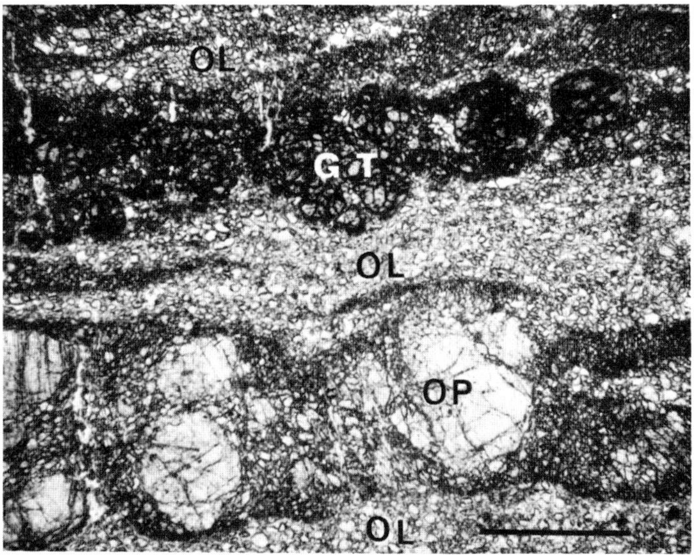

Fig. 47. Photomicrograph showing laminated and disrupted mosaic porphyroclastic texture in garnet lhzerzolite from the Bultfontein Mine. Laminae comprising mainly orthopyroxene *(OP)* (both porphyroclasts and neoblasts) or disrupted garnet *(GT)* alternate with monomineralic bands of olivine neoblasts *(OL)*; Scale bar = 1 mm; plane-polarized light

primarily upon the degree of recrystallization or regrowth of olivine, whereas subtypes (i.e. fluidal, laminated, disrupted) are based on the response of pyroxenes and garnet/spinel. Generally the *porphyroclastic* and *mosaic-porphyroclastic* are envisaged as being formed by increasing deformation and recrystallization of *coarse* rocks (Figs. 44, 45 and 46). Fluidal texture occurs when "tails" of neoblasts from recrystallizing pyroxenes extend into a matrix of olivine neoblasts. *Laminated* refers to the texture when pyroxenes (mainly orthopyroxene) have recrystallized and are drawn out into thin laminae composed of neoblasts, often with a "fluidal" tail (Fig. 47). *Disrupted* texture is present when former large garnets or spinel are broken up and strung out into chains of smaller grains (Fig. 47); this texture must indicate a considerable kinetic input since the garnets (or spinel) survive lower degrees of deformation. In a sequence from coarse to porphyroclastic-mosaic laminated and disrupted there is an overall decrease in grain size. It should be empahsized, however, that textures can vary widely over distances of as little as 1 cm, indicating high deformation gradients, perhaps connected in some cases with variations in hydrolytic weakening (Dawson et al. 1975). It should also be noted that the terms *porphyroclastic* and *porphyroclastic mosaic* cover earlier terms such as "sheared", "flaser" and "cataclastic"; and the "porphyritic peridotites" illustrated by Russian petrographers (e.g. Bobrievich et al. 1964; Figs. 17 and 18) show porphyroclastic mosaic texture. In the subsequent descriptions of rocks of the peridotite-pyroxenite suite, the broad term *deformed* will be used to denote rocks that have undergone varying degrees of recrystallization.

The term *granuloblastic* describes a rock containing few, if any, porphyroclasts and in which the grains of each mineral (excluding porphyroclasts) show a small size-range and granuloblastic texture (Fig. 48). They are interpreted as indicating extensive recrystallization of all minerals (except the porphyroclasts) with subsequent variable amounts of grain growth (annealing). In relatively rare examples, spinel, garnet and diopside may occur as inclusions within larger olivines and orthopyroxenes, as a result of deformation and recrystallization (as far as the disrupted stage where spinel and garnet are involved), followed by preferential grain growth of the olivine and orthopyroxene. Harte et al. (1975) have drawn up a complete deformation, recrystallization and grain growth cycle to embrace all those differing textural types. Of extreme importance is the fact that rocks with porphyroclastic or porphyroclastic-mosaic texture must have been in the process of deformation at the time they were entrained and brought up in the kimberlites (Boullier and Nicolas 1975); if they had been held in the upper-mantle thermal regime for any appreciable time, annealing would have eliminated their distinctive textures. Thus it appears that the formation of these particular textures was coeval, and possibly directly connected, with the kimberlite event.

The deformation giving rise to porphyroclastic and porphyroclastic mosaic rocks is usually accompanied by some degree of metasomatism, usually localized serpentinization or phlogopitization. Extreme metasomatism is found in some of the so-called polymict peridotites from the Bultfontein and De Beers Mines,

Fig. 48. Photomicrograph showing tabular granuloblastic texture in chromite harzburgite from the Letseng satellite pipe, Lesotho. Note strongly preferred orientation of long axes of both olivine and orthopyroxene crystals. Scale bar = 1 mm; plane-polarized light

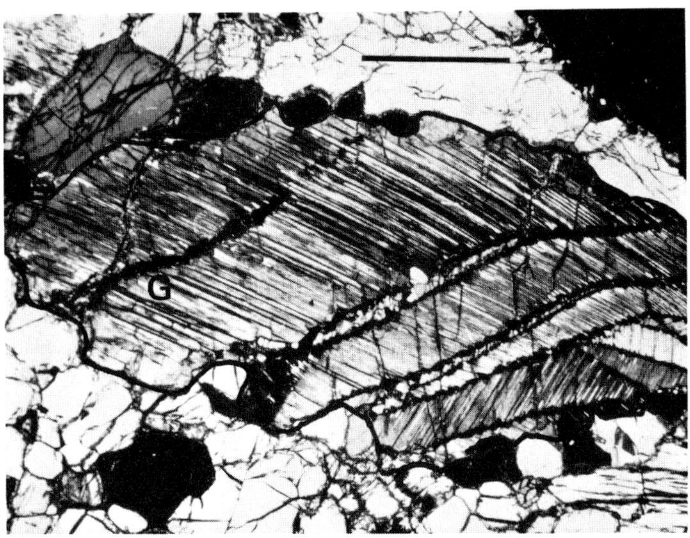

Fig. 49. Photomicrograph showing a large relict, deformed orthopyroxene within the granuloblastic matrix of a garnet lherzolite from the Monastery Mine. The margins of the orthopyroxene crystal have been inked in for clarity. The orthopyroxene is cut by numerous small microfractures along which garnet *(G)* has exsolved; clinopyroxene has exsolved (the light fine bands) parallel to the (100) crystallographic direction of the orthopyroxene; the pyroxene lamellae can be seen to be rotated between some microfractures. Exsolved garnet grains (isotropic) are seen around the margins of the orthopyroxene crystal. Scale bar = 0.5 mm; crossed polarizers

in which clasts and grains derived from deformed peridotite, pyroxenite and eclogite have been intermixed, and subsequently cemented by ilmenite, phlogopite and sulphides (Lawless et al. 1979).

In some relatively-rare granuloblastic textured rocks, quite large porphyroclasts of orthopyroxene have survived the earlier deformation event (Fig. 49). The porphyroclasts are invariable highly deformed (curved crystal-faces or kinks may be seen in hand specimen) and, in thin section the kinking on {100} [101] is invariably accompanied by the appearance of (100) exsolution lamellae of clinopyroxene. Other deformed enstatite crystals contain exsolution rods and blebs of pyrope garnet and chrome diopside, and in one particularly good example from the Monastery Mine, garnet blebs exsolving at kink-band junctions have coalesced to form continuous stringers of garnet (Dawson et al. 1980). It has been observed that under experimental conditions, kinking of orthopyroxene crystals occurs only at high temperatures and low deformation rates; under other conditions the deformation causes inversion to clino-enstatite (Carter 1971).

By contrast with the peridotites, pyroxenites generally are of coarse texture and show little, if any, deformation.

The second important group of textures are those resulting from sub-solidus reactions between constituent minerals. In many garnet-bearing peridotites, the garnets are surrounded by rims of "kelyphite" and, in some instances, the kelyphite completely replaces the garnet. The "kelyphyte" consists of a mineral aggregate in which spinel and mica are the main phases that can be recognized optically; amphibole, clinopyroxene, plagioclase and chlorite have been detected by X-ray diffraction. It is apparent that some of the kelyphite minerals are due to metasomatic addition of alkalies and volatiles from the host kimberlite, the metasomatism having obscured the original minerals of the coronae with the exception of spinel. However, in relatively unmetasomatised lherzolite xenoliths from some Lesotho diatremes (e.g. Letseng Le Terai 2) the garnets are surrounded by rims of spinel, orthopyroxene and clinopyroxene (Figs. 50 and 51); these reaction rims are similar to those described by Reid and Dawson (1972) in lherzolite xenoliths from the Lashaine volcano, Tanzania, in which the coronae are attributed to isochemical, subsolidus reaction between garnet and olivine as a result of changes in PT conditions from garnet-lherzolite facies conditions to spinel-lherzolite facies conditions. The reverse, however, has been noted in lherzolite from the Obnazhennaya pipe (Sobolev and Sobolev 1964); in olivine pyroxenite from the Slyudanka pipe (Frantsesson 1970); in lherzolite from Green Knobs, New Mexico (Smith and Levy 1976); and in olivine pyroxenite from the Frank Smith Mine (Dawson and Smith, unpublished data). In these rocks the spinel is overgrown and replaced by garnet which, in the case of the Slyudanka and Green Knobs examples, forms at the interface between spinel and enstatite.

Another possible sub-solidus texture is chromite grains in whorl-like filaments intergrown with other phases in some coarse-textured peridotites (Fig. 52). These "fingerprint" chromites have been found intergrown with olivine, orthopyroxene,

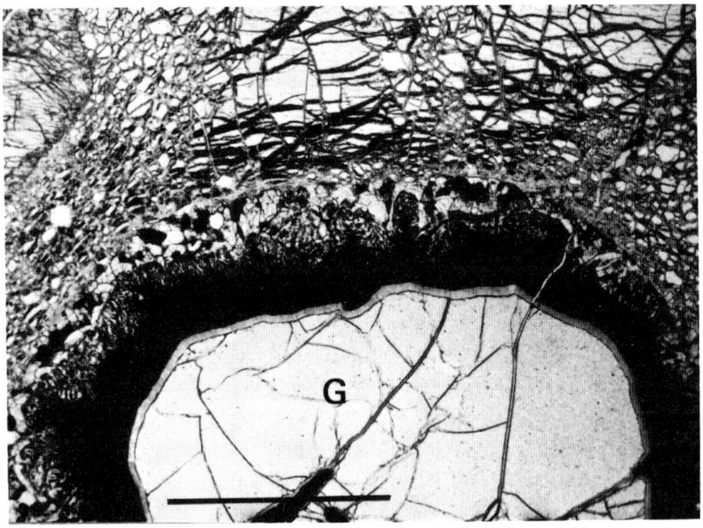

Fig. 50. Photomicrograph of a reaction corona around garnet *(G)* in garnet lherzolite from the Letseng satellite pipe, Lesotho. The fine-grained dark inner reaction zone is mantled outwards by a zone of orthopyroxene, clinopyroxene and intergrown aluminous spinel. Scale bar = 1 mm; plane-polarized light

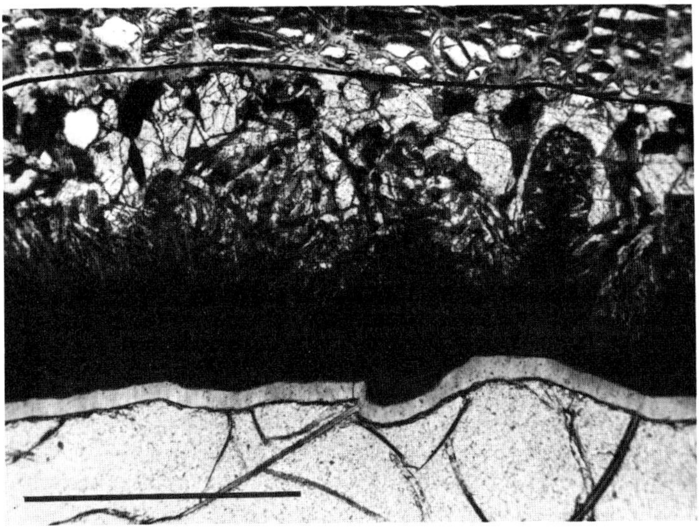

Fig. 51. Enlargement of part of the reaction zone shown in Fig. 50. The contact between the reaction zone and the surrounding porpyroclastic olivine is inked in for clarity. Note the intimate intergrowth of dark spinel and the pyroxenes in the outer part of the reaction zone. Scale bar = 0,5 mm; plane-polarized light

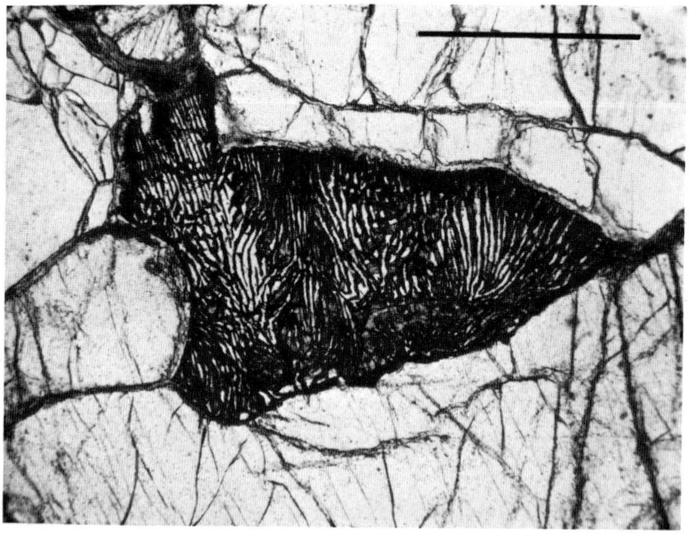

Fig. 52. Photomicrograph of a chromite-pargasite symplectite in chromite harzburgite from the Wesselton Mine. Surrounding grains are olivine. Scale bar = 0.5 mm; plane-polarized light

clinopyroxene and pargasite; the intergrowths are referred to as simple when the chromite is intergrown with one silicate grain, or complex if it is intergrown with two or more silicate grains, not necessarily of the same species (Dawson and Smith 1975a). The "fingerprints" often occur at triple-junctions of host-silicate grains, perhaps suggesting redistribution of chromite during recrystallization; however, within rocks containing fingerprint chromites, there are invariably chromite grains of normal morphology, and the reason for these contrasting morphologies cannot be explained at present.

Whereas the deformation and recrystallization textures mentioned above result from an external stress-field, Carstens (1971) has observed localized areas of plastic strain around inclusions of rutile and olivine in pyrope from lherzolite xenolith from South Africa and Yakutia; these localized stress fields are due to differences in the coefficients of thermal expansion and compressibilities between the inclusions and the pyrope host, and are manifested in birefringent haloes and dislocation tangles in the pyrope immediately adjacent to the inclusions.

4. Bulk Chemistry

a) Major Elements. Representative analyses of rocks of the peridotite-pyroxenite suite may be set in a sequential series (Table 18). This series is based upon the premise that, starting with a parental garnet lherzolite, more iron-rich liquids may be generated by small amounts of partial melting of garnet lherzolite and will

Table 17. Representative analyses of rocks of the peridotite-pyroxenite suite

	1	2	3	4	5	6	7	8
SiO_2	46.53	44.66	44.99	41.80–47.85	45.37	44.36	51.61	47.0
TiO_2	0.09	0.09	0.06	0.02– 0.16	0.01	0.01	0.11	0.01
Al_2O_3	1.84	1.22	1.40	0.53– 3.26	1.29	1.14	4.24	2.0
Cr_2O_3	0.38	0.35	0.32	0.21– 0.62	0.42	0.34	0.57	0.6
FeO^T	6.70	7.87	7.89	6.70– 9.17	5.50	5.90	7.17	4.0
MnO	0.11	0.11	0.11	0.09– 0.15	0.10	0.09	0.18	0.08
MgO	41.98	42.94	42.60	37.6 –45.7	42.5	44.3	27.65	45.8
NiO	0.32	0.37	0.26	0.25– 0.43	0.27	0.30	0.12	0.26
CaO	1.47	0.81	0.82	0.35– 3.50	0.65	0.53	6.61	0.04
Na_2O	0.16	0.15	0.11	0.02– 0.60	0.12	0.08	0.63	0.0
K_2O	0.15	0.12	0.04	0 – 0.35	0.06	0.06	0.06	

1. Average of 9 garnet lherzolites (Carswell and Dawson 1970)
2. Average of 41 S. African garnet lherzolites (Chen 1971)
3. Average of 202 garnet lherzolites (Maaløe and Aoki 1977)
4. Range of analyses used in averages in 1, 2 and 3
5. Garnet harzburgite (O'Hara et al. 1975)
6. Harzburgite (O'Hara et al. 1975)
7. Websterite 1042 (O'Hara et al. 1975)
9. Estimated composition of garnet harzburgite formed by pro-grade metamorphism of aluminous harzburgite (Dawson et al. 1978)

Table 18. Analyses of rocks of the peridotite-pyroxenite suite from Matsoku, Lesotho (Cox et al. 1973)

	Dunite LBM 20	Harzburgite LBM 21	Garnet lherzolite LBM 9	Garnet pyroxenite LBM 18
SiO_2	39.99	45.57	47.49	51.50
TiO_2	0.04	0.01	0.03	0.17
Al_2O_3	0.01	0.85	2.45	6.27
Cr_2O_3	0.05	0.36	0.52	0.40
Fe_2O_3	1.69	1.34	1.07	1.74
FeO	8.02	4.40	5.18	7.82
MnO	0.12	0.10	0.12	0.18
MgO	47.26	44.40	38.84	25.18
CaO	0.12	0.44	1.36	4.85
Na_2O	0.00	0.01	0.04	0.12
K_2O	0.03	0.05	0.16	0.06

crystallize to form garnet websterite and garnet pyroxenite; the complementary residual rocks in this melting sequence are garnet harzburgite, harzburgite and dunite. Three average analyses are given for garnet lherzolite in Table 17; there is reasonably good agreement between these three averages, the main differences being in Al_2O_3, CaO and K_2O. Compared with garnet lherzolite, garnet harzburgite and harzburgite is relatively depleted in FeO, TiO_2, CaO and alkalies, due to an increased proportion of modal olivine and particularly of orthopyroxene. The complementary garnet websterite is relatively enhanced in TiO_2, Al_2O_3, Cr_2O_3, total FeO, CaO and Na_2O. A gradational sequence from dunite to garnet websterite is shown in the specimens from the Matsoku pipe shown in Table 17, and in diagrams showing suites of rocks from the Matsoku and Bultfontein pipes (Fig. 53).

By contrast, a garnet harzburgite believed to have been derived by pro-grade metamorphism of aluminous harzburgite has an atypical chemistry in that extremely high magnesia values are combined with low CaO and Na_2O (Table 17).

b) Trace Element Chemistry. A summary of the trace element content of peridotites and pyroxenites is given in Table 19. The ranges for values of the elements for the whole suite are given, together with the relative concentration in the various phases and the maximum values found in the principal phase bearing the element; for more detailed information the interested reader can consult the references listed. The variations in the bulk-rock values reflect the modal proportions of the various phases and the substitution of trace elements for major elements. For example, NiO substituting for MgO, particularly in olivine, is more concentrated in dunite and harzburgite; whereas strontium substituting for CaO in diopside is relatively concentrated in websterites and pyroxenites. Rb, Li, Ba, Cs, substituting for K are found most concentrated in micas.

Rare-earth element patterns for bulk peridotites have either a flat chondrite-like pattern or a slightly negative slope (see Fig. 54). Haskin et al. (1968) and Philpotts et al. (1969) have shown that the heavy rare-earths are more concentrated in the garnets giving them a low Ce/Yb ratio, whereas the clinopyroxenes contain more light REE and have a high Ce/Yb ratio (Fig. 55). These REE patterns for garnets and clinopyroxenes have been confirmed by Mitchell and Carswell (1976) and Shimizu (1975a). The flat chondrite-like pattern for the bulk rocks indicates approximately equal amounts of garnet and clinopyroxene in the bulk rocks. Shimizu (1975a) has pointed out that deformed xenoliths apparently have flat or flat chondrite-like REE pattern of a very shallow negative slope, whereas coarse xenoliths, allegedly depleted in mobile elements (like the potassium and rubidium values, see p. 138) show relatively high values of total rare-earths and moreover have a much steeper negative slope. Again there is this apparent clash between major and trace element evidence, as to which type of xenolith is the more depleted.

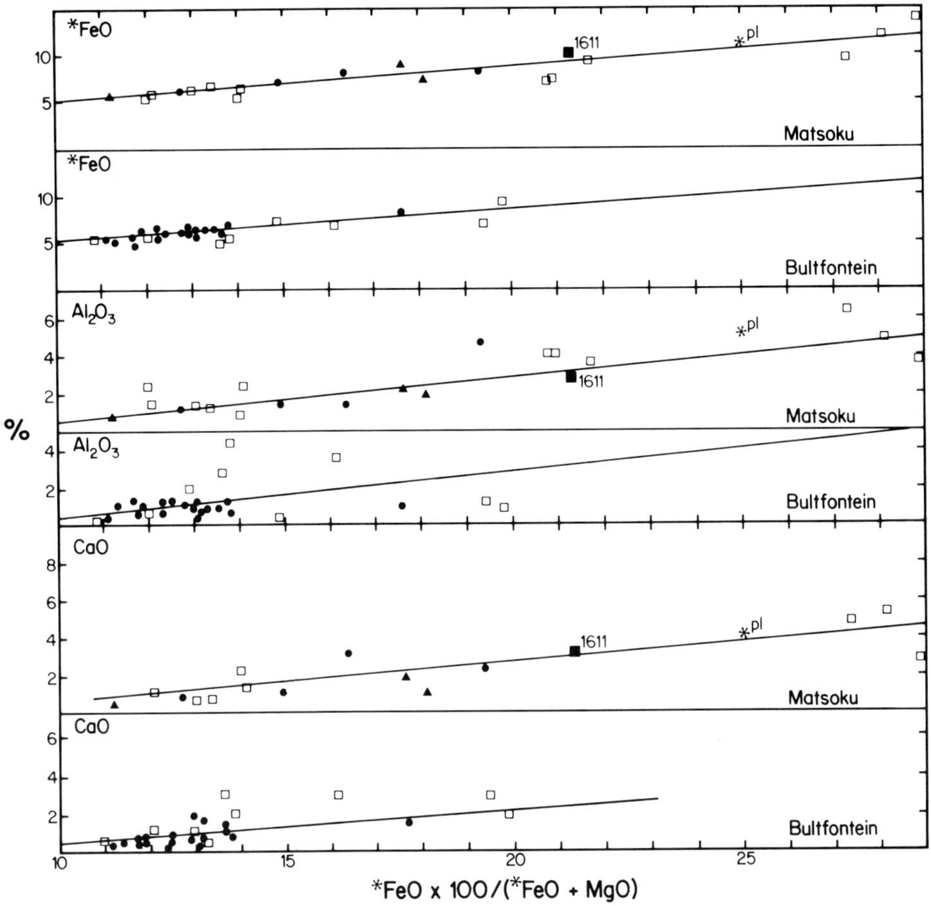

Fig. 53. Plots of CaO, total FeO and Al_2O_3 against 100 $FeO^T/(FeO^T + MgO)$ in xenolith suites from the Matsoku and Bultfontein kimberlites, southern Africa (after Gurney et al. 1975). *Open squares* coarse granular rocks; *solid circles* deformed rocks (porpyhroclastic and mosaic textures); *triangles* granuloblastic rocks; *1611* "fertile" garnet lherzolite PHN1611 (Boyd and Nixon 1973a); *pl* primary lherzolite (Kuno and Aoki 1970)

Because some garnet lherzolites contain high amounts of rare-earth elements and are enriched in light rare-earth elements related to heavy, Shimizu (1975b) has proposed that garnet lherzolites were originally cumulates consisting of olivine, orthopyroxene and clinopyroxene which have recrystallized to form garnet lherzolite; this hypothesis is supported by the textural evidence for the exsolution generation of some garnet lherzolites (Dawson et al. 1980).

Table 19. Trace element content of rocks and phases in the peridotite-pyroxenite suite

Element	Rock range ppm	Phase concentration order [a]	Concentration in main phase ppm	Reference
Li	0.5–3.0	m > ol = op = cp = gt	m 8	2, 13
Be	< 3			2
Sc	1–10	gt > cp	gt 90–110	2, 6
V	70–200	ct > gt	cp 400–600	1, 2
Co	90–120	cp > gt	cp 25–40	2, 6
Ni	90–350	ol > op > cp > gt	ol 2000	1, 2, 7
Cu	35–180			16
Rb	1–57	m > cp > op = gt > ol	m 200–700	4, 6, 8, 9, 12, 13, 14
Sr	50–70	cp > m > op = gt = ol	cp 90–550	2, 5, 8, 9, 12, 14
Y	30			2
Zr	?	cp > gt	cp 40–70	1, 2
Nb	2.8–14			17
Ag	0.05–3.6			16
Cs	0.1–5.2			11
Ba		m > cp > gt > op > ol	m 1300–1500	2, 8, 14
La	0.06–0.99			18
Ce		cp > gt	cp 24.3	19
Nd		cp > gt	cp 21.6	19
Sm	0.11–0.31	cp > gt	cp 5.33	18, 19
Eu		gt > cp	gt 1.04	18, 19
Cd		gt > cp	gt 3.82	19
Dy		gt > cp	gt 5.04	19
Er	0.03–0.29	gt > cp	gt 3.29	19
Yb		gt > cp	gt 2.96	18, 19
Au	0.06–0.81			15, 16
Pb	37.3	m > gt > cp	High in secondary m	2, 12
Bi	0.011–0.03			20
Th	0.2–1.7		High in secondary m	1, 3, 12
U	0.03–1.11	m > gt > cp > op = ol	High in secondary m	1, 3, 10, 12, 21

a m = mica, ol = olivine, cp = clinopyroxene, op = orthopyroxene, gt = garnet

References:
1. Holmes (1936); 2. Nixon et al. (1963); 3. Heier (1963); 4. Heier and Compston (1966); 5. Stueber and Murthy (1966); 7. Stueber and Goles (1967); 8. Griffin and Murthy (1968); 9. Allsopp et al. (1969); 10. Akimov et al. (1968); 11. Gurney and Berg (1968); 12. Manton and Tatsumoto (1971); 13. Stavrov and Ukhanov (1971); 14. Philpotts et al. (1972); 16. Ivanov et al. (1977); 17. Lutts et al. (1974); 18. Mitchell and Carswell (1976); 19. Shimizu (1975a); 20. Dunin-Barkovskaya and Ukhanov (1974); 21. Kramers (1977)

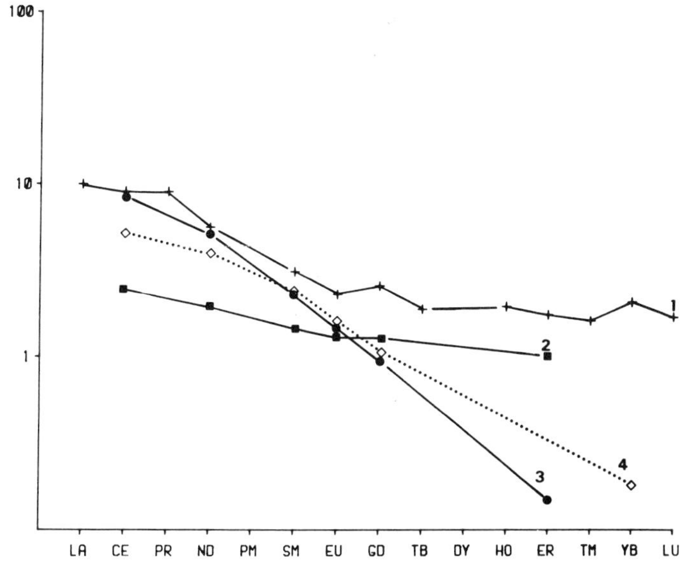

Fig. 54. Rare-earth element plots for bulk peridotites. Data were plotted after normalization against standard chondrite (Haskin et al. 1966). The specimens are *1* peridotite, Wesselton Mine (Haskin et al. 1966); *2* garnet lherzolite, Roberts Victor Mine; *3* lherzolite, Robert Victor Mine; *4* garnet lherzolite, Bultfontein Mine (*2–4* are from Phillpotts et al. 1972)

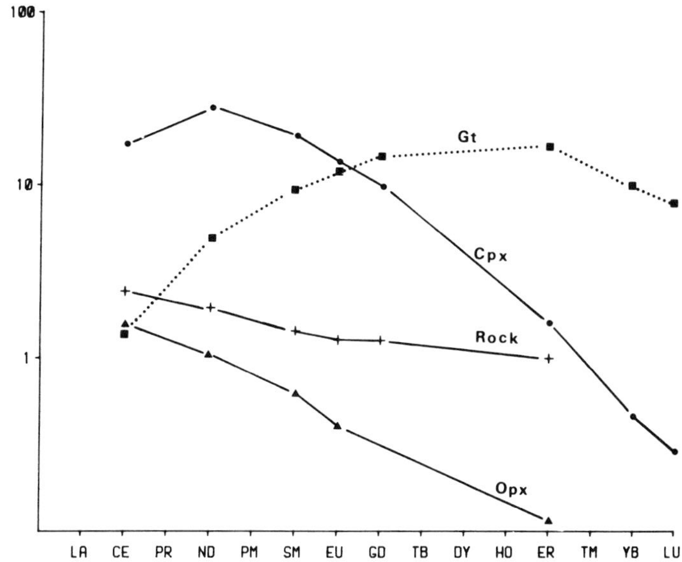

Fig. 55. Rare-earth element plot for bulk rock and phases in garnet lherzolite 187, Roberts Victor Mine (Phillpotts et al. 1972)

5. Phase Chemistry. Most analytical work has been aimed at establishing the compositions of phases in large numbers of these upper mantle rocks. Recently, however, the picture has become more complex when intensive studies on individual specimens has shown small but significant inter-grain differences within single specimens, associated either with localized deformation (e.g. Boyd and Finger 1975) or disequilibrium (Dawson et al. 1978b).

Olivine is the phase with the most simple chemistry and in most rocks ranges compositionally between 88% and 94% forsterite; some in more iron-rich, deformed (porphyroclastic or porphyroclastic-mosaic) dunites have compositions in the range 79%–83% forsterite (Dawson and Smith, unpublished); Nixon and Boyd (1973a) find that the olivines in deformed garnet lherzolites from Thaba Putsoa, Lesotho, are more iron-rich (Fo_{90}) and contain more Al_2O_3 and Cr_2O_3 than coarse garnet lherzolites (Fo_{93}), and this difference must be sought at other localities to determine whether it is a localized or more general phenomenon. NiO in olivines varies from 0.38–0.48 wt% and MnO from 0.07–0.89 wt%. High precision analyses for trace elements are Al_2O_3 0.006–0.020 wt%; Cr_2O_3 0.006–0.04 wt%, CaO 0.013–0.04 wt%; TiO_2 0–390 ppm (av. 130), Na_2O 40–130 (av. 90) ppm; $K_2O < 20$ ppm; P_2O_5 50–200 (av. 130) ppm (Bishop et al. 1978). Although the Na, P, and Ti contents are low they are nonetheless important in view of the high abundance of olivine in the upper mantle. There has been some debate, in view of relatively high amounts of Cr in lunar olivines, whether Cr in olivine in general is di- or tri-valent.

Orthopyroxene varies from about 87%–95% enstatite; in the following, the terminology of Stephens and Dawson (1977) is followed (for ranges in major oxides in these orthopyroxenes see Table 13). Table 20 gives representative analyses of 149 orthopyroxenes from the literature (Fig. 56), and shows *enstatite* to be the commonest orthopyroxene, particularly in coarse garnet-lherzolites; however, orthopyroxenes from other garnet lherzolites are richer in iron and titanium, though the fact that an equally large number of undeformed lherzolite orthopyroxenes are Ti-enstatites or Ti-bronzites indicates no overall correlation between texture and Fe/Ti contents in lherzolite orthopyroxenes. *Cr-Al enstatite* (containing appreciable garnet in solid solution) is less common and occurs principally in lherzolites and harzburgites in which chromite is present rather than an aluminous spinel. *Na-Ca enstatite* occurs mainly in deformed lherzolites, but also in coarse lherzolites. *Ti-enstatites* are present in deformed and coarse garnet lherzolites, and also is the orthopyroxene in most garnet-websterites. *High Ti-bronzite* is present in sheared and deformed garnet lherzolites and in garnet-websterites. From the histogram four main facts are evident: (1) enstatite is the commonest orthopyroxene in garnet lherzolite, (2) there is no overall evidence of enrichment of iron, sodium, calcium and titanium in deformed, as opposed to coarse, garnet lherzolites, as suggested by Nixon and Boyd (1973a) on the basis of studies on the suite of specimens from Thaba Putsoa, Lesotho; (3) garnet-websterites generally contain more iron- or titanium-rich orthopyroxenes; (4) the histograms underline the

Table 20. Representative analyses of ortho- and clino-pyroxenes in rocks of the peridotite-pyroxenite series. (Nomenclature after Stephens and Dawson 1977)

	1	2	3	4	5	6	7	8	9	10	11	12
SiO_2	57.55	57.51	56.6	56.8	57.20	56.8	55.72	55.32	55.92	52.65	55.54	55.68
TiO_2	0.03	0.00	0.00	0.00	0.18	0.11	0.22	0.00	0.14	0.82	0.05	0.09
Al_2O_3	0.75	0.97	2.43	2.11	0.97	0.74	0.88	1.39	1.75	5.92	2.43	2.48
Cr_2O_3	0.28	0.45	0.57	0.48	0.27	0.22	0.18	1.01	0.87	1.43	1.48	2.60
FeO^T	4.57	4.82	4.48	4.18	5.59	9.54	8.54	3.14	3.27	1.62	2.52	2.43
MnO	0.07	0.11	0.13	0.11	0.14	0.15	0.18	0.13	0.09	0.05	0.09	0.09
MgO	35.01	34.96	35.8	34.9	33.98	31.8	32.32	20.91	17.71	15.11	16.15	16.35
CaO	0.56	1.45	0.68	1.51	1.08	0.59	0.85	17.59	20.39	18.60	20.64	18.73
Na_2O	0.11	0.15	0.02	0.47	0.25	0.18	0.19	0.82	1.38	2.47	1.95	2.46
	98.92	100.42	100.71	100.56	99.66	100.17	99.08	100.31	101.52	99.92	100.84	100.91

1. Enstatite in coarse garnet websterite LBM9, Matsoku, Lesotho (Cox et al. 1973)
2. Enstatite in sheared garnet lherzolite 1591, Thaba Putsoa, Lesotho (Nixon and Boyd 1973a)
3. Cr-Al enstatite, coarse chrome-spinel harzburgite BD1733, Newlands Mine (Dawson and Smith 1975a)
4. Na-Ca enstatite, coarse chromite-pargasite harzburgite BD1368, Monastery Mine (Dawson and Smith 1975a)
5. Ti-enstatite, deformed garnet lherzolite 2001, Mothae, Lesotho (Nixon and Boyd 1973a)
6. High-Ti- bronzite, coarse garnet lherzolite, Letseng satellite pipe, Lesotho (Bishop et al. 1978a)
7. High Ti-bronzite, deformed garnet lherzolite 1582, Thaba Putsoa, Lesotho (Nixon and Boyd 1973a)
8. Sub-calcic diopside, deformed garnet lherzolite 1591, Thaba Putsoa, Lesotho (Nixon and Boyd 1973a)
9. Diopside, coarse garnet websterite LBM6, Matsoku (Cox et al. 1973)
10. Ti-Cr diopside, garnet lherzolite 433D/586, Obnazhennaya, U.S.S.R. (Milashev et al. 1963) (total includes small amounts of K_2O, H_2O)
11. Chrome-diopside, coarse garnet lherzolite JJG221, De Beers Mine, S. Africa (Cox et al. 1973)
12. Ureyitic-diopside, coarse garnet lherzolite LBM 11, Matsoku, Lesotho (Cox et al. 1973)

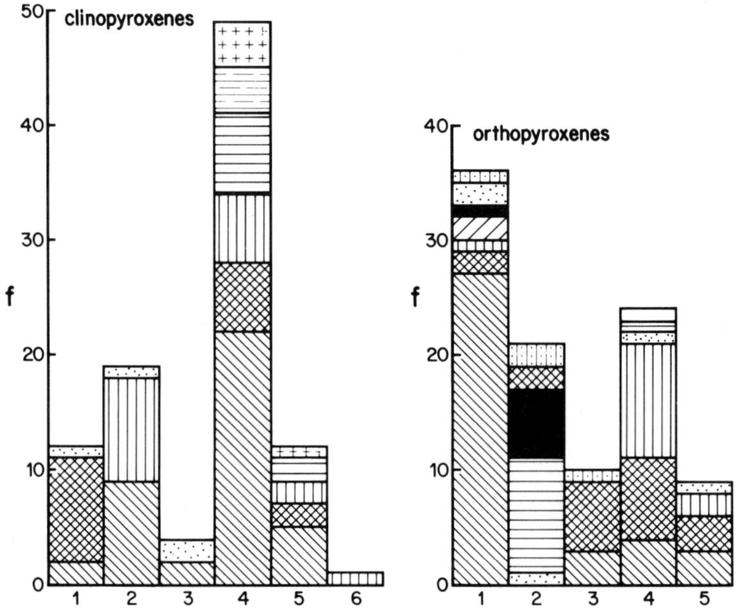

Fig. 56. Histograms of orthopyroxenes and clinopyroxenes in rocks of the peridotite-pyroxenite group. Numbers along the base refer to clinopyroxene and orthopyroxene nomenclature which are: clinopyroxenes – *1* subcalcic diopside; *2* diopside; *3* Ti-Cr diopside; *4* chrome diopside; *5* ureyitic diopside; *6* jaeditic diopside; orthopyroxenes – *1* enstatite; *2* Cr-Al enstatite; *3* Na-Ca enstatite; *4* Ti-enstatite; *5* Ti-bronzite; pyroxene nomenclature is after Stephens and Dawson (1977). Rock-type ornament is as in Fig. 57

generalization that orthopyroxenes of differing compositions (mainly in terms of MgO, FeO and TiO_2) occur in modally variable members of the peridotite-pyroxenite group.

From the point of view of minor and trace elements, with the exception of the Cr-Al enstatites, Al_2O_3 is < 2.5 wt% and, in the commonest orthopyroxene (enstatite) is usually < 1 wt%. In the enstatites, CaO is low ($< 1\%$) indicating only limited solution towards diopside. Dawson et al. (1980) divide harzburgite orthopyroxenes into two main groups, "barren" (containing relatively low amounts of Ca, Al, Cr and Na) and "fertile" (those containing relatively high amounts of those elements). Boyd (1975) has shown that orthopyroxene neoblasts in deformed garnet lherzolites are erratically but markedly enriched in CaO, Al_2O_3 and Cr_2O_3 relative to the enstatite porphyroclasts. Trace elements values for garnet lherzolite orthopyroxenes are Na_2O 530–1900 (1070); K_2O 10–110 (30); P_2O_5 0–90 (50); and TiO_2 10–1650 (480); values in ppm, averages in parentheses (Bishop et al. 1978).

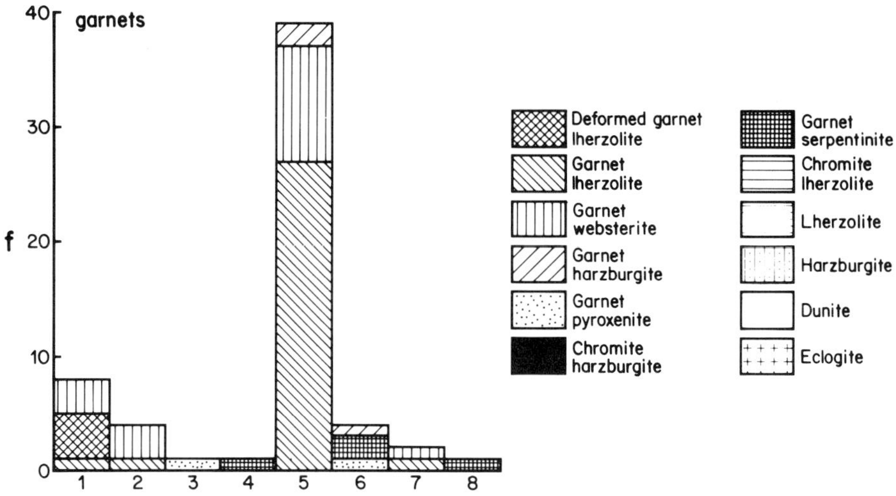

Fig. 57. Histogram of garnets in the peridotite-pyroxenite group. Types of garnet are: *1* titanian-pyrope; *2* calcic pyrope-almandine; *3* pyrope-grossular-almandine; *4* ferro-magnesian uvarovite grossular; *5* chrome pyrope; *6* low-calcium pyrope; *7* titanian uvarovite-pyrope; *8* knorringitic uvarovite pyrope; garnet terminology from Dawson and Stephens (1976)

Clinopyroxene nomenclature is that outlined in the chapter on kimberlite mineralogy. The clinopyroxenes show considerable compositional variation (Table 20), particularly with respect to the extent of enstatite solid solution. Ca/(Ca + Mg) ratios vary from 0.26 to 0.48 which can be experimentally correlated with equilibration temperature variation of 1400°–950°C (Davis and Boyd 1966), though pyroxene geothermometry is currently the subject of considerable debate. Those with low values (i.e. with most enstatite in solid solution) equilibrated at *relatively* high temperatures, though precise temperatures are subject to correction for other elements (Fe, Al, Cr, Na, Ti) and to bulk-rock compositions (Lindsley and Dixon 1977). Nixon and Boyd (1973a) correlated sub-calcic composition (high temperature of equilibration) with deformed texture in garnet lherzolites from Thaba Putsoa, but observations by Cox et al. (1973) and Dawson et al. (1975) from other localities indicate no overall correlation between texture of the host peridotites and temperature of equilibration of the clinopyroxenes. Nonetheless, whereas all coarse-grained and *some* deformed garnet lherzolites have equilibrated at 950°–1050°C, all rocks containing sub-calcic diopsides (implying high-temperature equilibration), are deformed. Apart from the major oxides, the clinopyroxenes contain appreciable amounts of Na_2O, Al_2O_3 and Cr_2O_3 (see Table 19), and also small, but significant amounts of K_2O (10–110, av. 30 ppm), P_2O_5 (90–630, av. 350 ppm) (Bishop et al. 1978), Rb (0.02–0.26 ppm), Sr (84–558 ppm), Ba (0.2–45 ppm) and Cs (0.1–10.2 ppm) (Shimizu 1975b). The presence of K_2O

has been attributed to high pressures (Sobolev 1977), an analogous effect being the presence of small amounts of Na_2O in solid solution in peridotite and eclogite garnets. Shimizu (1975b) demonstrates differing K/Rb ratios in clinopyroxenes from deformed peridotites (K/Rb 1100–2200) compared with those from coarse peridotites (200–1400); this relative rubidium depletion in the deformed peridotites is, like the bulk-rock REE evidence, contrary to the major element evidence (from FeO, CaO, Al_2O_3 and Na_2O values) that shows deformed peridotites to be *less* depleted than coarse-textured peridotites.

The clinopyroxene histogram (Fig. 56) shows chrome diopside to be the commonest type in rocks of the peridotite-pyroxenite group despite their variable modes and it is evident that both coarse and deformed garnet lherzolites contain clinopyroxenes of differing compositions.

Some chrome-diopsides in lherzolite xenoliths have cloudy margins or patches of altered diopside that is in optical continuity and has the same birefingence as unaltered pyroxene; relative to original pyroxene, these altered areas contain lower Na_2O and Al_2O_3, but similar Cr_2O_3 and Ca/(Ca + Mg) ratios (Carswell 1975).

Garnet of different compositions (Table 21) occurs in many rocks of the peridotite-pyroxenite group. They belong to eight of the twelve chemical groups established by Dawson and Stephens (1975, 1976) on the variability of TiO_2, Cr_2O_3, MgO, CaO and FeO^T in garnets from kimberlite, associated xenoliths and diamond inclusions (Fig. 57). FeO^T tends to be reasonably constant in peridotite garnets and variations are mainly in CaO vs. MgO, and in Cr_2O_3 vs. Al_2O_3. Chrome pyrope is by far the commonest type, occurring mainly in coarse garnet lherzolites, though also in websterites, pyroxenites and harzburgites. Garnet lherzoliths, though containing mainly chrome pyrope, also contain titanian pyrope (in both coarse and deformed lherzolites) and titanian uvarovite-pyrope. Diamond has been found in peridotites containing chrome pyrope, and in harzburgites and dunites containing low-calcium chrome pyrope and knorringitic uvarovite-pyrope (Sobolev 1977). It has been generally assumed that garnets are homogeneous in peridotites and pyroxenites, but recent studies by Boyd and Finger (1975) and Dawson et al. (1978a) show significant inter-grain and intra-grain variations in CaO, TiO_2, Cr_2O_3 and Al_2O_3. Garnet zoned mainly with respect to Cr_2O_3 and Al_2O_3 occurs in non-equilibrated harzburgite from Liqhobong, Lesotho (Dawson et al. 1978), and disequilibrium also occurs between garnet and spinel in lherzolites from Green Knobs, New Mexico (Smith and Levy 1976). Small amounts of Na_2O in peridotitic and eclogitic garnets were first reported by Sobolev and Lavrentyev (1971) who propose its presence is due to high pressures; subsequent high precision analyses by Bishop et al. (1978) show Na_2O 150–790 (340), together with P_2O_5 200–1040 (560), $K_2O < 20$, TiO_2 160–5190 (1470) (values in ppm, averages in parentheses). The latter authors show that (a) atomic Na usually exceeds P, (b) Na generally increases with Ti but usually is less than it, and (c) atomic Ti + P always exceeds atomic Na; they propose such coupled substitutions as $Na^{VIII} + P^{IV} \rightarrow Ca^{VIII} + Si^{IV}$ and $Na^{VIII} + Ti^{VI} \rightarrow Ca^{VIII} + Al^{VI}$, and see no need

Table 21. Representative analyses of garnets found in rocks of the pyrope-pyroxenite group

	1	2	3	4	5	6	7	8
SiO_2	42.68	41.19	41.7	40.6	43.0	44.4	41.68	41.5
TiO_2	0.80	0.31	0.05	0.52	0.04	0.0	0.80	0.19
Al_2O_3	21.15	21.97	21.2	20.5	17.6	19.7	16.58	12.5
Cr_2O_3	1.46	1.06	4.13	3.48	8.20	5.84	7.15	14.1
FeO^a	8.63	13.15	6.87	7.71	6.82	5.47	7.46	7.28
MnO	0.26	0.36	0.35	0.36	0.32	0.33	0.27	0.40
MgO	20.66	17.47	21.3	19.9	22.9	24.7	20.40	18.8
NiO	n.d.	n.d.	0.009	0.0	n.d.	0.0	n.d.	n.d.
CaO	4.30	4.41	5.83	4.43	1.93	0.42	5.90	6.24
Na_2O	0.07	n.d.	0.037	0.07	–	0.0	n.d.	n.d.
	100.01	99.93	100.54	97.61	100.8	100.8	100.4	101.0

a Total iron as FeO
1. Titanian pyrope, in deformed lherzolite, Thaba Putsoa, Lesotho (Nixon and Boyd 1973a)
2. Calcic pyrope-almandine, in garnet pyroxenite LBM18, Matsoku, Lesotho (Cox et al. 1973)
3. Chrome-pyrope, in garnet lherzolite 1149, Bultfontein, S. Africa (Bishop et al. 1978) (Total includes 0.069 wt% P_2O_5)
4. Chrome-pyrope, in garnet pyroxenite 1156, Bultfontein, S. Africa (Bishop et al. 1978) (Total includes 0.036 wt% P_2O_5)
5. Low-calcium chrome-pyrope, in garnet-serpentinite A-2a, Aikhal pipe, U.S.S.R. (Sobolev 1977)
6. Low-calcium chrome-pyrope, in garnet harzburgite 2093, N.W. pipe, Liqhobong, Lesotho (Dawson et al. 1978b)
7. Titanian uvarovite-pyrope, garnet "peridotite" Ud-5, Udachnaya pipe, U.S.S.R. (Sobolev 1970)
8. Knorringitic uvarovite-pyrope, diamondiferous garnet serpentinite A-4a, Aikhal pipe, U.S.S.R. (Sobolev et al. 1969)

to invoke substitution of six-fold coordinated Si as proposed by Sobolev and Lavrentyev (1971).

Amphibole. Two types of amphibole are known in peridotites: (1) chromiferous pargasite/pargasitic hornblende, occurring as a primary phase either as isolated grains or intergrown with fingerprint chromite (Fig. 58) (Table 22); and (2) potassic richterite that is a primary metasomatic phase occurring either as grains widespread throughout the metasomatized peridotites or in veins accompanied by phlogopite and diopside (see p. 183) of xenolith localities studied to date, richterite is most common in peridotites from the Bultfontein Mine. In most respects the primary metasomatic richterite is compositionally similar to K-richterite in rocks of the MARID-suite (see p. 186).

Mica. The chemistry of peridotite micas has been reviewed by Delaney et al. (1980) who recognize a correlation between the composition and textures of the micas. Primary micas are those which are in textural equilibrium with the

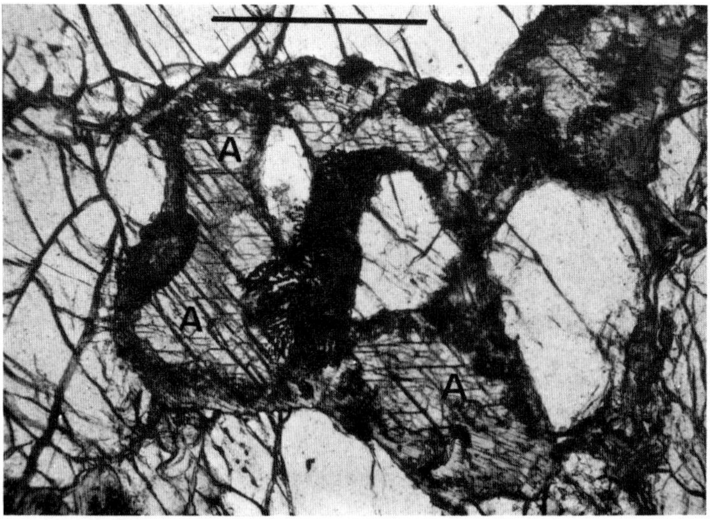

Fig. 58. Photomicrograph of chrome-pargasite *(A)* in amphibole harzburgite from the Monastery Mine. The other crystals are olivine. Scale bar = 0.5 mm, plane-polarized light

Table 22. Representative analyses of peridotite amphiboles

	1	2	3
SiO_2	45.5	45.1	55.4
TiO_2	0.01	0.0	0.46
Al_2O_3	11.1	11.1	0.99
Cr_2O_3	1.67	2.20	0.48
FeO^T	3.18	2.69	2.25
MnO	0.06	0.18	n.d.
MgO	20.0	20.3	22.9
NiO	0.10	0.36	n.d.
CaO	10.6	10.9	6.74
Na_2O	3.79	3.23	3.82
K_2O	0.60	1.34	4.63
	96.5	97.4	97.7

1. Pargasite, intergrown with chromite, in harzburgite, Wesselton Mine (Boyd 1971a)
2. Chromic pargasitic hornblende, separate grains and intergrown with chromite, in harzburgite, Monastery Mine (Dawson and Smith 1975a)
3. Average of 22 primary metasomatic richterites from peridotites, Bultfontein Mine (Erlank and Richard 1977)

surrounding silicate phases; they may occur as distinct isolated blades or else in aggregates or clots together with clinopyroxene. Secondary micas are those found in the so-called kelyphite rims around the garnet (often including secondary spinel) and also replacing orthopyroxene marginally or along cleavage traces. The chemistry of these various types of mica is shown in Table 23. The primary micas are phlogopites containing only moderate amounts of iron in the region 2.6 to 2.7 wt% FeO and only moderate amounts of TiO_2; they are fluor-phlogopites containing between 0.4 and 0.8 wt% F. Evidence that these are part of an equilibrium is showed by good correlation for *mg* and TiO_2 between the mica and co-existing orthopyroxene and clinopyroxene. Compared with the primary micas, the secondary micas contain greater amounts of TiO_2, Cr_2O_3, FeO, and Na_2O, together with lower amounts of K_2O and MgO; in addition, the primary micas contain approximately three times as much Cl. Rubidium values are approximately the same for both primary and secondary mica. The secondary micas show distinct departures from straight line correlation between values for *mg* and TiO_2 between the micas and coexisting orthopyroxene and clinopyroxene. The high Cr_2O_3 value of the

Table 23. Composition of peridotite and MARID-suite micas

	1	2	3	4	5	6
SiO_2	41.01	41.19	42.21	40.25	41.0 –42.7	42.23
TiO_2	0.18	0.46	0.66	1.79	1.07– 1.59	1.26
Al_2O_3	13.49	12.89	12.88	13.49	11.7 –13.2	10.21
Cr_2O_3	0.82	0.68	0.84	1.17	0.19– 0.81	0.22
FeO	2.60	2.76	3.28	3.84	2.79– 4.57	6.40
MnO	0.02	0.02	0.02	0.03	0.05– 0.05	0.03
MgO	26.03	25.67	25.91	24.15	23.9 –25.5	23.9
NiO	0.22	0.19	0.21	0.16	0.16– 0.24	0.19
CaO	0.01	0.04	0.02	0.03	0.05	–
Na_2O	0.31	0.48	0.35	0.51	0.17– 0.43	0.14
K_2O	10.00	10.06	9.47	9.74	9.27– 9.97	10.4
Cl	0.08	0.06	0.02	0.03		
F	0.43	0.76	0.50	0.68		
BaO	0.29	0.19	0.22	0.137		
Rb_2O	0.028	0.023	0.029	0.028		
mg	0.947	0.943	0.933	0.917		

1. Mean of 10 analyses of primary textured micas with no special feature
2. Mean of 6 analyses – primary micas associated with clinopyroxenes
3. Mean of 8 analyses – secondary textured micas in harzburgites, replacing orthopyroxene
4. Mean of 21 analyses – secondary textured micas in lherzolites, replacing garnet
   1–4 from Delaney et al. (1980)
5. "Primary metasomatic" phlogopite, range for 8 xenoliths, Matsoku, Lesotho (Harte and Gurney 1975b)
6. Mean of 8 analyses of MARID-suite phlogopites (Dawson and Smith 1977, Table 2, analyses 1–8)

micas replacing garnet is particularly apparent, the high Cr_2O_3 apparently being inherited from the replaced chrome pyrope.

Ilmenite is a comparatively rare mineral in peridotites and pyroxenites. It is magnesian with MgO contents similar to those in ilmenites from kimberlite but has a lower but variable oxidation state; for example, the two analyses in Table 24 for undeformed garnet lherzolites show the ilmenites to have very low Fe_2O_3; by contrast those in deformed lherzolite and dunite have a higher Fe_2O_3. All the rocks containing ilmenite are as a whole rich in titanium and contain garnets and clinopyroxenes more titanium-rich than is usually the case for lherzolites, and the presence of ilmenite is probably due to excess titanium over that which can be accommodated in the garnet and pyroxene lattices (Akella and Boyd 1972). Besides the low oxidation state these peridotite ilmenites differ from those occurring in the matrix of kimberlite in having a very low Al_2O_3 content (see Table 11). Ilmenite is also a reasonably common phase in mosaic- or porphyroblastic-textural iron-rich dunites.

Spinel. The chemistry of spinel in peridotite has been reviewed by Smith and Dawson (1975). Representative analyses are given in Table 25. Discrete grains occurring in the peridotitic rocks tend to have a fairly constant Mg/Fe ratio and the main variation is in the proportion of Cr_2O_3 to Al_2O_3. Broadly speaking the highest Cr/Al ratios are found in lherzolites, whereas the lowest ratios are found in harzburgites; spinels from garnet harzburgites and garnet lherzolites have intermediate values for Cr/Al. The Cr_2O_3 content of the spinel shown in analysis 1,

Table 24. Ilmenites occurring in peridotites

	1	2	3	4a	4b	5
SiO_2	0.01	0.07	0.19	0.03	0.03	0.11– 0.30
TiO_2	56.0	56.63	49.61	44.5	45.2	52.0 –56.1
Al_2O_3	0.54	0.47	0.69	0.04	0.08	0.13– 1.01
Cr_2O_3	5.45	1.40	2.17	1.8	2.0	1.70– 3.96
Fe_2O_3	–	2.93 [a]	8.97 [a]	8.1 [a]	7.8 [a]	–
FeO	21.8	26.65	25.70	34.9	33.2	22.9 –32.1
MnO	0.25	0.23	0.18	0.27	0.27	0.25– 0.48
MgO	15.6	13.52	10.49	7.3	9.0	11.9 –17.2
NiO	n.d.	n.d.	0.24	0.06	0.06	0.05– 0.19
	100.3	101.9	98.2	97.8	97.5	

a Calculated on basis of stoichiometry
1. Diamondiferous garnet lherzolite, Udachnaya (Pokhilenko et al. 1977)
2. Garnet lherzolite, Matsoku (Boyd and Nixon 1973b, Table 72)
3. Deformed garnet lherzolite, Matsoku (Boyd and Nixon 1973b, Table 72)
4. Porphyroclastic iron-rich dunite, Bultfontein, (a) core, (b) rim in contact with olivine (Smith and Dawson 1975)
5. Range of composition, in primary metasomatic veins, Matsoku (Harte and Gurney 1975b)

Table 25. Representative analyses of spinels in peridotites

	1	2	3	4	5 (a)	5 (b)	6	7
SiO_2	0.10	0.00	0.01	0.05	0.12	0.16	0.09	0.24
TiO_2	0.27	0.17	0.50	0.08	0.04	0.03	0.00	0.57
Al_2O_3	2.55	4.08	4.74	12.94	16.6	16.3	31.7	49.0
Cr_2O_3	68.7	65.3	61.7	54.83	48.2	49.8	38.3	15.0
Fe_2O_3	3.7	–	–	5.38	6.1	6.3	0.8	5.9
FeO	16.7	18.2 [a]	19.4 [a]	12.62	13.0	13.4	13.4	7.6
MnO	0.30	0.23	–	0.28	0.31	0.29	0.22	0.20
MgO	12.5	11.8	11.4	13.92	13.5	13.5	15.3	20.7
NiO	0.12	–	–	–	0.12	0.13	0.06	0.13
	101.2	99.8	98.0	100.1	97.8	100.0	99.9	99.1

a All iron as FeO
1. In chromite lherzolite, De Beers Mine, S. Africa (Smith and Dawson 1975)
2. In diamondiferous garnet dunite, ?Aykhal or ?Udachnaya pipe (Pohhilenko et al. 1976)
3. In diamondiferous garnet harzburgite, locality and reference as 2
4. In garnet-spinel lherzolite, Thaba Putsoa, Lesotho (Nixon and Boyd 1973a)
5. (a) separate grain, and (b) fingerprint spinel intergrown with pargasite, in pargasite harzburgite, Monastery (Smith and Dawson 1975)
6. In harzburgite, Newlands Mine, S. Africa (Smith and Dawson 1975)
7. In reaction corona, in garnet-spinel lherzolite, Letseng satellite pipe, Lesotho (Smith and Dawson 1975)

Table 25, is similar to that of chromite inclusions in diamond. The spinels in the symplectite rims around garnets are unlike the other spinels in having a relative high Mg/Fe ratio combined with relatively high amounts of Al_2O_3 (Table 25, analysis 8). Generally speaking, there is excellent agreement between the Cr/Cr + Al ratio between spinels and co-existing orthopyroxene and/or clinopyroxene, indicating that the spinel compositions are a reflection of bulk rock composition.

Rutile. The chemistry of rutiles in peridotitic rocks has been reviewed by Smith and Dawson (1975). Lherzolite rutiles contain appreciable amounts of Cr_2O_3 ranging from 1.6 to 7 wt%; the only other major impurity is Fe_2O_3 which varies from 0.17% to 2.6%. In a chromite rutile lherzolite from Bultfontein the rutile is zoned with respect to Fe_2O_3 from a core containing 0.7% to a rim with 2.6%; the rim contains 8.9% MgO which is otherwise very low. Rutile from a garnet pyroxenite from Matsoku has 0.25% Cr_2O_3 as its only significant impurity.

6. Isotope Chemistry

Carbon. Graphite from a garnet "peridotite" from the Mir pipe has $\delta^{13}C$ −5.9‰ and graphite from an "ultrabasic inclusion" in an Aldan kimberlite dyke has $\delta^{13}C$ −7.8‰, both relative to PDB (Kropotova and Fedorenko 1970). These values are within the range of −5.4 – −8.8‰ for most diamonds, possibly indicating a common source for the carbon.

Oxygen. $\delta^{18}O$ values for garnets and pyroxenes from five garnet lherzolites and two "peridotites" from the Roberts Victor and Bultfontein Mines range from 5.1‰ to 5.8‰ (Garlick et al., 1971); there is very close agreement between the values for coexisting garnet-pyroxene pairs.

Strontium. Strontium isotope data for some rocks and minerals of the peridotite-pyroxenite suite are given in Table 26. Allsop et al. (1969) have shown that, of the three Sr-bearing phases (diopside, garnet and secondary metasomatic mica), diopside contains the least strontium, and mica the most. The mica is secondary and, with its high amount of radiogenic strontium (together with high K and Rb), must elevate bulk-rock values above the true level for uncontaminated material. The garnets are contaminated to some extent by mica (accounting for their relatively high $^{87}Sr/^{86}Sr$ values) but the diopsides are relatively unaltered Sr isotope values, i.e. 0.705–0.707. Similar valures (in the range 0.7060–0.7075) have been found in diopsides from various "peridotites" from De Beers, Bultfontein and Dutoitspan by Barrett (1975) who reports primary micas with the same range of values. However, Shimizu (1975a) claims the diopsides from deformed lherzolites have lower $^{87}Sr/^{86}Sr$ values than those from granular ones – ~ 0.703 compared with 0.705; however it appears that both deformed and granular lherzolites from Thaba Putsoa have the same range of values, and Kramers (1977) suggests that each pipe has sampled different sectors of mantle that is isotopically heterogeneous. With the exception of the Thaba Putsoa samples, most peridotite values are still considerably higher than those of 0.701–0.703 for eclogite clinopyroxene (see p. 176) and point to real differences between the xenoliths of the peridotite-pyroxenite suite and those of the eclogite suite; they also indicate derivation of the peridotites from an area in the mantle that was relatively enriched in Rb and radiogenic Sr.

Lead. Lead isotope ratios for a garnet lherzolite from the Roberts Victor Mine are given by Manton and Tatsumoto (1971); the values are (a) $\frac{206}{204}$ 18.862, (b) $\frac{207}{204}$ 15.76, (c) $\frac{208}{204}$ 39.26. Compared with eclogites (see

Table 26. Summary of Sr isotope data on bulk rocks and phases in "peridotite" xenoliths

Material	Localities [a]	Rb/Sr (range)	$^{87}Sr/^{86}Sr$	Reference
Bulk "peridotite"	A B C D E F	0.035 –0.2355	0.7049–0.7073	1, 3, 4, 5
Diopsides	A B C D E F	0.0007–0.0089	0.703 –0.708	2, 5, 6
Garnet	C E F	0.0198–0.114	0.710 –0.712	2
Mica	C E	1.177 –3.577	0.712 –0.718	2

a A = Bultfontein, B = Wesselton, C = Dutoitspan, D = De Beers, E = Roberts Victor, F = Wilson's pipe, Tanzania

References: 1. Stueber and Murthy (1966), 2. Allsopp et al. (1969), 3. Manton and Tatsumoto (1971), 4. Mitchell and Crockett (1971a), 5. Barrett (1975), 6. Shimizu (1975a)

Table 32) the garnet lherzolite is relatively enriched in ^{206}Pb, ^{207}Pb and ^{208}Pb. Ratios from diopsides in the lherzolite suite at Bultfontein show a considerable range (a) 17.86–19.66, (b) 15.41–15.69, (c) 37.38–39.22, with no systematic variation with texture. By contrast, diopsides from banded lherzolites from Matsoku, regarded as cumulates, have very consistent values (a) 19.94–20.02, (b) 15.74–15.75, (c) 39.30–39.40 (Kramers 1977).

Sulphur. Bulk rock lherzolites and garnet pyroxenite from the Obnazhennaya pipe, Yakutia, and sulphide grains from the garnet pyroxenite give $\delta^{34}S$ values ranging from −0.5 to +2.1‰; by contrast peridotites with secondary amphibole and serpentine produce values ranging from +4.1 to +23.8, depending upon the degree of alteration (Grinenko and Ukhanov 1977). The values for the unaltered rocks resemble those for meteorites and unaltered eclogites in kimberlite.

Finally, Allegre et al. (1978) report equilibrium between Sr, Nd and Pb isotopes in coarse lherzolites from the Precambrian Premier Mine and Cretaceous S. African kimberlites; however, deformed garnet lherzolites show isotopic disequilibrium.

7. Conditions of Formation. The pressure and temperature limits of formation of rocks of the peridotite-pyroxenite suite may be deduced from the stability fields of individual minerals or assemblages, these fields being based upon current experimental data. Most methods are based on either element partitioning between coexisting phases or solubility of various components of one phase within another within the assemblage clinopyroxene-orthopyroxene-garnet. Briefly, methods most commonly used are the solubility of orthopyroxene in clinopyroxene coexisting with orthopyroxene (thermometry) and the solubility of Al_2O_3 in orthopyroxene coexisting with garnet (barometry) (see Boyd 1973a). The experimental results have been obtained on relatively simple systems and it should be appreciated that extrapolation from the simple systems to the naturally-occurring mineral assemblages is hedged about with many difficulties; some workers have tried to improve the accuracy by means of thermodynamic calculations to take account of small amounts of impurities existing in the natural systems but not in the synthetic systems (see reviews by Irving 1976; Harte 1978). A fundamental assumption is that both the experimental and the natural assemblages are equilibrium assemblages; this assumption of equilibrium in experimental systems has been questioned by Howells and O'Hara (1978) and there is some doubt as to whether all the rocks exhibiting deformational textures are in fact equilibrium assemblages. Nonetheless, some parameters are instructive for calculating *comparative* equilibrium conditions of differing nodules; these are calculated using the Ca/(Ca + Mg) ratio of the clinopyroxene and the Al_2O_3 content of orthopyroxene.

By using these parameters it is possible to estimate temperatures and pressures (depth) of formation of garnet-pyroxene-bearing assemblages. In the case of garnet lherzolites these have now been shown to have formed over a fairly wide PT regime (if it is assumed that the temperatures are in fact indicative of the ambient geothermal gradient, an important point when it is appreciated that the P factor is strongly T-dependent). Taken as a group, garnet lherzolites have formed at temperatures varying from approximately 900°C to almost 1400°C (Fig. 77) and have *apparently* been derived from depths of between 120 and 210 km. It should be stressed that inclusion suites from different kimberlite pipes show different PT conditions of equilibration. For example, those from the Thaba Putsoa pipe in Lesotho are interpreted as equilibrating over a wide PT inveral; by comparison the lherzolite suite from the Bultfontein Mine (although exhibiting a greater range of deformation textures) has equilibrated over a very narrow PT interval (see Fig. 59). In high pressure experiments on natural mineral mixes at 30–40 kb and 950°–1500°C, Mori and Green (1978) reproduced mineral compositions similar to those in some natural garnet lherzolites; although the phase equilibria can thus be reproduced experimentally, it was found that the experimental PT conditions varied considerably from PT estimates from pyroxene thermometry and barometry. Some rare garnet lherzolites contain either diamond or graphite (Boyd and Nixon 1975; Dawson and Smith 1975b; McCallum and Eggler 1976) demonstrating that kimberlites have sampled garnet lherzolites from depths on both sides of the graphite-diamond equilibrium curve.

Two further points are worthy of note: (a) there is no overall correlation between the pressure and temperature of formation of an individual xenolith and its overall texture (see Dawson et al. 1975); and (b) in the nodule suites from some kimberlite pipes (e.g. Thaba Putsoa, Udachnaya), which apparently equilibrated over a wide pressure-temperature interval, the curve outlined by the various equilibration points of individual xenoliths, a "fossil" geotherm, is interpreted as

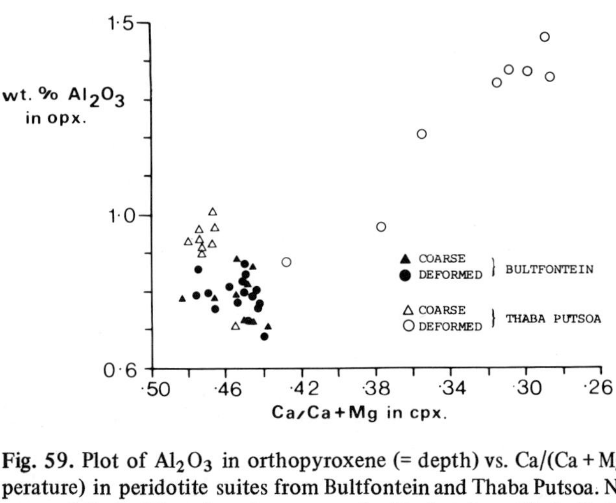

Fig. 59. Plot of Al_2O_3 in orthopyroxene (= depth) vs. Ca/(Ca + Mg) in clinopyroxene (= temperature) in peridotite suites from Bultfontein and Thaba Putsoa. Note the relatively restricted range of the Bultfontein suite compared with that from Thaba Putsoa

having an inflection or "kink". The tectonic implication of this will be discussed later (Chap. 9).

The rock series garnet lherzolite-garnet harzburgite-harzburgite-dunite has been linked together as being a sequential series: the garnet lherzolite is the most fertile and the dunite the most barren in the sense that dunite is the end member after complete extraction of basaltic liquids from parental garnet lherzolite (O'Hara et al. 1975; Carswell et al. 1979). In this sequence it is envisaged that elements such as calcium, aluminium, sodium and iron are removed from lherzolite in the form of a basaltic liquid leaving behind residua increasingly depleted in these elements; higher Mg/Fe and Cr/Al ratios are characteristic of the residua. Whilst this model is essentially correct (and has been proved experimentally), there are other alternative explanations for the genesis of some of these rocks. For example Dawson et al. (1980) have examined the relationship between garnet lherzolites and harzburgites, and have shown that some garnet lherzolites may be derived by exsolution of small amounts of garnet and diopside from harzburgites whose orthopyroxenes are rich in calcium, sodium and aluminium; some garnet harzburgites are the result of pro-grade metamorphism of orthopyroxene-olivine-chromite rock, and there are several instances of garnet-bearing rocks in which the garnet is seen to be replacing spinel (see p. 126); according to current experimental data (e.g. O'Hara et al. 1971), this replacement of spinel by garnet could have been produced by either increase in pressure at constant temperature or a fall in temperature at constant pressure, taking original spinel-bearing rocks into the garnet stability field.

The more iron-rich garnet websterites and garnet pyroxenites are believed to have originated by partial fusion of garnet lherzolite to form liquids relatively rich in iron; these liquids did not migrate far from their site of origin and consolidated under essentially the same pressure temperature conditions as the rocks from which they were derived (Gurney et al. 1975).

The very few spinel lherzolites presumably originated under lower pressure and/or higher temperature than garnet lherzolite, again according to current experimental data (e.g. O'Hara et al. 1971).

At present little is known about the origin of the iron-rich dunites, but in view of their invariable porphyroclastic texture, it is possible to speculate that they *may* have been deformed during a thermal event which did not affect the more magnesium-rich lherzolites and other peridotitic rocks. Furthermore, analyses of olivine in Kamfersdam and Bultfontein dunites (Boyd and Nixon 1978; Smith and Dawson, unpublished) indicate that they contain minor, but appreciable, amounts of TiO_2, Cr_2O_3, Al_2O_3 and CaO; the marked contrast with olivines from rocks of the peridotite suite (containing virtually none of these oxides) suggests one of two alternatives for the origin of these dunites: (a) that they are cumulates from a liquid precipitating olivine and a little ilmenite; or (b) they were originally normal peridotitic rocks but they have been subjected to a certain amount of infiltration metasomatism similar to that seen on a limited scale in other peridotites (see p. 183). The olivine chemistry and the common association with ilmenite certainly demonstrates that they are not residual dunites, i.e., the restites after considerable melting of a parental garnet lherzolite.

7.5.1.2 Eclogites

1. Introduction. Eclogites are coarse-grained rocks consisting principally of garnet and clinopyroxene, with a limited number of accessory minerals. Nodules of eclogite are much rarer in kimberlite than those of peridotite-suite rocks and may be absent in some intrusions; exceptionally, in some intrusions such as Roberts Victor (South Africa) and Zagadochnaya (U.S.S.R.), they may comprise a high proportion of the total xenolith suite.

The term eclogite was first applied by Hauy (1882) to a metamorphic rock from the Fichtelgebirge, Bavaria, consisting of green omphacite and pink pyrope. In the early days of diamond mining in South Africa, garnet-pyroxene nodules were found in the kimberlite intrusions and the name eclogite applied to them. However, to distinguish them from the eclogites occurring in metamorphic terrains, Beck (1899) proposed the name griquaite, stating at the same time his belief that these bimineralic aggregates were cognate segregations from the kimberlite host-rock. This term was not generally adopted though, in more recent years, distinctive mineralogical criteria have been recognized to permit these rocks to be distinguished from the eclogites occurring within gneissic terrains or accompanying blueschist-facies metasediments within Alpine-type fold belts. Smulikowski (1972) and Coleman et al. (1965) have shown that the garnets in the kimberlite eclogites are relatively rich in the pyrope molecule, and the clinopyroxenes have a relatively high diopside/jadeite ratio; in addition Banno (1970) has shown, on the basis of Fe-Mg partitioning between coexisting garnets and clinopyroxene, that the kimberlite eclogites have equilibrated at comparatively high temperatures.

Probably as a result of these distinguishing criteria, the term griquaite has been recently revived. Whereas some authors (e.g. Bosch 1971; Frick 1972; Smulikowski 1972) use the term as a synonym for "kimberlite eclogite", Nixon et al. (1963) and Nixon and Boyd (1973b) have used the term in its original sense in that they have recognized certain bimineralic aggregates of distinctive garnet and clinopyroxene that appear to have crystallized from the kimberlite magma. However, the relationship between the garnet-pyroxene rocks and their host kimberlites is difficult to ascertain in most instances, so the broad term eclogite will be retained here, on the understanding that it covers the terms eclogite, griquaite and garnet pyroxenite previously used in the literature.

2. Petrography. Eclogite xenoliths are usually ellipsoidal or discoidal in shape and are generally well-rounded, though preferential corrosion along monomineralic bands may give some xenoliths an irregular shape. Banding is often present, with garnet-rich bands alternating with pyroxene-rich bands and, in rarer instances,

Fig. 60. Eclogite with tightly interlocking fabric (Group II) from the Roberts Victor Mine

Fig. 61. Eclogite comprising mainly equigranular garnet and a little "interstitial" pyroxene (Group I eclogite) from the Roberts Victor Mine. Contrast this texture with that of the eclogite in Fig. 60

banding involves transition from bimineralic eclogite to kyanite- or corundum eclogite (Sobolev and Kuznetsova 1965; Dawson 1968; Mathias et al. 1970; Whitfield 1971; Lappin and Dawson 1975).

Although the eclogites are essentially holocrystalline, medium- or coarse-grained garnet/clinopyroxene aggregates, distinctive variations may be noted due to (a) differences in fabric; (b) differences in the colour and porportion of garnets and pyroxenes; and (c) more rarely, the presence of additional distinctive accessory minerals such as kyanite, corundum, graphite or diamond. Their fabric is generally a compact interlocking mosaic of irregular grains (Fig. 60), though in some eclogites well-shaped garnets are set in a matrix of pyroxene or pyroxene occurs between idiomorphic garnet grains, suggesting that garnet formed before pyroxene (Fig. 61). These variations in fabric have been used by MacGregor and Carter (1970) to propose subdivision of eclogite into two groups. Group I eclogites are composed of subhedral to rounded garnets set in an anhedral to interstitial matrix of pyroxene (occasionally with large pyroxenes surrounding garnet poikilitically), whereas Group II eclogites consist of irregular, anhedral crystals with a tightly interlocking fabric. MacGregor and Carter (1970) interpret these differences in fabric to terms of liquid-crystal equilibria processes, regarding the Group II eclogites as a set of cooled mafic liquids that have formed by continuous fractional crystallization through gravitative accumulation of garnet and clinopyroxene crystals – the Group I eclogites. An alternative explanation for these textures is that they result from subsolidus recrystallization or exsolution. For example Lappin and Dawson (1975) have described, within a single, small, layered hand-specimen from the Roberts Victor Mine, gradation from a "cumulate-textured" eclogite ("Group I") into kyanite eclogite with interlocking fabric ("Group II"). Evidence of strain and cataclasis in the kyanite and pyroxene within the kyanite-bearing portion, and replacement textures, are features incompatible with a crystallizing liquid and cumulate formation; moreover, some euhedral "cumulate" garnets in the "cumulate-textured" portion of the specimen have nucleii of strained kyanite that is in the process of being replaced by the garnet. The indications are that deformation within an anisotropic stress-field, followed by sub-solidus reaction between earlier phases, has given rise to small-scale variations in textural regime. Moreover, the rapid, gross variations in bulk composition between the kyanite-bearing and kyanite-free parts of this nodule, although perhaps reflecting an original igneous layering, are just as compatible with a metamorphic differentiation process. In other examples of eclogite and grospydite from the Roberts Victor Mine, garnet, in both lamellar and granular form, has exsolved from an earlier highly aluminous clinopyroxene (Harte and Gurney 1975a; Lappin 1978). In all these cases the texture, mineralogy and ?banding result from solid-state, subsolidus events rather than being original liquid-crystal equilibrium features.

In most eclogites the crystals of garnet and pyroxene are simple and the boundaries between adjacent crystals are uncomplicated (Fig. 62), though oriented inclusions of garnet in pyroxene are known (cf. Williams 1932, Plate 83). In some

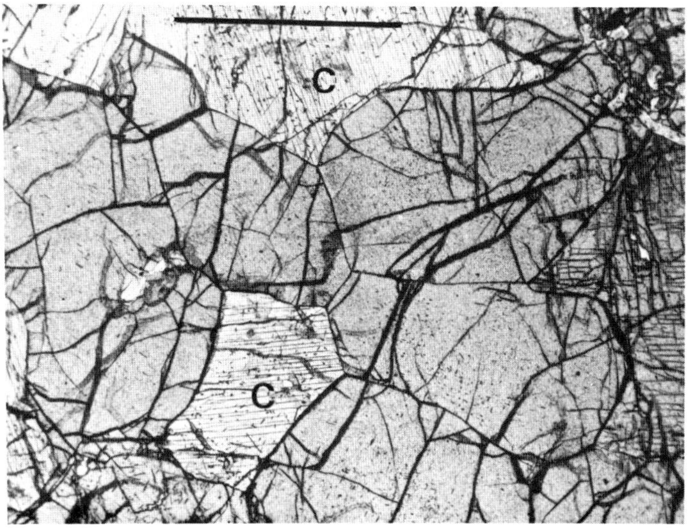

Fig. 62. Photomicrograph of an eclogite showing uncomplicated relationships between garnet and cleaved clinopyroxene *(C)*, in eclogite from Visser's pipe, Tanzania. Note the curved or straight grain boundaries and 120° angles at many triple grain junctions. Scale bar = 2 mm, plane-polarized light

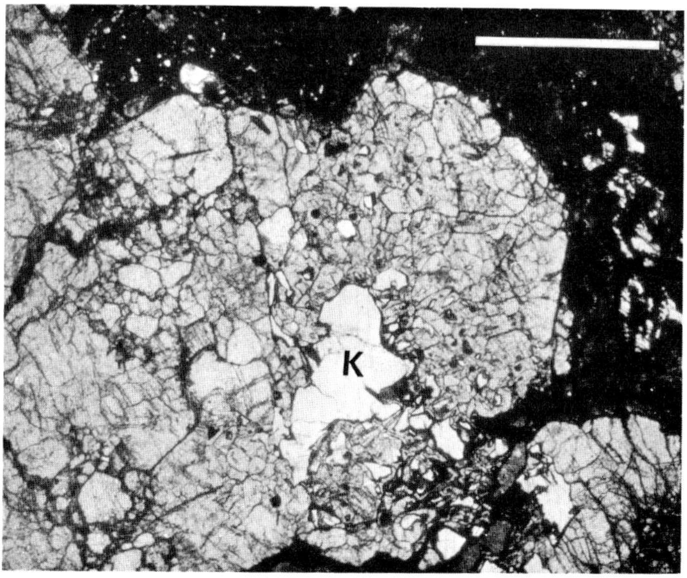

Fig. 63. Photomicrograph showing kyanite *(K)* being replaced by garnet in kyanite eclogite from the Roberts Victor Mine. This is the so-called "diablastic intergrowth" of Sobolev et al. (1968). Scale bar = 2 mm, plane-polarized light

kyanite eclogites, garnet replacing kyanite forms a corona on the kyanite (Fig. 63) (the so-called diablastic intergrowths of Sobolev et al. 1968), and garnet-kyanite-corundum and garnet-corundum intergrowths are recorded in xenoliths from the Zagadochnaya pipe (Sobolev et al. 1968). Many garnets and pyroxenes contain minute, oriented inclusions of an acicular highly-birefringent mineral (Fig. 64); in most cases these are too small to resolve but in one instance have proved to be rutile (Griffin et al. 1971).

In addition to the garnet and clinopyroxene that dominate the assemblages, the following minor or accessory minerals have been recorded in eclogites: coesite, plagioclase, sanidine, scapolite, orthopyroxene, olivine, amphibole, quartz, kyanite, sillimanite, corundum, phlogopite, rutile, ilmenite, sulphide, graphite and diamond. Of these minerals, there is some uncertainty as to whether scapolite, amphibole and phlogopite may be alteration products. In the case of the sulphides, pyrrhotite and pyrite are the most common, but Desborough and Czamanske (1973) and Frick (1973b) have reported polymineralic sulphide aggregates in eclogites from the Roberts Victor Mine. The aggregates described by Desborough and Czamanske are up to 1 mm in size and contain at least four sulphides distributed in a distinctive pattern; a central core of an unnamed $(FeNi)_9S_{11}$ sulphide containing 1.4%–3.2% Ni is surrounded by a more nickel-rich $(FeNi)_9S_{11}$ sulphide with ~20% Ni (akin to violarite) which also occurs as irregular flames and exsolution lamellae within the core sulphide, together with small grains of nickel-bearing pyrite.

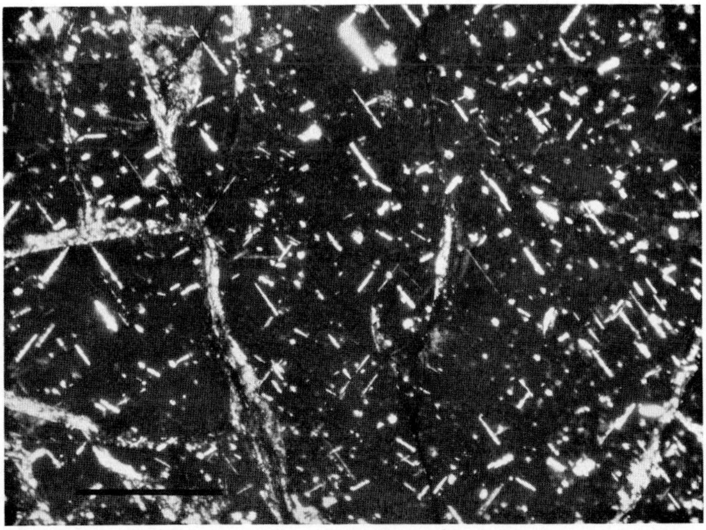

Fig. 64. Photomicrograph of oriented rutile needles in isotopic garnet in eclogite from the Visser's pipe, Tanzania. Scale bar = 0.25 mm, cross polarizers

The margins of the blebs consist of chalcopyrite. Because individual sulphide aggregates appear to have the same composition and occur within silicate crystals, it is suggested that they represent an immiscible sulphide liquid that solidified as a single phase and unmixed during cooling. Analogous sulphide aggregates are reported from eclogite in the Stockdale diatreme, Kansas, in which a narrow rim of chalcopyrite surrounds a phyrrhotite core in which occurs flame exsolution of pentlandite (Meyer and Brookins 1971).

3. Variants of Eclogite. Distinctive variants of eclogite are those containing additional primary minerals in more than accessory quantity (> 1.0 modal %). Known variants are plagioclase-bearing eclogite; diamond- and graphite-bearing varieties (carbonaceous eclogites); those containing kyanite and/or corundum and, very rarely, sillimanite (peraluminous eclogites), of which grospydites are a sub-group; amphibole eclogites; quartz eclogites; and orthopyroxene eclogites.

The plagioclase eclogites contain variable amounts of the feldspar and with increasing feldspar content grade into rocks that could be termed “granulite”. These have been recorded in the Zarnitsa, Udachnaya, Osenaya and Nievidimka pipes in Russia (Bobrievich et al. 1959) and in a number of the kimberlites in southern Africa (summarized by Rickwood et al. 1968).

Diamond-bearing eclogites are known from at least nine intrusions in South Africa, most being in the Barkly West area (Gurney et al. 1968; Rickwood and Mathias 1970; Reid et al. 1976), and also from the AK/1 pipe at Orapa in Botswana (Shee and Gurney 1979). In Russia, diamondiferous eclogites have so far been reported mainly from the Mir pipe (summarized by Sobolev 1977) and also from the Udachnaya pipe (Ponomarenko et al. 1973). In most cases the diamonds are the high-temperature octahedra, though in some instances the form is rhombic dodecahedral, indicating partial resorption before crystallization of the associated garnet and pyroxene (Robinson 1978). Graphite has been found accompanying diamond in the Newlands and Roberts Victor Mines, South Africa and in the Mir pipe, U.S.S.R., but has also been reported as the only carbon polymorph in eclogites from the Premier and Newlands Mines (Wagner 1911), Jagersfontein (Wagner 1914), Crown Mine (Rickwood et al. 1968), Roberts Victor Mine, and AK/1, Botswana (Robinson 1978); in the case of specimens from Roberts Victor in the author’s collection, the graphite does not pseudomorph diamond.

Peraluminous eclogites contain kyanite and/or corundum and, much more rarely, sillimanite. Both kyanite and corundum have been found in eclogites from a considerable number of kimberlite intrusions in southern Africa and Russia, and sometimes occur together in the same inclusion. Green chrome-kyanite is known from the Zagadochpaya pipe (Sobolev et al. 1968) and sapphire and ruby corundum are recorded in a considerable number of eclogites. By contrast, sillimanite is very rare; the “sillimanite” recorded by Williams (1932) in eclogite from the Wilsons pipe, Tanzania, has proved to be kyanite (Rickwood et al. 1968) and the only apparently authentic occurrence is in a kyanite-sillimanite eclogite from

Bellsbank, South Africa (Schutte 1967). Two phases intermediate in composition between the end-membergs Al_2O_3 and Cr_2O_3 occur in chrome-kyanite-bearing eclogite from the Zagadochnaya pipe (Sobolev 1977); the two phases contain 34.9 and 64.4 wt% Cr_2O_3. Diamond is reported in several kyanite and corundum eclogites from the Roberts Victor and Versluis Mines, S. Africa (Mathias and Rickwood 1970; Reid et al. 1976), and from the Udachnaya pipe, Yakutia (Ponomarenko et al. 1976). An almost constant feature of kyanite eclogites is the high degree of alteration of the clinopyroxene.

Grospydites are a special type of peraluminous eclogite, taking their name from the assemblage *gros*sularite-*py*roxene-*dis*thene (the old Continental name for kyanite). The principal difference between grospydite and kyanite eclogite is that the garnet in the grospydite contains > 50% of the grossularite molecule. First found in the Zarnitsa, Zagadochnaya and Dolgozhdannaya diatremes in Russia (Bobrievich et al. 1959), they have subsequently been found in South Africa in the Roberts Victor, Lovedale and Brakfontein pipes (MacGregor and Carter 1970; Frick 1972; Chinner and Cornell 1974; Lappin 1978) and the Bobbejaan dyke (Bishop et al. 1978), and from the Williamson Mine, Tanzania (Reid et al. 1976). Diamondiferous grospydite is reported from the Udachnaya and Sitikanskaya pipes (Koptil et al. 1976; Ponomarenko et al. 1976), and coesite and high-sanidine occur in grospydite from the Roberts Victor pipe (Smyth and Hatton 1977) and the Udachnaya pipe (Ponomarenko and Spetsius 1978).

Amphibole eclogites are known from kimberlites in South Africa and the Congo (Rickwood et al. 1968) from Russia, and from the U.S.A. (Meyer and Brookins 1971). In many instances the amphibole accompanies plagioclase in amphibole-plagioclase eclogite, the whole paragenesis apparently having equilibrated at lower pressures than plagioclase-free eclogites. Many of the amphiboles are doubtless of secondary origin, being associated with minerals such as chlorite, epidote and calcite in veinlets or along grain boundaries, and most amphiboles have been interpreted as alteration products. Nonetheless amphiboles in equilibrium with the two main phases are known in eclogites from the Stockdale kimberlite, U.S.A. (Meyer and Brookins 1971) and Roberts Victor (Bishop et al. 1978); the experimental work of Lambert and Wyllie (1968) indicates that amphibole could be stable in the upper parts of the upper mantle where at least some of the kimberlite eclogites equilibrated.

Quartz is a minor phase in several South African eclogites. In addition, Smyth (1977) has proposed that quartz aggregates in a Roberts Victor eclogite are pseudomorphs after coesite. Orthopyroxene comprises an appreciable proportion (up to 13%) rare *orthopyroxene-bearing eclogites* from the Newlands Mine (Beck 1907), Bellsbank and Beyersfontein (Mathias et al. 1970), and Mitchemanskraal (Schutte 1967), all these localities being in the Cape Province, S. Africa. Orthopyroxene eclogites reported from the Matsoku pipe, Lesotho (Mathias and Rickwood 1969) are in fact garnet pyroxenites (Cox et al. 1973).

Eclogites from the kimberlite pipes of the Colorado Plateau are unlike any others in that they contain garnets which are consistently high in iron and, furthermore, they contain the mineral *lawsonite* (Helmsteadt and Doig 1975). These are similar to the type C eclogites found in blue schist-facies metamorphic terrains (Coleman et al. 1965), and are interpreted as fragments of Mesozoic oceanic crust that has been subducted beneath the North American continent.

4. Alteration of the Xenoliths. Many eclogite xenoliths have been subjected to alteration by fluids from the water- and potassium-rich host kimberlite, and it is ironic that many of the eclogites from one of the world's main source of eclogites, the Roberts Victor Mine, are amongst the most extensively altered. Alteration is common for up to 2 cm inwards from the surface of the nodules extending further into the nodule along grain boundaries and minor fractures. The garnet is altered to phlogpite, amphibole, spinel, chlorite and plagioclase, whilst the pyroxene is altered to a turbid, microcrystalline aggregate of epidote/zoisite, calcite and a mineral that Berg (1968) has identified as an analcime-structured zeolite, possibly related to wairakite. MacGregor and Carter (1971) note that type I eclogites from the Roberts Victor kimberlite are markedly more altered than the type II. In some kyanite eclogites, the kyanite may be altered to zoisite, plagioclase, mullite and quartz. In addition, calcite, quartz and chalcedony may be present.

Whereas the alteration described above is metasomatic, more fundamental types of alteration have been described. High-temperature reaction, involving ionic diffusion from the host kimberlite, has been proposed to explain a virtually monomineralic garnet rind surrounding a grospydite core in a xenolith from Roberts Victor (Chinner and Cornell 1974). In some eclogites from the Roberts Victor Mine (including Chinner and Cornell's specimen) partial melting has taken place, these specimens being cut by veinlets and stringers of glass of basaltic composition that contain small crystals of spinel, microlites of plagioclase and augite, and analcime-filled vesicles (Switzer and Melson 1969); the glass is the same composition as the omphacite pyroxene which probably melted as a result of pressure-release. Switzer and Melson's (1969) first observations were on kyanite eclogites, but Whitfield (1971) and the author have observed these melting phenomena in diamondiferous, graphite-bearing and normal eclogites from this locality (Fig. 65). In addition, the author has observed examples of partial fusion in kyanite- and corundum-bearing eclogites from Bobbejaan, S. Africa, and Chicundo, Angola; kyanite and clinopyroxene have fused to form a brown glass that is precipitating plagioclase, and the kyanite is often replaced by secondary spindle-shaped corundum grains (Fig. 66). In a grospydite from Roberts Victor exhibiting partial melting, Lappin (1978) has identified nepheline in the melt patches derived from pyroxene.

5. Chemistry

a) Bulk Chemistry. Analyses of a compositional spectrum of eclogites, peraluminous eclogites and grospydites (mainly from the Roberts Victor Mine) are given in Table 27. They vary from the highly magnesian variety 1185, to others relatively

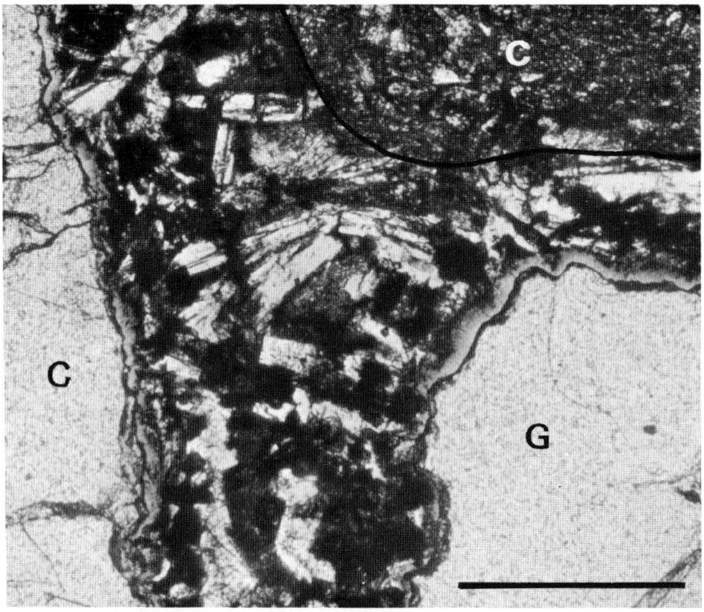

Fig. 65. Photomicrograph of veinlets of glassy material precipitating plagioclase, developing at the contact of garnet *(G)* and altered pyroxene *(C)*, in eclogite from the Roberts Victor Mine. Scale bar = 0.5 mm, plane-polarized light

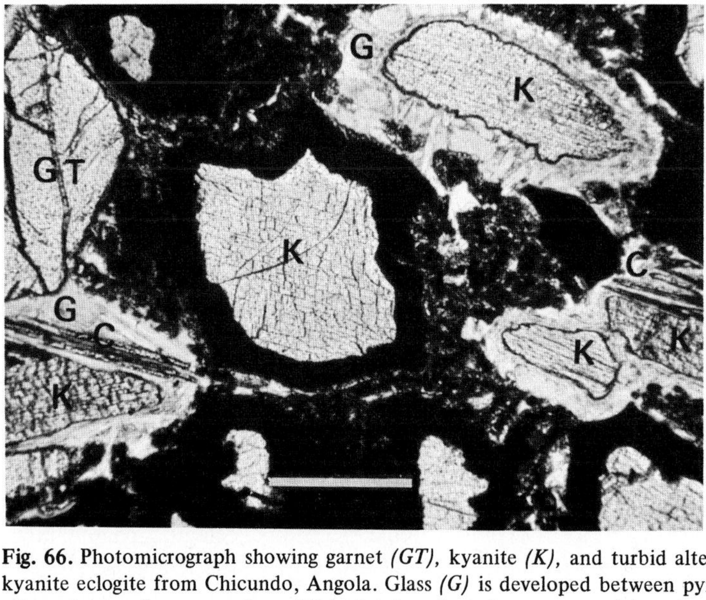

Fig. 66. Photomicrograph showing garnet *(GT)*, kyanite *(K)*, and turbid altered pyroxene in kyanite eclogite from Chicundo, Angola. Glass *(G)* is developed between pyroxene and kyanite and secondary corundum *(C)* is developed at the expense of kyanite. White scale bar = 0.5 mm, plane-polarized light

Table 27. Analyses of eclogites, peraluminous eclogites and grospydites. (From Roberts Victor Mine, unless stated otherwise)

	1	2	3	4	5	6	7	8	9	10	11	12	13	14	15	16
SiO_2	48.4	48.0	46.8	45.67	47.24	47.73	49.49	46.9	41.7	45.80	43.46	42.42	42.27	40.26	42.00	52.69
TiO_2	0.15	0.11	0.19	0.42	0.49	0.61	0.52	0.53	0.30	0.40	0.35	0.18	0.46	0.18	tr	0.09
Al_2O_3	10.6	13.2	16.5	17.85	15.05	11.97	8.46	17.7	29.7	23.86	24.11	24.60	22.16	29.43	31.04	21.43
Cr_2O_3	4.42	0.32	0.04	0.07	0.09	0.12	0.24	0.03	0.03	–	0.02	–	0.03	0.04	0.01	0.04
Fe_2O_3	–	–	–	2.88	2.25	3.79	3.20	1.65	1.74	–	1.99	7.39 [b]	1.18	0.23	1.40	–
FeO	4.24 [a]	5.1 [a]	11.9 [a]	8.46	8.21	9.56	5.80	8.09	5.82	7.22	4.82	–	6.63	2.50	2.49	3.76 [a]
MnO	0.20	0.15	0.22	0.17	0.28	0.33	0.24	0.20	0.18	0.07	0.17	0.10	0.25	0.06	0.07	0.08
MgO	19.0	19.1	9.7	11.90	13.39	13.28	16.23	9.20	5.68	8.28	9.05	6.55	6.51	9.48	9.54	5.62
NiO	–	–	0.01	–	–	–	–	–	–	–	–	–	–	–	–	–
CaO	12.3	12.2	11.5	7.35	8.68	8.76	10.60	8.89	9.14	11.34	10.55	13.91	9.75	14.02	7.98	11.77
Na_2O	1.17	1.03	2.90	2.01	1.74	2.18	1.67	3.56	2.19	0.33	1.87	2.41	4.31	1.63	1.22	4.08
K_2O	0.02	0.01	0.0	0.39	0.22	0.40	1.12	0.86	0.83	–	0.53	0.04	0.70	0.18	0.30	0.24
H_2O^+	–	–	–	2.01	1.38	1.28	1.43	1.46	1.83	1.34			2.12			–
H_2O^-	–	–	–	1.07	0.68	0.33	0.64	0.55	0.49	0.12	2.65	2.29	0.77	1.55	3.77	–
CO_2	–	–	–	–	–	–	–	–	0.52	0.44			0.39			–
P_2O_5	–	–	–	0.04	0.06	0.01	0.64	0.14	0.18	0.07	0.11	0.11	0.01	0.05	0.12	0.01
Total	–	–	–	100.29	99.76	100.34	99.68	99.8	100.3	99.27	99.92	100.0	99.56	99.61	99.94	99.56

a Total iron as FeO
b Total iron as Fe_2O_3

1. High Mg, Cr-rich eclogite 1185
2. High Mg, low Cr eclogite 1698
3. High Fe eclogite 1191

(1.–3.) Eclogites, calculated from phase composition data on basis of 50% garnet, 50% pyroxene. Phase data from Bishop et al. (1978)

4.–7. Eclogites (Kushiro and Aoki 1968)
8. Eclogite band and 9. kyanite eclogite band in banded eclogite (Lappin and Dawson 1975)
10. Kyanite eclogite (Williams 1932)
11. Kyanite eclogite, Zagadochnaya (Sobolev et al. 1968)
12. Kyanite eclogite, Rietfontein, S. Africa (Gurney et al. 1971)
13. Kyanite-corundum eclogite (Whitfield, 1971)
14. Corundum eclogite, Obnazhennaya, U.S.S.R. (Sobolev and Kuznetsova 1965)
15. Grospydite, Zagadochnaya (Sobolev et al. 1968)
16. Coesite-sanidine grospydite (calculated from phase data) (Smyth and Hatton 1977)

high in iron, and those (containing a discrete aluminous phase) which are extremely high in Al_2O_3; examples with corundum (e.g. analysis 14) are very low in silica (40%) by contrast with the coesite-bearing specimen which has greater than 52% SiO_2. It should be noted that specimens containing appreciable amounts of K_2O (e.g. analyses 7–9) are described as containing secondary phlogopite; by contrast analyses calculated on the basis of garnet and pyroxene alone (e.g. analyses 1–3) contain virtually no K_2O. Broadly speaking, on the basis of their silica content and their calcium, magnesium, iron and aluminium values, most eclogites can be matched in a broad sense with some of the numerous types of basalt and, following their work on the experimental conversion of basalt to eclogite, Yoder and Tilley (1962) proposed that "for every major basalt-type there is a chemically equivalent eclogite". The same stand-point has been adopted by Kushiro and Aoki (1968) who propose that they present crystallization in the upper mantle of liquids of olivine basalt or picritic basalt composition, a viewpoint supported by Smulikowski (1972). In addition, the experimental work of Green (1967) and Khitarov et al. (1971) assumes the broad chemical equivalence of high-alumina basalt with kyanite eclogite and grospydite.

There are, however, others who challenge the validity of the basalt = eclogite assumption. Angel (1957) has argued against this on the grounds that few eclogites contain both normative nepheline and normative olivine, but numerous eclogite norms do contain these two normative minerals, thereby invalidating Angel's argument on this particular basis. More significantly Forbes (1965) has outlined small, but important, differences between the bulk composition of most basalts and bimineralic eclogites; the basalts contain significant amounts of potassium which occurs in negligible quantities in the garnet-clinopyroxene assemblages.

In Figs. 67–70 are plotted the fields of various eclogite types and the same parameters for the basaltic rocks listed in Table 28. These plots show on the basis of $CaO:MgO:\Sigma FeO$ and $CaO:Al_2O_3:(\Sigma FeO + MgO)$ ratios, a broad equivalence between eclogites and most of the more common basalt types. Some points of difference should be noted, however. On the $CaO\text{-}MgO\text{–}\Sigma FeO$ plot (Fig. 69), these average basalts have a relatively constant $CaO/\Sigma FeO$ ratio with variable MgO; by contrast the eclogites show wider $CaO/\Sigma FeO$ ratios, and many of the peraluminous eclogites and grospydites are much more calcic than even the high-alumina basalt. The absence of true equivalents of the peraluminous eclogites and grospydites is also emphasized by the $CaO\text{-}Al_2O_3\text{–}(\Sigma FeO + MgO)$ plot (Fig. 70). Although these plots indicate a *broad* equivalence between some eclogite and the commoner basaltic types, they do *not* show the two main discrepancies, that of the virtual absence of K_2O in the eclogites (as pointed out by Forbes 1965), and also of titanium. The paucity of potassium in eclogites was highlighted by the experimental work of Erlank and Kushiro (1970) who show that, even up to pressures of 32 kb, synthetic clinopyroxenes accept less than 150 ppm K into their lattice; garnets synthesized in the same experiment contain less than 100 ppm. Bishop et al. (1978) report 50–1600 (av. 370) ppm K_2O in eclogite clinopyrox-

Table 28. Analyses of basic igneous rocks

	1	2	3	4	5	6	7	8	9	10	11	12	13
SiO_2	49.36	44.10	44.33	46.41	46.46	49.61	51.8	49.60	48.33	47.37	52.22	45.94	46.0
TiO_2	2.50	2.74	2.65	1.98	3.01	1.43	1.13	1.53	0.80	0.46	0.56	0.34	0.61
Al_2O_3	13.94	11.20	12.80	8.53	14.64	16.01	14.8	16.58	18.40	6.79	5.42	2.98	10.86
Fe_2O_3	3.03	2.84	3.38	2.47	3.27	–	3.92	3.00	0.93	1.18	0.98	6.23	3.98
FeO	8.53	9.91	9.14	9.82	9.11	11.49 [a]	7.26	7.62	8.13	8.08	8.88	4.80	6.86
MnO	0.16	0.16	0.15	0.15	0.14	0.18	0.17	0.17	0.15	0.19	0.22	0.18	0.14
MgO	8.44	15.13	11.05	20.81	8.19	7.84	7.1	10.45	9.30	20.39	15.25	33.79	22.15
CaO	10.33	10.69	10.52	7.38	10.33	11.32	10.57	7.18	11.45	8.31	12.83	4.73	7.88
Na_2O	2.13	1.66	3.60	1.58	2.92	2.76	2.40	3.56	2.54	0.39	1.21	0.15	1.07
K_2O	0.38	0.54	0.99	0.32	0.84	0.22	0.74	0.06	0.12	0.06	0.09	0.03	0.04
P_2O_5	0.26	0.30	0.43	0.20	0.37	0.14	0.13	0.14	0.05	–	–	–	–

a Total Fe as FeO
1. Average of 181 Hawaiian tholeiites and olivine tholeiites (MacDonald and Katsura 1964)
2. Average of 7 Hawaiian ankaramitic picrite basalts (MacDonald and Katsura 1964)
3. Average of 10 Hawaiian nepheline basanites (MacDonald and Katsura 1964)
4. Average of 14 Hawaiian oceanitic picrite basalts (MacDonald and Katsura 1964)
5. Average of 28 Hawaiian alkali basalts (MacDonald and Katsura 1964)
6. Average ocean-floor basalt (Cann 1971)
7. Average Karroo basalt (Cox and Hornung 1966)
8. Average of 4 metabasalts from the Mid-Atlantic Ridge, calculated water-free (Thompson and Melson 1972)
9. Average of 3 analyses of Warner high-alumina basalt, California (Yoder and Tilley 1962)
10. Average basaltic komatiite (Geluk type), Barberton, S. Africa (Viljoen and Viljoen 1969)
11. Average basaltic komatiite (Badplaas type), Barberton, S. Africa (Viljoen and Viljoen 1969)
12. Average peridotitic komatiite (Onverwacht Group), Barberton, S. Africa (Viljoen and Viljoen 1969)
13. Chilled margin of picrite dyke, Skye (Drever and Johnston 1967)

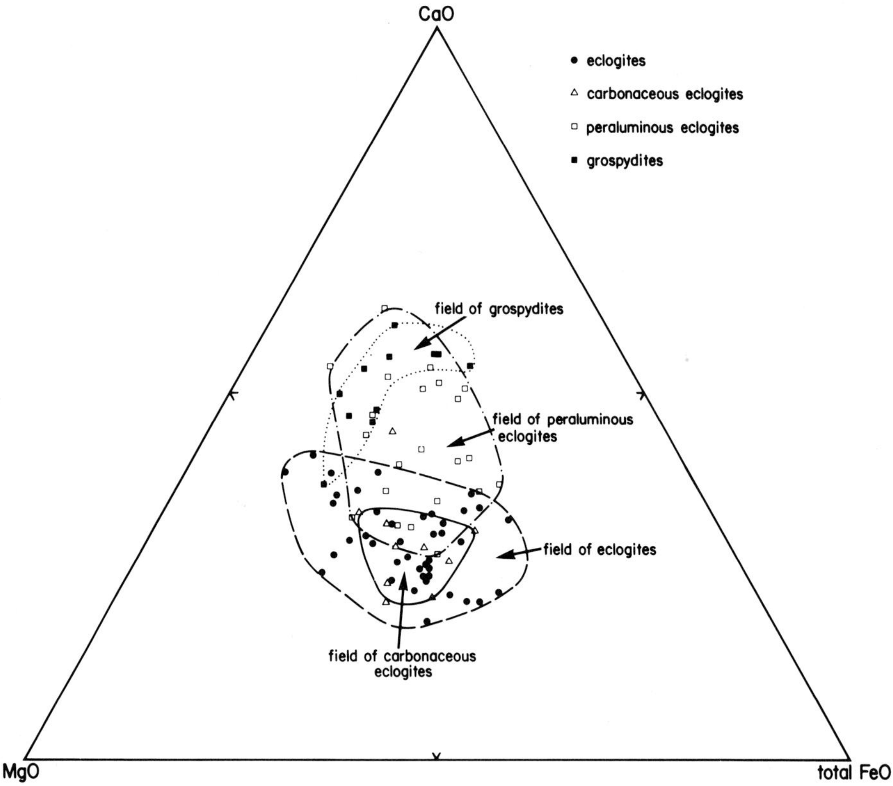

Fig. 67. Plot of CaO, MgO, total FeO in bulk eclogites. Data in this plot are derived from: Bobrievich et al. (1964), Frantsesson (1970), Gurney et al. (1971), Holmes (1936), Kushiro and Aoki (1968), Lappin and Dawson (1975), Meyer and Brookins (1971), Mikheyehko et al. (1970), Nixon et al. (1963), O'Hara et al. (1975), Sobolev and Kuznetsova (1965), Sobolev et al. (1968, 1969, 1970), Wagner (1914), Whitfield (1971), and Williams (1932)

enes, again underlining the trivial amounts in these rocks. It must be reiterated that, in the absence of a proven *primary* potassic phase, the potassium must be assumed to be in secondary phlogopite, amphibole and zeolites sited along fractures and grain boundaries, or in sub-microscopic alteration. Hence, although some lower-pressure Alpine-type and migmatite-terrain eclogites are chemically equivalent to basalt (due to the presence of equilibrium-assemblage micas and amphiboles), most unaltered kimberlite eclogites are quite distinct in this particular aspect.

There are, to be sure, some rare, basaltic rock-types that contain low amounts of potassium (analyses 6 and 8–13, Table 29); some of these are more ultrabasic in composition and are generally regarded as being "primitive" melts, not having been subjected to pre-eruption differentiation. Nonetheless, even these low-potas-

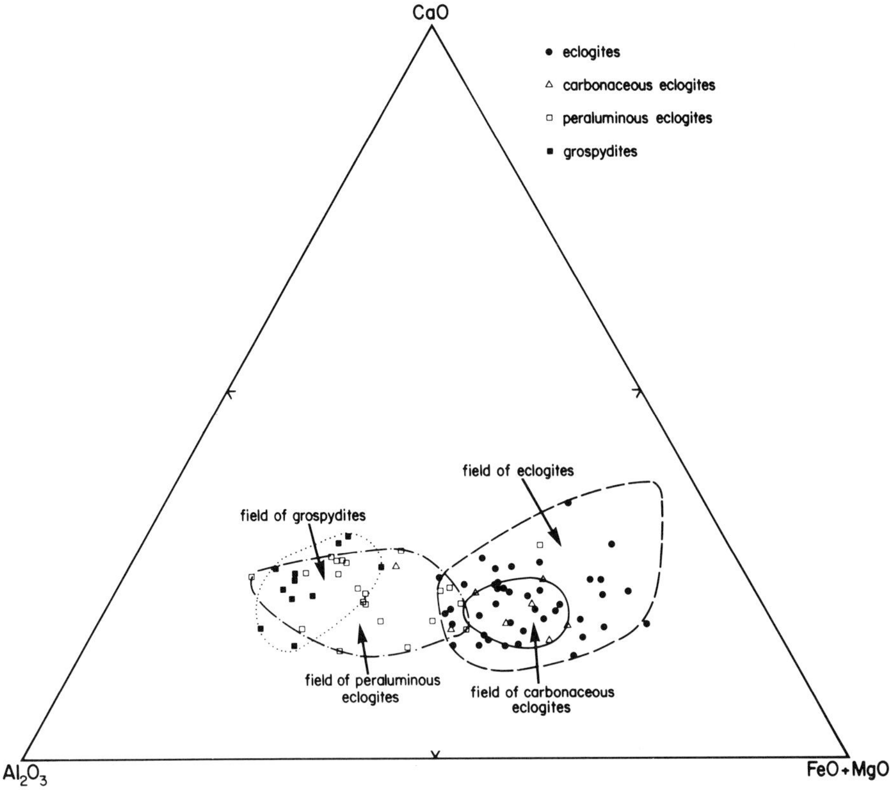

Fig. 68. Plot of CaO, Al_2O_3, and total FeO + MgO in bulk rock eclogites. Data sources as in Fig. 67

sium basalts differ from most eclogites in certain respects. The more ultrabasic komatiites owe their low potassium content to a very high proportion of modal olivine, and this is likewise reflected in a very low sodium content – quite unlike eclogite. Again on Figs. 69 and 70, where the low-potassium basalts are marked in solid black squares, it is apparent that most have higher MgO/CaO and MgO/Al_2O_3 ratios than eclogites; only the average Badplaas komatiite plots within the eclogite field, and this compositional overlap is tenuous when examined in detail, since the overlap applies only to certain varieties of eclogite that have a high pyroxene/garnet ratio (see Viljoen and Viljoen 1969).

In summary, eclogites containing ?primary amphiboles and/or phlogopite, and others that are visibly metasomatised may be chemically equivalent to certain basaltic rocks. However, very few unaltered kimberlite eclogites are chemically similar to the vast majority of basaltic rocks. To explain this there are several possibilities, of which three are (a) eclogites originally may have crystallized from

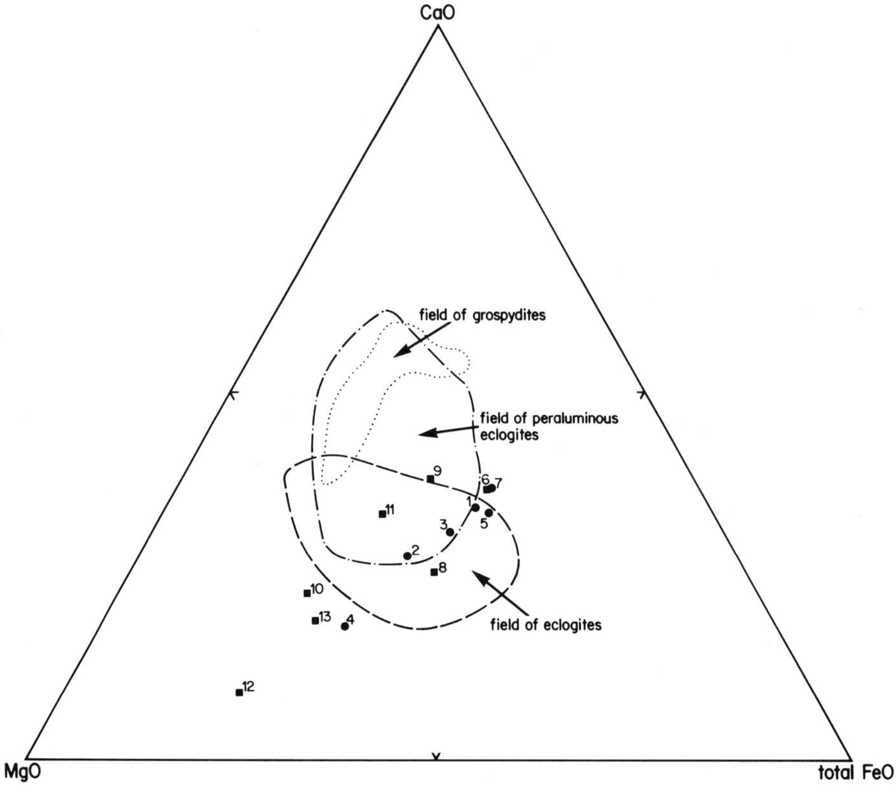

Fig. 69. Plot of CaO, MgO, total FeO in eclogites from Fig. 68 and the average basalts in Table 28. *Solid squares* (basalts 8–13) are low-potassium basalts

melts of basaltic composition but during the primary crystallization the pressures were sufficiently high to preclude incorporation of "basaltic" elements such as potassium and titanium in the eclogite phases (this raises the problem of where these elements migrated and in what form); (b) basalts and eclogites originate from melts of similar initial composition but, during their ascent, most basalts become enriched in potassium and titanium, having undergone some degree of fractionation; in this context, it is relevant that "primitive" lavas contain the smallest amounts of potassium; (c) eclogites and basalts may result from chemically distinct melts, perhaps reflecting small differences in the mineralogy of the upper mantle where these melts originate; a possibility is that most basalts may arise in areas of the mantle containing potassic amphibole or phlogopite-bearing peridotite, whereas low-potassium basalts and eclogite melts originate in areas of peridotite lacking these phases.

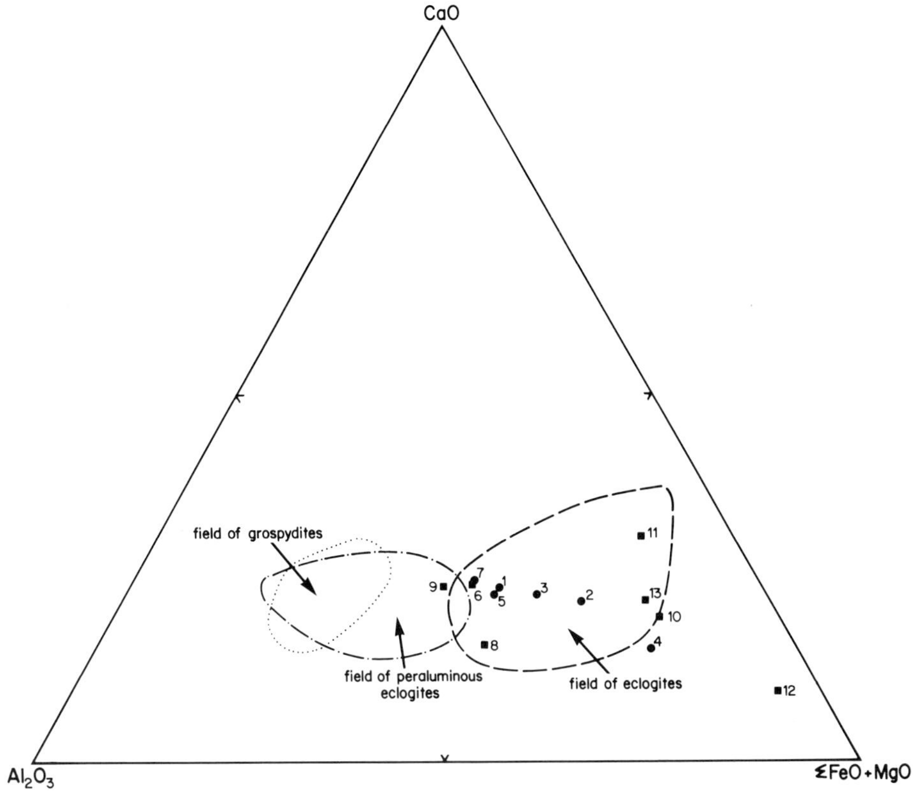

Fig. 70. Plot of CaO, Al_2O_3, total iron + MgO for eclogites from Fig. 67 and average basalts from Table 28

b) Trace-Element Chemistry. The trace element content of analysed eclogites is given in Table 29; (Ti, Cr, Mn, Ni and P are included with the major-element analyses, Table 27). It will be appreciated that these trace-element data are strongly weighted by specimens from the Roberts Victor Mine and the highly metasomatised nature of some Roberts Victor eclogites should be re-iterated; both absolute values for bulk-rock analyses and certain geochemical parameters (e.g. K/Rb ratios) for these specimens must be viewed with circumspection.

V, Co, Ni, Sr and Ba are the more abundant trace elements, with Li, Sc, Rb, Cs, Th and U amongst the less abundant. Attempts to balance some bulk-rock and mineral concentrations in many Roberts Victor samples leads to a noticeable disparity in the case of Rb, Cs and Ba; bulk-rock values are greater than can be attributed to any combination of the principal constituent minerals and must be ascribed to secondary minerals in cracks, or along grain boundaries. There is a strong cor-

Table 29. Trace element content of eclogites and eclogite minerals

Element	Locality	Concentration in ppm Rock [a]	Cpx	Gt	Mica	Reference
Li	Kaalvallei		<10	<3		2
	Kao		<10	<3, 10		2
	Obnazhennaya	3.0	4.5	4.5		10
Be	Kaalvallei		<3			2
	Kao		<3			2
F	Roberts Victor	200				1
Sc	Kaalvallei		10	75		2
	Kao		10	40, 120		2
V	Roberts Victor	260				1
	Jagersfontein	210				1
	Kaalvallei		450	80		2
	Kao		450	50, 13		2
Co	Kaalvallei		45	65		2
	Kao		45	13, 70		2
Ni	Kallvallei		1000	140		2
	Kao		1000	30,9		2
	Roberts Victor	220				1
	Jagersfontein	310				1
Ga	Kaalvallei		15	4		2
	Kao		15	15.8		2
Rb	Roberts Victor	9.1–82.8 (7)				4
	Jagersfontein	3.8				4
	"Kimberley"	3.5–57.0 (2)				4
	Bultfontein	3.0				4
	Wilson, Tanzania	6.2				4
	Roberts Victor	2.9–23.7 (16)				5
	Jagersfontein	2.5–3.3 (2)				5
	Wilson, Tanzania	2.1, 3.8				5
	Bultfontein	2.4	0.88, 0.82	1.84		7
	Roberts Victor	5.5	1.3, 1.1	1.5		7
	Roberts Victor	6.0	0.51, 0.68	1.0		8
	Roberts Victor	19.5	0.81, 0,97	1.0		8
	Roberts Victor				561.2	8
Rb	Roberts Victor	2.1–28.0 (5)				10
	Roberts Victor	2.99				13
	Roberts Victor		0.56	1.14		13
	Obnazhennaya	2.8	2.4	1.3		11
	Obnazhennaya	5.2	3.4	3.0		11

Table 29 (continued)

Element	Locality	Concentration in ppm Rock	Cpx	Gt	Mica	Reference
Sr	Kaalvallei		150			2
	Kao		150			2
	Roberts Victor	159	227	6.4		8
	Roberts Victor	150	191	13.2		8
	Roberts Victor				77	8
	Roberts Victor	180	220, 238	7.5		7
	Bultfontein	36	58, 59	4.2		7
	Roberts Victor	133–370 (5)				10
	Roberts Victor	277				13
	Roberts Victor		249	8.25		13
Y	Kao			30		2
Zr	Kaalvallei		25	50		2
	Kao		25	3,25		2
Cs	Roberts Victor	0.6–12.8 (16)				5
Ba	Kaalvallei		<10	<10		2
	Kao		<10	<10		2
	Bultfontein	71.8	28.6, 31.6	46.9		7
	Roberts Victor	136	33.0, 41.7	28.6		7
	Roberts Victor	1440				13
	Roberts Victor	327	178	52.6		13
	Roberts Victor		134	43.7		13
	Jagersfontein	302				13
Au	Obnazhennaya	0.08–0.45 (8)				12
Pb	Kaalvallei		<3	<2		2
	Kao		<3	10, <2		2
	Roberts Victor	0.95				6
	Roberts Victor	0.79–5.50 (5)				10
Th	Jagersfontein	0.5				1
	Wesselton	2.1				1
	Roberts Victor	1.3				1
	Roberts Victor	0.29–1.25 (7)				3
	Jagersfontein	0.17				3
	Bultfontein	0.31				3
	Wilson's, Tanzania	0.45				3
	Roberts Victor	10.7, 11.0, 11.8				6
	Roberts Victor	0.35–1.1 (5)				10

Table 29 (continued)

Element	Locality	Concentration in ppm Rock	Cpx	Gt	Mica	Reference
U	Jagersfontein	0.66				1
	Wesselton	1.47				1
	Roberts Victor	1.26				1
	Roberts Victor	0.04–0.80 (8)				3
	Roberts Victor	0.04, 0.05				6
	Roberts Victor	0.09–0.26 (5)				10
	Obnazhennaya	0.097	0.062	0.119		9
	Zagadochnaya	0.024 [b]				9

a Some figures denotes ranges in concentration; the figures in brackets indicate the number of samples
b Grospydite
References: 1. Holmes (1936); 2. Nixon et al. (1963); 3. Heier (1963); 4. Heier and Compston (1966); 5. Gurney et al. (1966); 6. Lovering and Tatsumoto (1968); 7. Griffin and Murthy (1968); 8. Allsopp et al. (1969); 9. Akimov et al. (1968); 10. Manton and Tatsumoto (1971); 11. Stavrov and Ukhanov (1971); 12. Ukhanov and Pchelintseva (1972); 13. Philpotts et al. (1972)

correlation between the absolute amounts of K and Rb, but not between K and Cs, in the Roberts Victor samples; the siting of these large-radius cations is believed to be in secondary phlogopite or small-scale intergrowths of secondary amphibole (Erlank 1970) or in submicroscopic alteration products of the clinopyroxene (Gurney et al. 1966; Berg 1968). The wide range in K/Rb ratios of rocks from this locality (176–349 – Heier and Compston 1966; 257–534 – Gurney et al. 1966; 210–407 – Griffin and Murthy 1968) is also due to varying degrees of alkali metasomatism. By contrast, eclogites from other localities, such as Jagersfontein, Bultfontein, Wilson's pipe (Tanzania) and Obnazhennaya have relatively low absolute values for K and Rb, and they have a reasonably limited K/Rb ratio. For example, K/Rb ratios of coexisting minerals from a Bultfontein eclogite are: garnet 239, pyroxene 255–265 (Griffin and Murthy 1968), and from an Obnazhennaya eclogite: garnet 270, pyroxene 250 (Stavrov and Ukhanov 1971); furthermore the K/Rb ratios for the minerals and the bulk rock are not very different, in the case of the Obnazhennaya eclogite the bulk-rock ratio being 286. In the case of U, Akimov et al. (1968) found 0.1 ppm or less in both bulk-rocks and minerals, and challenge the upper-mantle estimates of U proposed by Heier (1963) on the basis of his analyses of Roberts Victor eclogites (0.04–0.80 ppm). Their contention is supported to some extent by the work of Manton and Tatsumoto (1971), who show that the eclogites they examined from the Roberts Victor Mine have suffered some U, Th, Pb, Rb and Sr contamination from the host kimberlite; in the case of Pb, from 38%–96% of the eclogite lead was shown to be introduced from the host rock.

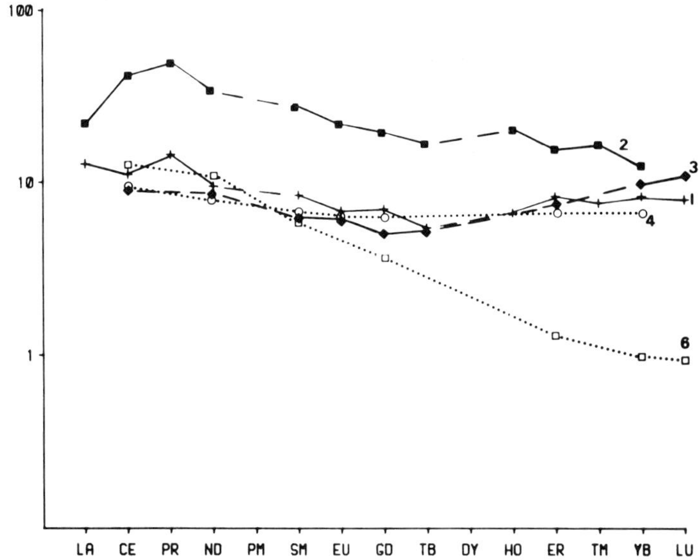

Fig. 71. Rare-earth element plots of bulk eclogites (normalized against standard chondrite; Haskin et al. 1966). *1* Roberts Victor; *2* Dutoitspan eclogites (Haskin et al. 1966); *3* Roberts Victor eclogite 151; *4* Roberts Victor eclogite 155; *6* Jagersfontein eclogite 152; *3, 4* and *6* from Philpotts et al. (1972)

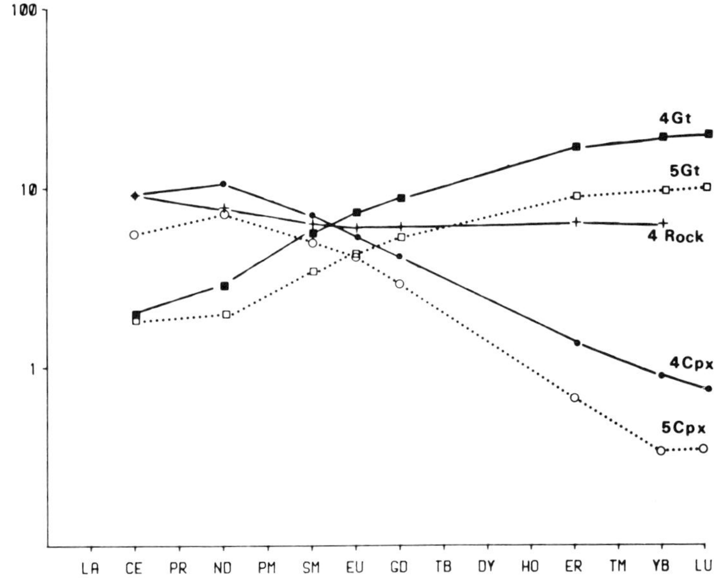

Fig. 72. Rare-earth element plots for bulk eclogite *4* and phases in eclogites *4* and *5*; *4* Rovic eclogite 155; *5* Rovic eclogite 250 (Philpotts et al. 1972)

Values for rare-earth elements in eclogites and their principal phases are plotted (after normalization against the average chondrite of Haskin et al. 1968) in Figs. 71 and 72. It is evident that the rare-earth concentrations in the phases are complementary, the garnet containing relatively high amounts of the heavy RE elements and the clinopyroxene high amounts of the light RE elements (Fig. 72). The normalized RE plots of some whole rocks, such as eclogites 1, 3 and 4 from the Roberts Victor Mine are remarkably consistent and the concentrations can be accounted for in terms of the garnet and pyroxene concentrations, assuming that the minerals are present in roughly equal porportions. Eclogite 6, from the Jagersfontein Mine, has a decided negative slope (Fig. 71), suggesting either that pyroxene may be the dominant mineral in this particular rock, or it contains a contaminating phase rich in the light rare-earths. In the case of the Roberts Victor eclogites, is appears that only the Ce and, to a lesser extent, Nd concentrations in the garnet have been enhanced due to contamination, and "the best trace-element model appears to be that the Roberts Victor eclogite omphacites and garnets were in equilibrium with a liquid similar, at least in relative rare-earth concentrations, to (host) kimberlite" (Philpotts et al. 1972).

c) Phase Chemistry. There is now a tremendous wealth of data on the chemistry of the phases in eclogites and their variants. It is impossible to survey all the data, so this section will be mainly illustrated by analyses of phases in eclogites and their variants from the Roberts Victor pipe, South Africa; these cover a very wide range of compositions and examples from other kimberlite intrusions will only be mentioned to illustrate certain features.

Garnet. Eclogite garnets show considerable variation in their calcium, magnesium and iron contents (Table 30). In a survey of xenolith garnets in general, Dawson and Stephens (1976) showed that eclogites contain garnets of six distinct compositions on the basis of their TiO_2, Cr_2O_3, FeO, MgO and CaO contents. Furthermore, they showed that garnets from eclogites in individual intrusions vary widely in composition; an exception to this is the eclogites from the intrusions of the Colorado plateau described by O'Hara and Mercy (1966) which are consis tently high-iron types. Referring specifically to the Roberts Victor eclogites, the Ca-Mg-Fe plot (Fig. 73) illustrates the wide variation in garnet composition; some are highly magnesian, others reasonably high in iron, whilst others are more calcic; there is no real trend and they are unlike garnets of the peridotite-pyroxenite group in which Fe is relatively constant, and variations are mainly in Ca relative to Mg. Garnets from kyanite-bearing eclogites are more calcic and less magnesian than many other types (Fig. 74) and grospydite by definition contains $>50\%$ of the grossular molecule; the other eclogites, including kyanite-bearing types, show a variation in calcium molecule from 5% to 48%. Note that some of the more magnesian garnets both in their magnesium and chromium contents are very similar to garnets in lherzolites. Reid et al. (1976) report that diamondiferous eclogites apparently have no different compositions in their garnets compared with non-diamondiferous eclogites. Trace elements in the garnets are TiO_2 (210–7800 ppm

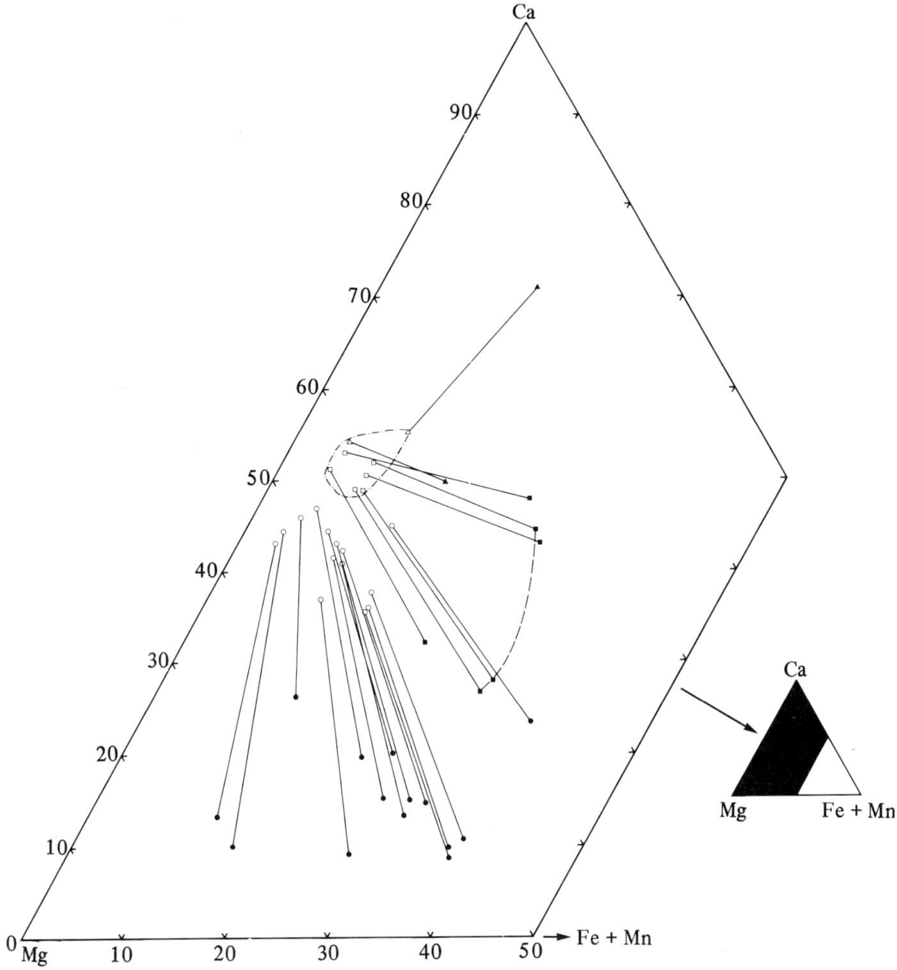

Fig. 73. Ca-Mg-Fe plot for coexisting garnets and clinopyroxenes from Roberts Victor eclogites. *Open circle* clinopyroxenes, *closed circle* garnets; *squares* kyanite-bearing eclogites; *triangles* for grospydites. Data from Bishop et al. (1978); Kushiro and Aoki (1968); Lappin (1978); Lappin and Dawson (1975); Sobolev (1977, Table 25); Smyth and Hatton (1977)

av. 2000), P_2O_5 (160–940 ppm av. 530), and Na_2O (100–1420 ppm av. 610); K_2O was found to be < 20 ppm (Bishop et al. 1978). The small but significant amounts of Na_2O in eclogite garnets were first recognized by Sobolev and Lavrentyev (1972), who proposed that this high sodium was specially significant in diamond-bearing rocks and thus attributed the sodium content to high pressures. However a later study by Bishop et al. (1978) shows that garnets from a large

Table 30. Representative analyses of garnets in eclogites, kyanite eclogites and grospydite from

	1	2	3	4	5	6	7	8	9	10	11
SiO_2	40.4	40.4	41.6	39.8	41.4	41.7	39.4	40.2	41.7	41.34	40.40
TiO_2	0.12	0.31	0.21	0.30	0.08	0.34	0.17	0.39	0.12	0.42	0.46
Al_2O_3	23.1	22.5	19.2	22.8	23.1	23.5	22.5	22.7	23.3	22.45	21.47
Cr_2O_3	0.25	0.08	6.4	0.12	0.27	0.49	0.04	0.16	0.42	0.24	0.24
F_2O_3	–	–	–	–	–	–	–	–	–	1.09	1.68
FeO	11.7[a]	14.8[a]	6.6[a]	18.6[a]	9.6[a]	9.41[a]	18.2[a]	15.3	8.09[a]	13.01	16.57
MnO	0.22	0.46	0.32	0.71	0.24	0.20	0.41	0.58	0.25	0.54	0.99
MgO	15.5	14.7	20.8	14.9	14.5	15.7	10.3	14.4	21.2	17.54	14.56
NiO	0.01	0.01	0	0	0	0.02	0	0.02	0.01	–	–
CaO	7.88	5.88	5.4	3.43	10.0	10.7	8.89	5.93	4.02	3.39	3.64
Na_2O	0.05	0.14	0.04	0.09	0.04	0.11	0.05	0.14	0.03	–	–
K_2O	0	0	0	0	0	0	0	0	0	–	–
P_2O_5	0.05	0.05	0.04	0.05	0.02	0.08	0.02	0.05	0.02	–	–
Total	99.28	99.33	100.61	100.81	99.25	102.26	99.99	99.88	99.17	100.02	100.01

a Total iron as FeO

1–9. Eclogites (Bishop et al. 1978)

10–14. Eclogites (Kushiro and Aoki 1968)

15. Eclogites (Lappin and Dawson 1975); 16, 17, kyanite eclogite (Lappin and Dawson 1975)

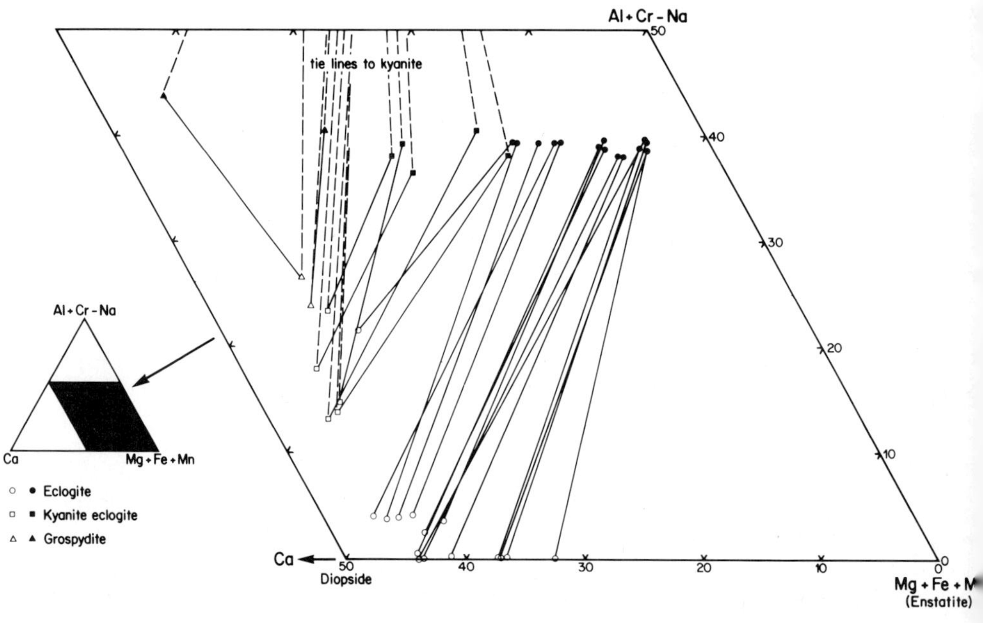

Fig. 74. C-M-A plot of coexisting garnet and clinopyroxene from Roberts Victor eclogites, kyanite eclogites and grospydites. Data sources as for Fig. 73; symbols as in Fig. 73

Roberts Victor kimberlite, South Africa

12	13	14	15	16	17	18	19	20	21	22
40.31	40.98	41.04	39.7	40.2	40.7	41.6	40.16	40.23	40.31	40.04
0.57	0.35	0.38	0.22	0.25	0.19	0.3	0.35	0.35	0.17	0.08
21.69	23.01	22.69	23.4	22.1	21.9	24.0	22.70	22.34	21.94	21.58
0.12	0.12	0.07	–	–	–	–	–	0.03	0.14	–
1.22	0.70	0.61	–	–	–	3.3	2.17	2.44	–	0.04
17.66	13.23	12.64	15.2[a]	15.9[a]	13.8[a]	8.48	9 36	11.43	10.25[a]	9.11[a]
0.58	0.43	0.24	0.35	0.29	0.19	0.1	0.19	0.21	0.29	0.14
14.02	15.50	14.58	11.5	11.1	7.44	12.0	7.17	7.95	7.57	3.51
–	–	–	–	–	–	–	–	–	–	–
3.91	5.63	7.77	10.6	11.1	17.3	12.3	18.12	15.95	18.90	–
–	–	–	0.15	0.13	0.13	0.1	0.12	0.16	0.07	–
–	–	–	–	–	–	–	–	–	–	–
–	–	–	–	–	–	–	–	–	0.05	–
100.06	99.95	100.02	101.1	101.1	101.6	102.2	100.34	100.39	99.69	100.88

18–20. Kyanite eclogites (Sobolev 1977, Table 25)
21. Grospydite (Smyth and Hatton 1977)
22. Grospydite (Lappin 1978)

number of non-diamond-bearing eclogites contain similar amounts of sodium, and the presence of sodium can be explained by coupled substitution with phosphorus and titanium. There is no need to invole six-fold coordination of silicon (as postulated by Sobolev and Lavrentyev (1972), and on a plot of atomic Na – P v. atomic Ti all the garnets from the Roberts Victor eclogites lie near a line Ti = 1.5 (Na – P), suggesting that a common factor controls the distribution of these elements. Furthermore, the data for diamondiferous eclogites (Reid et al. 1976) also fall close to the same line; as these diamondiferous eclogites come from several localities, the trend is not confined solely to specimens from the Roberts Victor pipe.

Garnet compositions may vary widely within individual hand specimens. For example, in a banded specimen from Roberts Victor which contains bands of bimineralic eclogite grading into kyanite eclogite, garnets vary from grossular 25, pyrope 42, almandine 33 in the garnet-pyroxene layer to grossular 44, pyrope 28, almandine 28 in the kyanite-bearing portion of the xenolith (Lappin and Dawson 1975). Also, Chinner and Cornell (1974) report a range of garnet composition from pyrope 22% to 53% in a concentrically zoned Roberts Victor grospydite. In another Roberts Victor xenolith in which the garnet is exsolving from clinopyroxen, variation in composition of the garnets is linked with the texture; granular garnet contains 36% grossular molecule which increases in coarse lamellae and reaches 55% grossular molecule in very fine exsolution lamellae (Harte and Gurney 1975a). Generally within the eclogites, garnets are homogeneous; however, slight zoning has been recorded in some garnets from kyanite eclogite (Lappin

Table 31. Representative analyses of clinoproxenes in eclogites, kyanite eclogites and grospydites

	1	2	3	4	5	6	7	8	9	10	11
SiO_2	55.3	55.3	55.3	55.0	56.4	55.1	54.2	54.7	54.3	53.74	54.29
TiO_2	0.20	0.36	0.10	0.36	0.09	0.36	0.21	0.38	0.09	0.31	0.34
Al_2O_3	9.26	9.88	2.12	6.96	9.40	8.92	10.5	8.82	3.04	5.06	4.40
Fe_2O_3	–	–	–	–	–	–	–	–	–	2.65	2.92
Cr_2O_3	0.32	0.08	2.45	0.15	0.24	0.89	0.05	0.18	0.22	0.34	0.11
FeO	2.32[a]	4.02[a]	1.88[a]	7.24[a]	1.6[a]	2.12[a]	5.55[a]	4.36[a]	2.12[a]	3.46	5.29
MnO	0.02	0.06	0.08	0.20	0	0.02	0.04	0.07	0.05	0.16	0.28
MgO	11.5	10.7	17.2	12.7	11.9	12.1	9.14	11.0	17.0	15.60	13.66
CaO	16.0	13.7	19.2	13.0	16.7	16.0	14.1	13.8	20.4	15.48	14.20
Na_2O	4.95	5.70	2.31	4.66	4.37	5.03	5.76	5.47	2.04	2.62	2.90
K_2O	0.06	0.15	0.04	0.11	0.01	0.06	0.01	0.16	0.03	0.17	0.20
P_2O_5	0.04	0.03	0.04	0.04	0.02	0.02	0.02	0.04	0.04	–	–
Total	99.9	99.9	100.7	100.4	100.7	100.7	99.6	99.9	99.3	99.59	96.68

a Total iron as FeO
1–9. Eclogites (Bishop et al. 1978)
10–14. Eclogites (Kushiro and Aoki 1968)
15. Eclogites (Lappin and Dawson 1975); 16, 17, kyanite eclogite (Lappin and Dawson 1975)

and Dawson 1975). Some extremely rare kyanite eclogites and grospydites from the Zagadochnaya pipe contain chromium-bearing grossular-rich garnets containing 1.5%–7.2% Cr_2O_3; the grossular molecule varies from 27% to 62%; uvarovite molecule is up to 23%, and the colour varies from pale violet to green (Sobolev 1977). Compositionally these rare garnets can be matched with the isolated garnets found in some South African kimberlites (e.g. Hornung and Nixon 1973; Clarke and Carswell 1977).

Clinopyroxenes. Stephens and Dawson (1977) distinguish seven chemically distinct varieties of clinopyroxene in eclogites and grospydites; they vary from chromiferous diopsides, indistinguishable from those in garnet lherzolites, to jadeites and jadeitic diopsides with high sodium and aluminium contents. The composition of clinopyroxene from Roberts Victor eclogites, kyanite eclogites and grospydites (Table 31) are shown in Figs. 73 and 74 together with the coexisting garnets; on these plots it can be seen that the clinopyroxenes have a relatively small range in composition compared with the coexisting garnets, the inference being that variations in the bulk rock composition are accommodated mainly by variation in the garnet. The clinopyroxene coexisting with the most magnesian garnets contains appreciable amounts of Cr_2O_3, a feature not readily recognized from the diagrams. The main variations may be seen in terms of varying jadeite to diopside ratios (Fig. 75); departures from the jadeite-diopside join are evident in varieties relatively high in iron. In most instances the tschermackite molecule comprises not more than 2% but one exceptional example from the Roberts Victor from the Roberts Victor kimberlites

12	13	14	15	16	17	18	19	20	21	22
54.21	54.72	54.73	56.1	56.5	57.9	52.1	53.3	56.3	56.63	56.51
0.55	0.41	0.41	0.24	0.30	0.12	0.3	0.3	0.29	0.07	0.09
6.76	10.51	14.10	18.1	15.9	17.5	17.5	17.9	17.3	17.04	20.72
3.10	1.69	1.17	–	–	–	–	–	0.04	0.04	–
0.15	0.16	0.10	–	–	–	–	–	–	–	0.03
4.44	2.56	1.74	2.19 [a]	2.49 [a]	2.01 [a]	1.5 [a]	1.8 [a]	1.95 [a]	1.56 [a]	1.98 [a]
0.4	0.07	0.04	0.02	0.01	tr	–	–	0.04	0	0.01
12.39	11.02	8.84	6.23	6.42	5.40	7.2	6.4	5.30	6.24	3.47
13.68	13.36	11.62	9.90	10.5	9.94	11.9	12.1	9.26	11.57	7.96
4.47	5.05	6.61	7.90	7.76	7.40	6.8	6.4	8.51	7.23	8.90
0.21	0.17	0.24	–	–	–	0.3	0.2	0.05	0.20	0.02
–	–	–	–	–	–	–	–	–	0	–
100.19	99.72	99.60	100.7	99.9	100.3	97.6	98.4	99.04	100.6	99.71

18–20. Kyanite eclogites (Sobolev 1977, Table 25)
21. Grospydite (Smyth and Hatton 1977)
22. Grospydite (Lappin 1978)

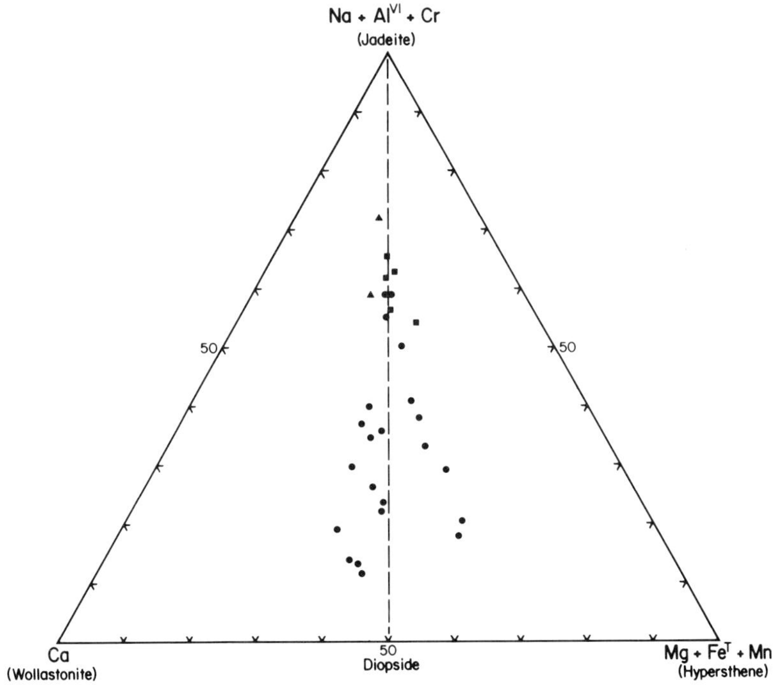

Fig. 75. Plot of Ca –(Mg + Fe^{T} + Mn) – (Na + Al^{vi} + Cr) for eclogite, kyanite eclogite and grospydite clinopyroxenes. Data sources as for Fig. 73; symbols as in Fig. 73

Mine contains over 30% tschermackite (Harte and Gurney 1975a). The pyroxenes in kyanite eclogite and grospydite are very similar in composition (Table 31 and Fig. 74) and are characterized by high amounts of calcium, high amounts of sodium and aluminium and relatively small amounts of iron and low MgFe/Mg ratios. A consistent feature is the presence of excess amounts of Al^{VI} (sometimes 20% of the Al^{VI}) which may be regarded as kyanite in solid solution (Lappin 1978; Sobolev et al. 1968). Chrome omphacites, containing 0.92%–2.89% Cr_2O_3, occur in rare kyanite eclogite from the Zagadochnaya pipe (Sobolev et al. 1968); they coexist with chromium-bearing grossular-rich garnets and chrome-kyanite.

Clinopyroxenes from diamond-bearing eclogites have very little, if any, Al^{VI} which contrasts with those in other eclogites, and iron is mainly Fe^{2+} (Sobolev 1977). Regarding trace elements, Bishop et al. (1978) report 110–520 (av. 300) ppm P_2O_5, 50–1620 (av. 370) ppm K_2O and 330–3800 (av. 2000) ppm TiO_2 in S. African eclogite clinopyroxenes, including some from Roberts Victor.

Amphiboles are mainly secondary. However, some are in textural equilibrium with the major phases; analyses of two examples (Table 32) show them to be of

Table 32. Analyses of amphiboles, micas and ilmenite from eclogites

	1	2	3	4	5	6
SiO_2	40.7	40.1	42.37	36.9	40.1	–
TiO_2	3.61	0.19	2.40	2.6	0.93	55.7
Al_2O_3	14.0	18.9	11.06	16.6	14.5	0.17
Cr_2O_3	0.05	0.16	n.d.	0.11	0.40	0.10
Fe_2O_3	–	–	1.61	–	–	–
FeO	10.7 [a]	7.8 [a]	8.00	10.8 [a]	4.41 [a]	34.2 [a]
MnO	0.12	0.18	0.11	0.11	0.03	0.48
MgO	13.8	14.2	20.00	17.6	24.6	10.0
NiO	n.d.	0.02	n.d.	0.01	0.07	–
CaO	10.6	10.7	0.71	0.02	0.0	0.07
Na_2O	3.59	3.0	0.72	0.42	0.21	–
K_2O	1.09	1.3	9.08	9.99	10.5	–
H_2O^+	–	–	3.42	–	–	–
H_2O^-	–	–	0.34	–	–	–
P_2O_5	–	–	n.d.	–	–	–
Total	98.3	96.55	99.82	95.16	95.75	100.7

a Total iron as FeO
1. Ti-rich amphibole in eclogite 1128 g, Stockdale kimberlite, Kansas, U.S.A. (Meyer and Brookins 1971)
2. Amphibole, Roberts Victor eclogite 1188 (Bishop et al. 1978)
3. Sedondary phlogopite, Roberts Victor eclogite (Kushiro and Aoki 1968)
4, 5. Secondary phlogopites, Roberts Victor eclogites (Bishop et al. 1978)
6. Ilmenite from eclogite, Stockdale, Kansas, U.S.A. (Meyer and Boctor 1975)

variable composition; compared with peridotite amphiboles (Table 21) they are low in MgO and Cr_2O_3, and compared with MARID-suite potassic richterites are low in K_2O but high in Al_2O_3.

Micas are secondary; analyses of three from Roberts Victor eclogites (Table 32) show them to be highly variable with respect to TiO_2, Al_2O_3, MgO and FeO (hence *mg*) and K/Na. Unlike the secondary micas in peridotites, they are low in Cr_2O_3 (cf. Table 23).

Feldspar Sanidine ($Or_{98}Ab_2$) containing no detectable CaO (despite high CaO in coexisting garnet and pyroxenes) occurs in coesite-grospydite from Roberts Victor (Smyth and Hatton 1977). ?Potassic feldspar blebs ($K_2O \sim 19\%$, Na_2O 0.37%–0.72%, CaO 0.02%–0.12%) occur in omphacite in a diamondiferous eclogite from Roberts Victor (Reid et al. 1976).

Kyanite. Most kyanites are reasonably pure with only a little Fe_2O_3 (<0.5 wt%) in solid solution (Gurney et al. 1971; Bishop et al. 1978). Exceptionally, kyanites from eclogites in the Zagadochnaya pipe contain up to 16.8 wt% Cr_2O_3, generally occurring in paragenesis with Cr_2O_3-rich garnets and clinopyroxenes (Sobolev 1977, p. 65); there may be considerable compositional variation in individual grains.

Ilmenite is rare. A sample from a Stockdale, Kansas, eclogite (Table 32) is a magnesian variety but, compared with most kimberlite ilmenites, is low in Al_2O_3.

Corundum. Very few analyses are available. One from Bobbejaan, S. Africa, is very pure with only a little Fe_2O_3 (0.16 wt%) (Bishop et al. 1978).

Rutile. Smith and Dawson (1975) have reviewed the chemistry of rutiles in eclogites. By contrast with rutiles from rocks of the peridotite-pyroxenite group, rutiles in eclogites contain virtually no Cr_2O_3, the principal additional oxide being Fe_2O_3; they may contain up to 2.3% Fe_2O_3 and, with respect to this oxide, the composition of differing grains within a single eclogite may be variable. An exception to the rule that these rutiles contain low Cr_2O_3 is shown by rutile from a kyanite eclogite from the Zagadochnaya pipe from Yakutia; this has the unusually high Cr_2O_3 value of 9.8% (Sobolev et al. 1972).

Other Oxides. Sobolev (1977) reports two phases from kyanite-bearing eclogite, intermediate in composition between the end-members Al_2O_3 and Cr_2O_3, the two phases containing 34.9 and 64.4 wt% Cr_2O_3. In experiments on the system grossular-uvarovite at one atmosphere pressure, Huckenholz and Knittel (1975) found that Al_2O_3-Cr_2O_3 solid solution exists only at temperatures in excess of 1280°C.

d) Isotope Chemistry

Carbon. Diamond in an eclogite from the Mir pipe, U.S.S.R., has $\delta^{13}C = -8.5‰$ (relative to PDB) and graphite in the same sample $-8.7‰$ (Kropotova and Fedorenko 1970). These values are similar to the range known for colourless diamonds and to the $-7.8‰$ value for graphite in an ultramafic inclusion in a kimberlite dyke in the Aldan area; they are, however, lower than the -5.9% for graphite in a pyrope peridotite xenolith from the Mir pipe. Kropotova and Fedorenko (1970) interpret their results as indicating a common source for the carbon in the diamonds occurring both in eclogite xenoliths and in kimberlite.

Oxygen. The oxygen isotope compositions ($\delta^{18}O$) of eclogites from the Roberts Victor Mine range from 2‰ to 8‰ relative to SMOW (Garlick et al. 1971). The eclogites interpreted as crystal cumulates (Type I) have $\delta^{18}O$ value of 5.4‰–8.0‰ and are mainly within the range of 5‰–7‰ for the great majority of ultramafic and mafic igneous rocks (Taylor 1968). By contrast, the Type II eclogites (interpreted as late-crystallizing fractionated liquids) have values mainly in the range 2.2‰–4.3‰; this variation is interpreted as due to fractional crystallization, and as being pressure-dependent. However, in view of the highly variable metasomatic alteration of eclogites from this locality, oxygen isotope data on the host kimberlite would be desirable. Garlick et al. (1971) also give values of 6.0‰ and 6.3‰ for garnets from the two Russian grospydites.

Sulphur. $\delta^{34}S$ in bulk eclogites and sulphide grains from the Obnazhennaya kimberlite ranges from +0.5‰ to +1.8‰; values for altered eclogites range up to +16.0‰ (Grinenko and Ukhanov 1977). Other eclogites from the Premier and Roberts Victor intrusions yield values of $\delta^{34}S$ +0.2‰ – +2.1‰ (Tsai et al. 1979b) which, together with the peridotite data, suggest the $\delta^{34}S$ value of upper-mantle sulphur is close to zero.

Strontium. Allsopp et al. (1969) give $^{87}Sr/^{86}Sr$ data for eclogites and their phases from the Roberts Victor Mine. Whereas the fresh omphacites have low ratios of 0.701–0.703, the bulk-rock values (0.707–0.709) are enhanced by the higher values obtained on garnet (0.708–0.712) which is coated by a more radiogenic rind containing phlogopite with $^{87}Sr/^{86}Sr$ 0.738. The low strontium isotope ratios in the omphacite, which are regarded as the most reliable data, indicate that the formation of the eclogites was a Precambrian event, assuming that primary upper mantle has Rb/Sr $\geqslant$ 0.023. Combined with the Cretaceous (105 ± 20 m.y.) age for the Roberts Victor kimberlite (Allsopp et al. 1969), the evidence is that these eclogites are accidental xenoliths not genetically connected with the formation of the kimberlite.

Further $^{87}Sr/^{86}Sr$ ratios for five eclogites from the Roberts Victor Mine are given by Manton and Tatsumoto (1971), who report values varying from 0.7039 to 0.7096, the highest ones containing secondary mica. As the two least altered specimens have the lowest ratios (0.7039 and 0.7040) and also contain residual lead with the least radiogenic content (see below), it is argued that (on the assumption that at least some strontium has been added as a contaminant from the host kimberlite) the unaltered rocks had an even lower value than 0.7039, agreeing well with the conclusions of Allsopp et al. (1969). Two other eclogites have ratios close to the 0.7077 of the host kimberlite. Hand-picked omphatices from Roberts Victor eclogites give values ranging between 0.70328 and 0.70702 (Kramers (1979).

Lead. Lead isotope data in eclogites from the Roberts Victor Mine have been given by Lovering and Tatsumoto (1968) and Manton and Tatsumoto (1971). The earlier authors compare the isotopic ratios of lead in an omphacite from a Roberts Victor eclogite with those in kimberlite from the Kimberley Mine, 80 km distant, and conclude that the isotopic compositions were so different that the eclogites must be of accidental origin, on the assumption that the leads at the Kimberley and Roberts Victor Mines are identical and that complete extraction of lead had been obtained. The first assumption has been invalidated by Manton and Tatsumoto (1971) whose data on the Roberts Victor kimberlite show isotopic differences from the Kimberley kimberlite (Table 33); they also showed that the lead in five eclogites has two components, a lead that is pyroextractable at 1100°–1200°C, and a residual lead. Each individual eclogite appeared to have an independent history since, in three of the eclogites, the pyroextractable lead may be a contaminant from the host kimberlite; whereas this is not the case for the other two, one of them containing a much higher proportion of radiogenic ^{206}Pb. Calculated isotopic compositions for the residual leads vary greatly. Clearly much more data are needed, including verification of the primary nature of the residual lead. Analyses of hand-picked omphacites from Roberts Victor eclogites show a wide range of Pb-isotope and $^{87}Sr/^{86}Sr$ ratios, and there is no correlation between $^{87}Sr/^{86}Sr$, $^{87}Rb/^{86}Sr$ and Pb isotope ratios; on a $^{207}Pb/^{204}Pb$ vs. $^{206}Pb/^{204}Pb$ diagram, the Roberts Victor data define a linear array which, *if* interpreted as an isochron, give an age of 2465 ± 200 m.y. (Kramers 1979).

e) Conditions of Formation. In recent years, two main hypotheses have been postulated for the formation of eclogites (summarized by Dawson 1971). The first is that eclogites are accidental xenoliths derived from the upper mantle, their origin having nothing to do with the kimberlite which has transported them to the surface. The second hypothesis is that eclogites are cumulates of phases precipitating from the kimberlite liquid, and are thereby related to the kimberlite which brought them to the surface. Although textural and chemical data do indicate that eclogites are cumulates (e.g. Hatton and Gurney 1977), it is more likely that they originate during cooling of very much older basic magmas since the isotope and age data indicate that they are of Precambrian age and not cogenetic with their host kimberlite.

Eclogites and their variants occur within the earth's upper mantle over a considerable PT range. The high-level PT boundary is governed by the PT conditions at which plagioclase disappears from rocks of broadly basic composition and is governed by two broad reactions: (1) forsterite + anorthite → garnet; and (2) the reaction albite → jadeite + quartz. The deeper-level PT boundary is given by a more poorly recognized transition – the one at which pyroxene starts to go into solution within the garnet structure (Ringwood 1975). These two boundaries have been investigated experimentally in both simple and naturally-occurring systems of basalt composition (summarized by Ringwood 1975). Assuming that eclogites

Table 33. Isotopic composition of lead in kimberlites and eclogites

Sample	Location	Total Pb ppm	206/204	207/204	208/204		Reference
Kimberlite	Kimberley Mine	7.46	19.77	15.83	39.66		1
Kimberlite R17	Roberts Victor	37.3	17.305	15.64	38.16		2
			17.275	15.62	38.10		
Extracted leads						*% Pb extracted*	
Omphacite from eclogite	Roberts Victor	0.95	19.08	15.47	37.41		1
Eclogite R7	Roberts Victor	0.79	17.602	15.58	38.05	79	2
			17.612	15.62	38.11	79	2
Eclogite 14	Roberts Victor	0.84	18.878	15.65	38.35	95	2
Eclogite 22	Roberts Victor	1.86	17.449	15.63	38.22	90	2
			17.456	15.63	38.21	96	2
			17.423	15.58	38.09	98	2
Eclogite 64	Roberts Victor	5.50	15.573	15.13	36.01	58	2
			15.630	15.14	36.02	62	2
			15.622	15.12	36.00	63	2
			15.688	15.14	36.18	66	2
Eclogite 71	Roberts Victor	3.62	16.308	14.95	35.40	38	2
			16.998	15.07	35.87	52	2
Calculated residual leads						*% Pb residual*	
R7 [a]			17.88	16.20	40.12	Not given	2
14 [a]			12.68	14.94	35.70	Not given	2
64			13.59	14.63	33.98	38%	2
71			14.42	15.00	34.93	62%	2
71			14.64	15.00	35.13	48%	2

a The lead content of these rocks is so low that the error in determining the % extraction (and hence the calculation of the composition of the residual lead) is large

References: 1. Lovering and Tatsumoto (1968); 2. Manton and Tatsumoto (1971)

are chemically equivalent to various types of basalt (however see earlier), it has been shown that the bulk chemical composition does have some effect upon the basic granulite to eclogite transition. For example, according to Ringwood and Green (1966), at a temperature of 1100°C the transition from garnet granulite (garnet + pyroxene + placioclase) to eclogite (garnet + pyroxene ± quartz) occurs at 15 kb for rocks of alkali-poor olivine tholeite composition, at 18–20 kb for alkali olivine basalts and quartz tholeiite, and at about 24 kb for high-alumina basalt. This broad zone of transition is shown in Fig. 76, together with other reactions which assist in breaking down the very large eclogite stability field into narrower distinctive fields. There is for example, the quartz/coesite transition curve which has been experimentally determined by Boyd and England (1960); the

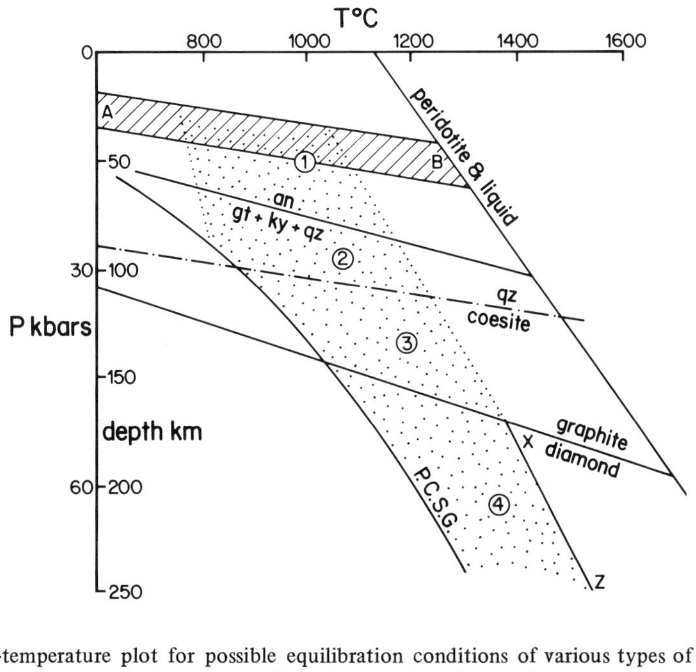

Fig. 76. Pressure-temperature plot for possible equilibration conditions of various types of eclogite. *1* zone of ordinary eclogites; *2* zone of kyanite eclogites; *3* zone of coesite eclogites; *4* zone of diamondiferous eclogites. *Shaded area A-B* represents the granulite → eclogite transition zone, according to the bulk chemistry. Reaction curve An = Gt + Ky + qz is from Boettcher (1970); the quartz/coesite curve is from Boyd and England (1960); the graphite/diamond equilibrium curve is from Kennedy and Kennedy (1976); *P.C.S.G.* is the Precambrian shield geothermal gradient (after Clark and Ringwood 1964); the peridotite solidus is from Ito and Kennedy (1967); the *stippled area* is based on the extrapolation of temperature estimates from the data given by Bishop et al. (1978) and Hatton and Gurney (1977a) (see text); the *line X-Z* gives the highest estimated temperatures for diamondiferous eclogites (Reid et al. 1976). Ordinary eclogites *may* occur in, e.g., diamond-eclogite zone, if no carbon is present in their composition

transition passes through 700°C at 27.3 kb and 1400°C at 35.2 kb. There is also the diamond-graphite equilibrium curve; however, it will be appreciated that this curve and that for the quartz/coesite transition have only limited application in view of the very limited number of specimens which contain these mineral species. Another experimentally determined curve is that of anorthite → garnet + kyanite + quartz in the system CaO-Al_2O_3–SiO_2-H_2O (Boettcher 1970) which is used by some authors to give the lower stability limit for kyanite eclogites and grospydites; however, O'Hara and Yoder (1967) have pointed out that some high-temperature aluminous clinopyroxenes formed in experimental runs have kyanite in solid solution, and this "occult" kyanite may be capable of exsolving as a subsolidus phase at lower temperatures. In addition, in all the experimental systems the presence of garnet has been a factor in classifying the experimentally determined assemblage as "eclogitic". Nonetheless, there is increasing evidence that some garnets in rocks which are *now* eclogite may have evolved by exsolution of the garnet from previously-occurring high-alumina clinopyroxenites (e.g. Harte and Gurney 1975a; Howells et al. 1975; Lappin 1978) or by reaction between preexisting kyanite and high-alumina clinopyroxene (Lappin and Dawson 1975). The actual pressures and temperatures of equilibration of eclogites are hard to estimate. Råheim and Green (1974) have shown that the distribution co-efficient K_D = (FeO/MgO) ga/(FeO/MgO) cpx for coexisting garnets and clinopyroxenes may be used as a geothermometer for eclogites. Their experiments were carried out over a wide range of basaltic compositions and in a series of glasses with the range $6.2 < 100 \frac{Mg}{Mg + Fe} < 93$. For *mg*-values ranging from 6.2 to 85 the bulk composition does not perceptibly affect the K_D values and, for a given pressure, the K_D values decrease with increasing temperature; for example, at 30 kb the K_D value ranges for 18.0 at 600°C to 1.45 at 1400°C. K_D values of eclogites from the Roberts Victor Mine reported by Bishop et al. (1978) range from 2.19 to 4.92; at an assumed pressure of 30 kb these values give a range of temperatures varying from 800° to 1200°C, or at an assumed pressure of 20 kb the temperature range is from 800° to roughly 1100°C; the temperature range in both instances is in the order of 300°–350°C. Similarly Hatton and Gurney (1977a) record a temperature range of some 280°C for another suite of nodules from the Roberts Victor Mine.

In the case of diamondiferous eclogites, some from Roberts Victor, these obviously crystallize within the stability field of diamond, and Reid et al. (1976) show that a suite of diamondiferous eclogites examined by them splits into two groupings. One group yields K_D values varying from 2.2 to 2.8, whereas the other group has K_D values ranging from 3.28 to 3.96. If the first group are to fall within the stability of diamond, then pressures in excess of 50 kb and temperatures above 1250°C are required; the other group would indicate pressures in excess of 45 kb and temperatures over 1050°C. It has been suggested that these data may indicate that these diamondiferous eclogites come from at least two regions within the mantle where different temperatures appertain.

However, a note of caution must be struck against accepting such temperatures as firm estimates, in view of current progress in geothermometry. For example, if the calcium content of the garnets is taken into consideration, as proposed by Ellis and Green (1979), the temperature range of most Roberts Victor eclogites is very restricted – 1017° to 1160°C (D.A. Carswell, personal communication); this assumes P = 45 kb and all iron as Fe^{2+}, and does not include the disequilibrium assemblages of Harte and Gurney (1975) and Lappin (1978). This newer more restricted range, if correct, suggests that the Roberts Victor eclogites, including the diamondiferous ones, originate from a restricted upper-mantle zone. The equilibration range of published eclogite garnet/pyroxene pairs, using the Råheim and Green (1974) K_D values, is shown with the restricted Roberts Victor range referred to above, on Fig. 77; the equilibration ranges of peridotites, inclusions in diamond and the megacryst suite are shown for comparison.

In most eclogites the relatively homogeneous crystals indicate that equilibration had taken place. However in some kyanite eclogites and grospydites (Lappin and Dawson 1975; Lappin 1978), there is evidence that the outer parts of both garnets and clinopyroxenes are zoned. When K_D values are calculated it is apparent that these outer zones indicate that the original eclogites after initial equilibration actually moved into a higher pressure regime before incorporation into kimberlite and transportation to the surface (Fig. 78). These examples, together with the evidence of formation of garnet harzburgite from original spinel harzburgite (see p. 147) are further indicators of pro-grade metamorphic processes within the upper mantle.

The kyanite eclogites and grospydites from Zagadochnaya pipe, Yakutia, provide a good example of the inter-relationship of the various oxides in the stabilizing of different phases. Chinner et al. (1960) have experimentally obtained a continuous solid-solution series between pyrope and grossular at 32.6 kb. However, the natural pyrope-grossular garnets had for a long time been regarded as having

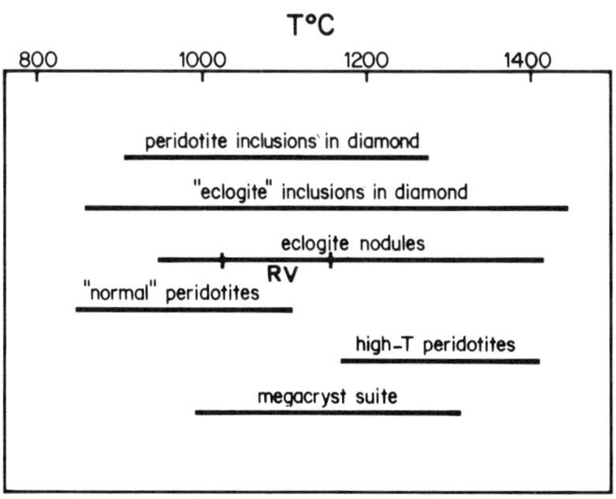

Fig. 77. Temperature ranges of equilibration of peridotite and eclogite xenoliths, peridotite- and eclogite-suite inclusions in diamond, and the megacryst suite. Compiled from ranges cited in the text. *Vertical bars* and *RV* on eclogite line refer to the restricted range for Roberts Victor eclogites when calcium content of garnets is considered (see text)

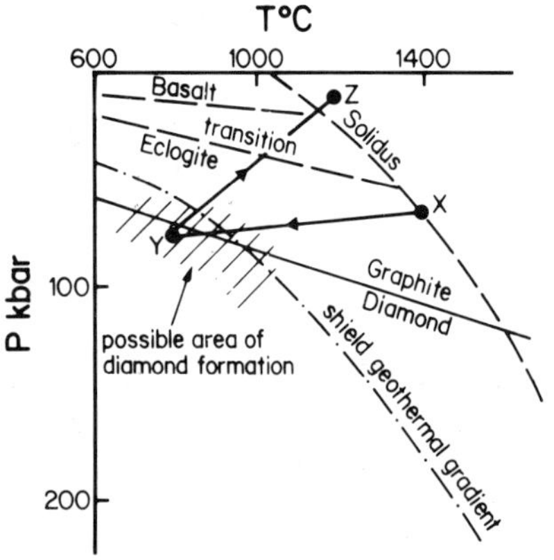

Fig. 78. Possible cooling and ascent paths of a kyanite eclogite from the Roberts Victor Mine (modified from Lappin and Dawson 1975). $X \rightarrow Y$ is the cooling path, and $Y \rightarrow Z$ is the ascent path. The kyanite eclogite solidus is from T. Green (1967); the shield geothermal gradient is from Clarke and Ringwood (1964)

a compositional break between 40% and 60% grossular, and in a soda-free system garnets of similar composition to the natural garnets have been reproduced experimentally at 30 kb by Boyd (1970) together with the expected break in composition. However, it must be recognized that the natural garnets contain small amounts of iron and, in eclogites, also co-exist with sodium-bearing pyroxenes both of which features will tend to lower the pressures below those found experimentally. The kyanite eclogites and grospydites from the Zagadochnaya pipe contain a series of pyrope-grossular garnets which fill in the previously-determined break between grossular 40 and grossular 60 (Sobolev et al. 1968). Furthermore a clear relationship exists between the sodium content of pyroxene and the calcium content of the garnets; for garnets of identical iron content, for garnets containing less than 50% of the grossular molecule, the Ca-content of the garnet decreases with increase in the amount of sodium in the pyroxene; and in the true grospydites (i.e. with garnets containing greater than 50% of the grossular molecule) the Ca-molecule of the garnet increases with increasing sodium content of the co-existing pyroxene. In short, the relatively low sodium content of the bulk rock enables the garnets of intermediate composition to stabilize. In most other garnet series it appears that the natural break is probably due to the increase in the chemical potential of sodium which leads to the disassociation of garnet of intermediate composition to form clinopyroxene plus kyanite. Nonetheless, it should be stressed that grospydites only develop from bulk rocks which have extremely high amounts of both calcium and aluminium.

It has been noted above that some of the eclogites at least are interpreted as being cumulates from basic igneous melts. The mineralogical and compositional banding has been referred to earlier; this may have arisen in various ways. It may be due to metamorphic differentiation or alternatively it may be an original effect; the changes are so rapid that it appears unlikely that the banding would have been caused by an in situ differentiation process, and Hatton and Gurney (1977a) have proposed that these rapid variations are due to multiple magma intrusions within the original magma chambers. They point out that small-scale inhomogenities are more common in Group 2 eclogites and propose that these are formed from relatively small volumes of magma. whereas the more homogeneous Group 1 eclogites may have crystallized from larger pulses of injected magma in which the effects of magma mixing will be less marked. In addition Hatton and Gurney (1977b) propose that some inhomogeneous kyanite eclogites may have arisen from other eclogites through liquid immiscibility; however, this hypothesis does not account for the observed reaction relationship between kyanite eclogite and bimineralic eclogite in which kyanite is replaced by garnet.

7.5.1.3 Metasomatised Peridotites

Many of the xenoliths within kimberlite show evidence of penetration by fluids which have developed new minerals by interaction between pre-existing minerals and introduced fluids; it is evident that they have been modfied by metasomatic processes. In addition to the late stage metasomatism which has been imposed upon the xenoliths by infiltration of fluids from the host kimberlite (which has resulted in the formation of serpentine from olivine and late stage micas around existing garnet), there is evidence of earlier mantle metasomatism which occurred prior to the incorporation of the xenoliths into their kimberlite hosts. Gurney and Harte (1980) have distinguished two types of metasomatism: (1) infiltration metasomatism and (b) diffusion metasomatism.

a) Infiltration Metasomatism. In the case of infiltration metasomatism, movement of a fluid appears to be involved and the xenoliths have been altered by the introduction of elements leading to the formation of new minerals although there may also be changes in the chemistry of pre-existing phases (the secondary alteration from the kimberlite host comes into this particular category although it is of course a late, high-level event).

The polymict peridotites described by Lawless et al. (1979) in which there is strong deformation associated with cementation of the clastic rocks by ilmenite, phlogopite, rutile and sulphide, are highly metasomatised rocks; and infiltration metasomatism is commonly associated with deformation in many peridotites, including the so-called "fertile" peridotites of Nixon and Boyd (1973a).

The most commonly-developed minerals from infiltration metasomatism are phlogopite, ilmenite, rutile and sulphides, together with rarer richterite and carbonate. The primary (mantle-derived) nature of the phlogopite was documented first on textural criteria (Dawson 1972) and confirmed subsequently on chemical criteria (Carswell 1975; Boyd and Nixon 1978; Delaney et al. 1980). The most extensive studies on metasomatised peridotites are those on the veined xenoliths from the Matsoku pipe (Gurney et al. 1975; Harte and Gurney 1975b; Harte et al. 1975) and richterite-bearing nodules mainly from the Bultfontein Mine (Erlank and Rickard 1977). At Matsoku the metasomatic minerals include phlogopite, ilmenite, rutile and sulphides (mainly pyrrhotite, pentlandite and chalcopyrite); both ilmenite and rutile show spinel exsolution. The minerals are disseminated through the matrix of deformed and recrystallized nodules but occur most distinctly in veinlets, narrow zones and patches in coarse peridotites and pyroxenites, and it is possible that the veinlets and narrow zones represent original planar fractures. Chemical as well as textural features suggest that the Matsoku metasomatic minerals are not just the product of high-level contamination of the xenoliths by host kimberlites. The metasomatic minerals show restricted but distinctive mineral chemistry; the phlogopites are similar in composition to the primary phlogopites distinguished by Delaney et al. (1980; see Table 23) and ilmenites fall within the compositional range for other peridotite ilmenites (Table 24). In addition the phases in the peridotite host cut by the veinlets (olivine, orthopyroxene, clinopyroxene and garnet) have recrystallized together with the metasomatic veining, and show close similar compositions to those found in rocks not associated with metasomatic effects. In summary, it is apparent that the metasomatic veins and rocks consisting predominantly of metasomatic minerals were in reasonable chemical and textural equilibrium at the same PT conditions as peridotites unaffected by metasomatism. Gurney and Harte (1980) summarize evidence suggesting that the metasomatic mineralisation at Matsoku may be connected to a magmatic event associated with the intrusion of clinopyroxene-rich sheets, which in itself may be related in age to the Karoo magmatism and hence quite distinct from the kimberlite event itself.

In the case of peridotite nodules containing potassic richterite, Erlank and Rickard (1977) describe the formation of richterite together with phlogopites, opaques and calcite by metasomatism which includes the development of poikoblastic and replacement textures as well as the occurrence of these minerals within narrow veinlets. It is proposed that phlogopite develops principally at the expense of pre-existing garnet, and richterite at the expense of pre-existing diopside which in many instances can be seen to be replaced by the richterite. The phlogopite compositions are very similar to those in the MARID xenoliths (Table 34), whereas the richterite is highly enriched in Si, Ti, K_2O and Na_2O compared with peridotite pargasites but poor in Cr (Table 22). Truncation of the metasomatised zones and veins by the xenolith margins suggests that metasomatism did not take place within the kimberlite host, and Sr-isotope data (Erlank and Shimizu 1977) indicate that the metasomatic event was older than the age of the host kimberlite and is isotopically completed unrelated to the kimberlite event. In addition the writer has observed, in nodules also from the Bultfontein Mine, evidence of development of richterite and phlogopite at the expense of pre-existing enstatite and hence Erlank and Rickard's contention, that the minerals developed at the expense of garnet and clinopyroxene, must be modified.

The available evidence shows that the metasomatic minerals are extremely rich in elements which are depleted in most kimberlite xenoliths; for example, some of the richterite-bearing peridotites contain > 2% K_2O and rutile-rich primary veins from Matsoku contain > 18% TiO_2. The phases themselves are rich in incompatible elements; for example, rutile from Matsoku is rich in Zr and Nb, and Erlank and Rickard (1977) report an opaque phase containing 9.1% BaO, 7.4% ZrO_2, 2.4% SrO and 0.5% Nb_2O_5. In short, the infiltration metasomatic effects indicate the presence and migration within the upper mantle of fluids whose chemistry and movement is capable of leading to the formation of minerals richer in Fe, K, Ti, H_2O, S, Nb, Zr, Hf, Sr, Cu, Rb, Ba, Na and CO_2. In view of the observed association between veinlets consisting of richterite, mica, ilmenite, rutile veins (chemically similar to the liquids proposed as parents for the MARID xenoliths see p. 187), which have metasomatised adjacent wallrocks in hand-specimens of veined peridotite in the Bultfontein Mine, it is possible that the metasomatic fluids in general are closely related to, and derived from, adjacent magma bodies.

b) Xenoliths Showing Diffusion Metasomatism. In the case of diffusion metasomatism, diffusive transfer has taken place across boundaries between upper mantle rocks leading to modifications in earlier mineral chemistry. Diffusive transfer of elements is an important factor in achieving local chemical equilibrium within rocks in general; the efficiency of diffusion in mantle rocks is testified to by the homogeneity of the mineral phases in most xenoliths derived from kimberlite, specially in peridotites, and this homogeneity is even apparent when the rocks show strong modal layering (Gurney et al. 1975). However, in the xenolith suite from Matsoku, there are contacts between sheets rich in clinopyroxene which have been magmatically intruded into the host rock of coarse-grained, common garnet lherzolite. Despite the obvious intrusion relationships between the pyroxene sheets and host rock, the mineral composition across the contact is virtually the same for most elements with the particular exception of chromium; moreover, the host-rock phases are noticeably higher in iron and titanium than is usual for common peridotites (Harte et al. 1977). These chemical features, with the host rock and intrusive sheets apparently trying to reach equilibrium, are attributed to diffusion metasomatism between the sheets and the host peridotite; the bulk chemical composition of the peridotites shows the elevation of iron and titanium which is particularly characteristic of the high-temperature, so-called fertile peridotites of Nixon and Boyd (1973a), again raising the question as to whether the high-temperature peridotites are "fertile" or have in fact been "fertilized" (see section (a) on infiltration metasomatism).

Diffusion metasomatism has been noted between units of differing mineral chemistry in xenoliths of eclogite in which, for example, single xenoliths of eclogite have been diffusively metasomatised by host-rock kimberlite (Chinner and Cornell 1974); and also in the case of xenoliths where there are strong chemical gradients between bands of distinctly different chemical composition (Lappin and Dawson 1975; Hatton 1978).

7.5.1.4 Glimmerites and MARID-Suite Rocks

Nodules mainly dominated by phlogopite occur in several South African and Yakutian kimberlite pipes; these nodules, often described loosely as "glimmerites" are usually dominated by phlogopite, but amphibole, clinopyroxene, ilmenite, rutile, apatite and olivine also occur in various amounts and combinations. Although at first these rocks might appear to form a coherent suite, Aoki (1974) and Dawson and Smith (1977) found considerable variations in the chemistry of minerals from mica-rich nodules in South African kimberlites, and Dawson and Smith recognize the MARID-suite that consists of varying proportions of compositionally distinct phlogopite, potassic-richterite, rutile, ilmenite and diopside (Table 34). The nodules are rounded or discoidal up to 15 cm across and foliation of lineation or both were observed, resulting from alignment of phlogopite and amphibole crystals; banding in some specimens is due to different concentrations of the various phases and is interpreted as cumulate layering. All specimens are deformed, this deformation being apparent in bending of amphibole crystals and undulose extinction of phlogopites. The MARID micas are phlogopite with *mg* ranging from 0.88–0.82 which are distinguished from micas in textural equilibrium with other silicates in peridotite nodules by their low NiO, Cr_2O_3 and FeO and in having relatively low Al_2O_3 but (calculated) high Fe_2O_3 (see Table 22). The amphibole is a potassic richterite (Table 34) with *mg* varying from 0.87 to 0.91, K/(K + Na) varying from 0.47 to 0.58, Na + K > Ca and considerable Fe^{3+} in

Table 34. Analyses of representative phases in MARID-suite rocks. (Dawson and Smith 1977)

	Phlogopite	K-richterite	Rutile	Ilmenite	Diopside
SiO_2	42.2	54.2	0.1	0	54.0
TiO_2	1.48	0.64	95.9	53.1	0.54
Al_2O_3	9.55	1.05	0.0	0.02	0.61
Cr_2O_3	0.11	0.11	0.7	1.54	0.21
Fe_2O_3	n.d.	n.d.	2.7 [c]	2.2 [a]	n.d.
FeO	6.90 [b]	4.18 [b]	0	30.2	4.40 [b]
MnO	0.02	0.05	0	0.41	0.11
MgO	23.5	21.5	0.3	11.4	17.9
NiO	0.16	0.05	0	0.18	0.06
CaO	0.01	6.22	0	0.03	20.0
Na_2O	0.13	3.91	0	–	1.24
K_2O	9.89	5.28	0	–	0.02
Total	94.0	97.3	99.7	99.2	99.1

a Fe_2O_3 calculated on basis of stoichiometry
b Total iron as FeO
c Total iron as Fe_2O_3

tetrahedral sites. They are most similar chemically to potassic richterite occurring in the groundmass of kimberlites. The rutile, like that from peridotites and eclogites, has appreciable chrome and also appreciable Fe_2O_3. The ilmenites from MARID-suite xenoliths have TiO_2, Cr_2O_3 and FeO values found within the wide ranges reported for ilmenite megacrysts in kimberlites; but the MARID ilmenites are distinctive in containing exceptionally low Al_2O_3. The diopsides are fairly pure diopsides showing little solid solution towards enstatite and hedenburgite; compared with lherzolite diopside they contain little Al_2O_3, Cr_2O_3 and Na_2O but are fairly high in TiO_2. Zircon was considered to be an additional mineral of the MARID-suite by Dawson and Smith (1977), and has subsequently been found in other MARID-suite xenoliths (J.J. Gurney, personal communication). As a suite the MARID phases are lower in Al_2O_3 and Cr_2O_3, but richer in total iron, than minerals in peridotite xenoliths.

D/H ratios of coexisting phlogopite and richterite in mica-rich nodules from the Dutoitspan and Wesselton kimberlites vary from δD -43.2‰ to -77.4‰ for phlogopites and -103.1‰ to -132.5‰ for richterites (Kuroda et al. 1975); the δD values for phlogopites are very similar to those found for phlogopite megacrysts in kimberlite (Sheppard and Dawson 1975; see Fig. 22). Rare gas analyses indicate that the rare gas isotopic compositions are atmospheric except for ^{4}He and ^{40}Ar, and no excess ^{129}Xe was found (Kaneoka et al. 1977); a large excess of ^{40}Ar indicates that the phlogopite did not crystallize at shallow depths (Kaneoka and Aoki 1978).

Dawson and Smith (1977) postulate that the MARID rocks crystallized under high oxidation conditions from a magma chemically similar to kimberlite, within the higher parts of the upper mantle; the presence of amphibole restricts the depths to less than approximately 100 km. However, other mica and amphibole-rich xenoliths containing olivine and orthopyroxene, as well as minerals of similar, but not the same, composition as MARID-types are possibly metasomites, perhaps representing wallrock of the magma body from which MARID-suite rocks crystallized; these are possibly transitional to the metasomatic peridotites. A metasomatic origin for richterite-bearing peridotites is favoured by Aoki (1975) and Erlank and Rickard (1977); this is discussed more fully in Chap. 9.

7.5.1.5 Miscellaneous Xenoliths

1. Alkremite Xenoliths. Xenoliths comprising coarse-grained garnet and dark green spinel are found in kimberlites of the Udachnaya pipe, U.S.S.R. (Ponomarenko 1975), and in the Bellsbank and Jagersfontein kimberlites, South Africa (Nixon et al. 1978). The name is a Russian mnemonic derived from the preponderance of Al, Si and Mg in the rock analyses given by Ponomarenko (1975). The xenoliths comprise approximately two-thirds of a chromium-poor titanium-poor pyrope garnet; in their low chromium and titanium contents (approximately 0.18% and 0.16% respectively) this pyrope is unlike both the titanian pyrope megacrysts found in kimberlite and the chrome pyropes derived from or found within rocks of the peridotite-pyroxenite group. There is a range in the MgO contents of the garnets from approximately 14.8% to 19% MgO and this varies inversely with the CaO content which varies approximately from 5% to 10.5% CaO. The spinel is a hercynitic variety containing in the order of 60%–65% Al_2O_3, 1%–5% Cr_2O_3, 11%–14% total FeO, and 18%–22% MgO. In addition to garnet and spinel, corundum is present in an alkremite from the Bobbejaan kimberlite, Bellsbank (J.B. Dawson, personal observation). The intercumulus nature of the spinel and current experimental data suggest that the alkremites formed as early dense cumulates from an aluminous mantle melt under pressure conditions equivalent to about 70 km depth; at greater pressures perhaps more pyrope-rich garnets would be capable of being formed. The alkremite xenoliths are considered by Nixon et al. (1978) to have become separated from the magma prior to the crystallization of clinopyroxene, whereas more complete fractional crystallization would have produced garnet-clinopyroxene assemblages (i.e. eclogites) with elimination of early-formed spinel. This interpretation is supported by the evidence that the compositional range of garnets in the Bellsbank alkremites is very similar to that found in the ordinary and corundum-kyanite eclogites from Bellsbank (Bobbejaan), reported by Bishop et al. (1978).

In their paper on alkremites, Nixon et al. (1978) include a large (1.5 kg) spinel nodule BD 1988 from the Roberts Victor Mine (Lawless 1974); in addition to spinel and small amounts of garnet this nodule also contains olivine. The nodule BD 1988 differs from the alkremites in having a more chrome-rich garnet (0.65 Cr_2O_3) and more chromiferous hercynite spinel (10.03 wt% Cr_2O_3).

2. **Quenched Orthopyroxene-Ilmenite Rock.** A unique non-equilibrated pyroxene-ilmenite nodule has been described from the kimberlite of the Weltvreden Mine, South Africa, by Rawlinson and Dawson (1979); the xenolith comprises mainly acicular high-Ti bronzite (terminology of Stephens and Dawson 1977, see Table 13) irregularly intergrown with magnesian ilmenite in macro-spherulites and comb-layered units. Some macro-spherulites are nucleated upon larger crystals of high-Ti bronzite. In addition the rocks contain small amounts of high-Ti pyrope and titanian phlogopite, the latter occupying interstices between acicular orthopyroxene, and low-chrome diopside forming regular intergrowths with ilmenite. The composition of these phases, and the morphology of the pyroxene-ilmenite intergrowths, can be matched with those in the megacryst suite (see Chap. 8). In the rock are small patches of fine-grained calcite, serpentine, ilmenite, phlogopite and diopside, together resembling a kimberlite groundmass assemblage, which are interpreted as the final product of an essentially closed-system crystallization of an inferred high-titanium liquid.

Phases of similar composition to those in the quenched orthopyroxene-ilmenite rock occur in a coarse, large single crystal of enstatite, also from the Weltvreden Mine (Meyer et al. 1979); in this large (approximately 20 cm) single crystal of titanian bronzite are inclusions of magnesian ilmenite and titanian pyrope; the pyrope grain has a core of a pink chrome-pyrope garnet which itself contains small crystals of chrome diopside, chromite and ilmenite. Furthermore within the enstatite host and closely associated with the titanian garnet are calcite, titanian phlogopite and serpentine. It is a possibility that it may have formed from a liquid similar to the one proposed by Rawlinson and Dawson (see above) with the liquidus phase (i.e. titanian pyrope) nucleating upon a pre-existing xenocryst of chrome pyrope with associated high chrome phases (possibly a micro-xenolith). However, the conditions of crystallization of such a proposed liquid would have been drastically different from those of the quenched orthopyroxene-ilmenite rock.

Chapter 8 The Megacryst Suite

8.1 Petrography and Phase Chemistry

Single crystals of large size (commonly > 2.0 cm) are widespread and sometimes abundant as inclusions in kimberlite; they have been referred to both as megacrysts (the terminology adopted here) and discrete nodules (Nixon and Boyd 1973b). In considering this mineral suite, the assumption is that the minerals are so large that they could not have been derived by fragmentation of upper-mantle xenolith rock types, the grain-size of which is usually in the range 2–4 mm. Two series of megacrysts are known, those which are rich in chromium and those which are chrome-poor; the chrome-rich ones are by far the rarer and have so far been described from the Excelsior Mine, South Africa, and from some Colorado kimberlite pipes. A chrome-rich (8.5 wt% Cr_2O_3) and low-lime (2.7% CaO) garnet, 2 cm in diameter, has been described from the Excelsior Mine, S. Africa (Boyd and Dawson 1972); it is compositionally similar than the high-Cr, low-Ca pyropes found as diamond inclusions, and occurring as small isolated crystals in the kimberlite of the Finsch Mine (Gurney and Switzer 1973). The garnet contains small altered crystals of ?olivine or ?orthopyroxene.

The megacrysts of the Sloan kimberlite, Wyoming (Eggler and McCallum 1976) provide the best-described example of the Cr-rich and Cr-poor megacrysts occurring within the same intrusion. The main compositional differences between garnets and pyroxenes in the two suites are shown in Table 35; the main features are higher Mg and Cr in one suite, compared with higher Ti and Fe in the other; in addition, chromite and olivine occur as inclusions in the high-Cr megacrysts.

As in Colorado, the chrome-poor megacryst group is by far the commoner in the S. African and Lesotho kimberlites. It includes large crystals of olivine, orthopyroxene, clinopyroxene, garnet and ilmenite; in many cases small inclusions of one species are found within a single large crystal of another and in some cases regular intergrowths occur, especially of clinopyroxene and ilmenite and to a lesser extent of orthopyroxene and ilmenite. In the most extensive study of abundant megacrysts from a single location, the Monastery Mine, South Africa (Jakob 1977; Gurney et al. 1979) the megacrysts show a substantial range of chemical variation particularly in Mg/Fe (Fig. 79). The detailed compositions are different from those of most peridotite or eclogite xenoliths but some of the more magnesian megacrysts do show certain similarities to the phases in high-iron high-titanium deformed peridotites described from the Thaba Putsoa intrusion, Lesotho

Table 35. Ranges of compositions of phases in two megacryst assemblages from the Sloan 2 kimberlite pipe, Wyoming (Eggler and McCallum 1976)

	Cr-rich Megacrysts	Cr-poor Megacrysts
Garnets		
TiO_2	0.6– 0.9	0.7– 1.1
Cr_2O_3	6.5–12.7	1.4– 4.3
CaO	7 – 9	4.4– 5.1
Mg/(Mg + Fe)	82 –84	78 –83
Clinopyroxenes		
TiO_2	0.1– 0.2	0.2– 0.4
Cr_2O_3	1.0– 2.3	0.4– 0.8
Al_2O_3	1.1– 1.4	1.9– 2.2
Mg/(Mg + Fe)	92 –93	87 –90
Orthopyroxenes		
Cr_2O_3	0.4– 0.6	0.1– 0.3
Mg/(Mg + Fe)	92 –93	88 –91

(Nixon and Boyd 1973a). The olivines fall into two groupings, those of composition Fo_{84-88} with a nickel oxide content of ~ 0.3%; the other group are olivines Fo_{72-78}, NiO < 0.12 wt%. Inclusions of kimberlite have been found in some of the more magnesian high-Ni olivines (Haggerty and Boyd 1975); and one of these included kimberlites has a bulk composition very similar to that of the host kimberlite from the Monastery Mine. Orthopyroxene forms brown glassy, homogeneous megacrysts which define a trend of varying Mg vs. Fe at constant Ca (see Fig. 79) with the more Fe-rich sometimes forming irregular intergrowths with ilmenite. The clinopyroxenes vary from subcalcic diopsides through titanian chrome diopsides to low chrome diopsides (terminology of Stephens and Dawson 1977) defining a trend from high-Mg, low-Ca low-Fe megacrysts to low-Mg, high-Ca and slightly increased Fe (Fig. 79); the more calcic ones tend to form regular lamellar intergrowths with ilmenite. The garnets define the trend of changing Mg/Fe ratio from highly magnesian megacrysts to iron-rich garnet, the more iron-rich ones sometimes containing inclusions of ilmenite. In all cases they are low in chromium and the more magnesian ones may be termed high-titanium pyrope or titanian pyrope (terminology of Dawson and Stephens 1975; see Table 12). The ilmenites show a wide range in compositions particularly with respect to Fe and Mg but where co-existing with clinopyroxene, orthopyroxene and garnet these are all similar to each other and within the compositional range for discrete ilmenite megacrysts.

Other well-investigated megacryst suites are those from the kimberlites of northern Lesotho (Nixon and Boyd 1973b; Boyd and Nixon 1975), with particular

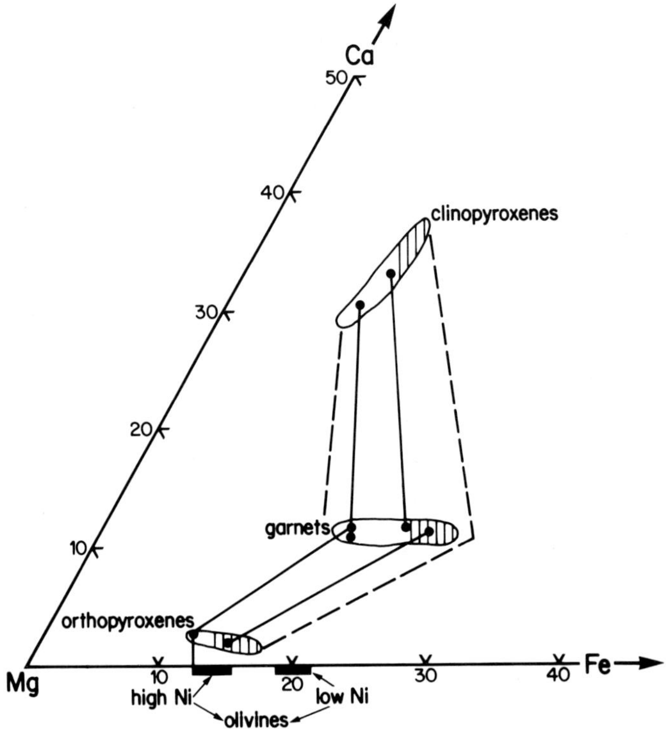

Fig. 79. Ca-Mg-Fe diagram for compositional fields of megacryst-suite olivines, pyroxenes and garnets. *Tie-lines* join intergrown phases. Phases with compositions lying in the *shaded areas* of the compositional fields (i.e. those richer in Fe) are intergrown with magnesian ilmenite. (Modified from Gurney et al. 1979)

emphasis on the megacryst suite from the Thaba Putsoa kimberlite. The phases and ranges of composition are extremely similar to those from the Monastery Mine, but there is one additional point of extreme interest. From the Thaba Putsoa intrusion, Nixon and Boyd (1973b) describe composite nodules comprising subcalcic low-chrome diopsides and titanian pyropes, the compositions of which can be matched with individual megacrysts; they are interpreted as being aggregates of megacrysts which have crystallized out from the kimberlite magma and as such conform to the original "griquaite" of Beck (1899); and indeed these aggregates are referred to as "griquaites" by Nixon and Boyd (1973b). The phases in these griquaites are compositionally unlike the phases found in other kimberlite eclogites, although the term griquaite and eclogite are often used synonymously (see introduction to eclogites, Chap. 7.5.1.B).

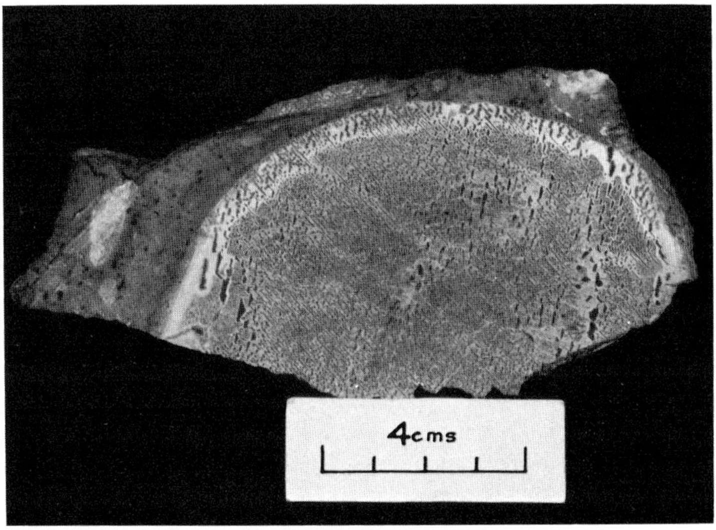

Fig. 80. Photograph of a single large crystal of low-Cr diopside intergrown with magnesian ilmenite, from the Monastery Mine, South Africa. The nodule, which is cut normal to the *c*-axis of the diopside, is calcitized around its periphery *(white rim)*

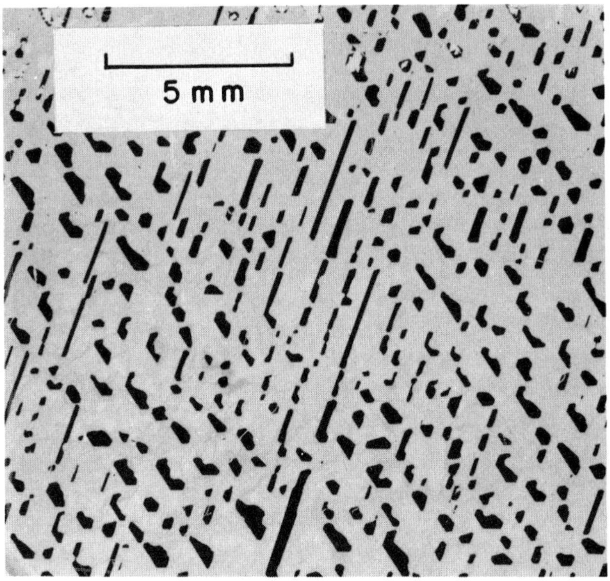

Fig. 81. Photomicrograph of a thin-section cut from the specimen illustrated in Fig. 80, showing the regular arrangement of the ilmenite lamellae

Of the silicate-oxide intergrowths, the most spectacular are the lamellar intergrowths of low-chromium diopside and ilmenite. Fine bars of ilmenite grow parallel to the *c* axis of the host clinopyroxenes which are large single crystals up to 20 cm in length (Dawson and Reid 1970); the ilmenite lamellae may intersect at angles of 60° or 120° (Figs. 80 and 81), thereby resembling the Widmanstätten pattern of intergrowths between phases in meteorites and alloys. The ratio of clinopyroxene to ilmenite varies very widely from nodule to nodule, the clinopyroxene content varying from 10%–90%. The nodules are generally strongly calcitized around their periphery and in the case of smaller nodules the pyroxene has been entirely replaced by calcite and epidote. Although developed most spectacularly at the Monastery Mine, other clinopyroxene-ilmenite lamellar intergrowths have been described from the Frank Smith, Sonop, Uintjes Berg kimberlites, South Africa; from the Kao, Thaba Putsoa, Solane, Matsoku, Mothae and Pipe 200 kimberlite intrusions, Lesotho; from the Hanaus II, and Mukorob kimberlites, S.W. Africa; from the Colossus and Wessels kimberlites, Rhodesia; from the Artur de Paiva kimberlite, Angola; from Mir, Dalnaya, and Kharakhtakh kimberlites, U.S.S.R.; and from the Iron Mountain, Stockdale, and Ison Creek kimberlites, U.S.A. The intergrowth nature of these diopside-ilmenite nodules has given rise to considerable debate as to their origin. Briefly, they are thought to be either (a) formed by exsolution from a pre-existing high temperature phase, or (b) a eutectic intergrowth (see review by Gurney et al. 1973). Gurney et al. (1973) on geochemical evidence favour the eutectic origon and indeed the lamellar intergrowth pattern has been reproduced experimentally by Wyatt (1977). Furthermore, diopside-ilmenite intergrowths have been shown to take place from a crystallizing magma in the case of the non-equilibrated ilmenite pyroxene rock from the Weltvreden Mine (Rawlinson and Dawson 1979; see p. 188).

Within the Monastery ilmenite suite as a whole, exsolution of oxides is widespread; the main phases to have exsolved are magnesian titanohaematite, magnesian titanomagnetite and aluminous spinel; in addition there is evidence for subsolidus reduction in the crystallization trends of both the discrete nodules and of the intergrowth ilmenites (Haggerty et al. 1979). Other ilmenite nodules from a variety of localities contain intergrowths of titanian chromite, rutile and a Cr-Ti-spinel (summarised by Wyatt 1979), and rods of pleonaste and lamellae of titanomagnetite are found in the nodules of ilmenite from the Frank Smith kimberlite (Pasteris et al. 1979).

Two clinopyroxene-ilmenite intergrowth nodules from the Monastery Mine contain abundant tubes containing pyrrotite and pentlandite which are preferentially located at the terminal contacts of elongate ilmenite laths (Haggerty et al. 1979); in some instances the tubes pierce both the ilmenite and the pyroxene and it is considered that the only reasonable explanation to account for their presence is to invoke an immiscible sulphide liquid which was present during the formation of the intergrowths. Alteration assemblages developed at clinopyroxene-ilmenite contacts are Fe-rich phlogopite plus orthopyroxene plus perovskite; the phlogopite has developed at the expense of both the ilmenite and the pyroxene, but tends to be concentrated in the ilmenite. Pyrrhotite-pentlandite intergrowths are also found in a clinopyroxene-ilmenite intergrowth from the Frank Smith Mine where they are accompanied by djerfisherite – a potassium- and chlorine-bearing iron sulphide [structural formula $K_6(Fe, Ni, Cu)_{25}, S_{26}Cl_1$] and two unnamed potassic sulphides which are intermediate in composition between iron sulphide and djerfisherite (Clarke 1979); it is suggested that the sulphide assemblage results from original immiscible mono-sulphide droplets enclosed in the ilmenite-clinopyroxene intergrowths which gave rise to pyrrhotite and exsolved pentlandite when the temperature dropped below 610°C; this was accomapnied by metasomatic introduction of potassium, chlorine and possibly copper from metasomatic fluids which have also started the incipient breakdown of the clinopyroxene.

The less spectacular, irregular intergrowths of orthopyroxene and ilmenite comprise only a small proportion ($< 10\%$) of the pyroxene-ilmenite intergrowths. They are known from the Monastery, Frank Smith and Premier kimberlites (S. Africa), Colossus (Rhodesia) and Kharakhtakh (U.S.S.R.). Compositionally, the orthopyroxenes are Ti-rich enstatites or titanian bronzites (terminology of Stephens and Dawson 1977), and the ilmenites are similar to those in the clinopyroxene-ilmenite intergrowths. In general, these nodules are less strongly altered than those containing clinopyroxene, though Haggerty et al. (1979) report interface assemblages of olivine + rutile + Ti-rich phlogpite at orthopyroxene-ilmenite contacts. Irregular intergrowths of orthopyroxene and ilmenite have been produced experimentally from a melt (Rawlinson and Dawson 1979), and hence, like the diopside-ilmenite intergrowths, may be interpreted as being the result of cotectic crystallization from a melt.

Three other species of megacrysts do not fall readily into either the high-chrome or low-chrome suites outlined above:

a) Diopside megacrysts of a distinctive apple-green colour have been reported from the Kamfersdam mine (Boyd and Nixon 1978) and also occur in the Bultfontein Mine (author's own observation); the nodules, which are up to 5 or 6 cm in size, are generally rounded and may be mono-mineralic or (in the Bultfontein specimens) coated with (or veined by) ilmenite. The nodules are often strongly deformed, with a strained core surrounded by a zone of smaller, strain-free neoblasts; and in some instances the ilmenite is interspersed between the new neoblasts. Compositionally the pyroxene is a fairly pure, chrome-poor diopside showing very little solid solution towards enstatite; they most closely resemble the diopsides found either in the kimberlite matrix or in the MARID-suite xenoliths.

b) Mica megacrysts, which are not generally greater than 1–2 cm in size, but in some exceptional instances, as at the Monastery Mine, are up to 15 cm. Compared with the groundmass micas, the megacrysts (1) are greater in size, (2) are generally corroded and have a bleached alteration rim of hydrophlogopite + magnetite, (3) have a dark-brown, rather than bronze, colour, (4) show buckling with attendant undulose extinction (in some, but not all, specimens); (5) do not contain inclusions of groundmass phases, such as perovskite or magnetite. The megacrysts are all highly magnesian (> 20% MgO) but have variable amounts of FeO, TiO_2, Cr_2O_3 and Al_2O_3. Dawson and Smith (1975b) have recognized two main groups of megacryst in this respect resembling the two groups of megacrysts (high-Cr and low-Cr) mentioned above. The first contains > 0.5 wt% Cr_2O_3, < 1.0 wt% TiO_2 and < 3.7 wt% FeO; these are chemically very similar to the primary micas in lherzolite xenoliths (p. 139) and may possibly have been derived from disrupted lherzolites. The other group comprises megacrysts containing < 0.5% Cr_2O_3, > 0.60% TiO_2 (but varying up to 2% TiO_2), < 12% Al_2O_3 and > 3.7% FeO. Although very few of these "non-peridotite" megacrysts are chemically similar to the micas in the MARID-suite xenoliths, plots of NiO vs. Mg/(Mg + Fe) indicate that they are chemically distinct and Dawson and Smith (1975b) suggest that the "non-peridotite" megacrysts are probably phenocrysts crystallizing from the kimberlite magma. A phlogopite exceptionally high in TiO_2 (6.1 wt%) from the Murfresboro' kimberlite, Arkansas (Giardini et al. 1974), and phlogopite inclusions in zircons from the Monastery Mine (Kresten 1973c) have the same chemical characteristics as the MARID-suite (glimmerite) micas; all have such a low Al content that the presence of appreciable amounts of ferric iron in the Y site must be inferred.

c) Zircon megacrysts. Zircon is another megacryst phase found within many kimberlite intrusions; it is this type or zircon, containing small amounts of uranium and lead, which is the source of the zircon dates for many kimberlite intrusions (Davis 1977). Although generally occurring as monomineralic grains, it has also been found with inclusions of phlogopite (Kresten 1973c) and forms intergrowths with ilmenite and/or rutile (Raber and Haggerty 1979); the study on zircons from the Monastery, Mothae and Kimberley kimberlites (Raber and Haggerty 1979) shows that the ilmenites associated with zircons have a very narrow range in MgO content (8–10 wt%) and in Cr_2O_3 (0.02–0.85 wt%): for a given Cr_2O_3 value they are usually more magnesian than ilmenites not occurring in association with zircon. Subsolidus reactions in these crystals at the zircon-ilmenite interfaces, yield an exotic mineral assemblages consisting of zirconolite ($Ca\,Zr\,Ti_2O_7$), titanium-rich baddeleyite [$(Ti\,Zr)O_2$], and diopside in association with calcite; in the ilmenite-rutile-zircon assemblages, aggregates of zirconolite + baddeleyite + diopside are found at ilmenite-zircon interfaces; and baddeleyite + sphene ($Ca\,Ta\,SiO_5$) at rutile-zircon interfaces, both assemblages being in association with calcite. In the case of rutile-zircon assemblages, the reaction products are zirconium-armalcolite and ilmenite, these assemblages having been produced in the absence of calcite. Raber and Haggerty (1979) suggest that these exotic assemblages along calcite-associated interfaces are the results of reactions being due to the availability of a carbonatitic fluid which induced an intercrystalline reaction between zircon and either ilmenite or rutile; this report of zirconolite is the first in kimberlite and the zirconium-armalcolite from Mothae is the second reported occurrence of armalcolite in kimberlite.

8.2 Trace Element and Isotope Chemistry

To date there are relatively few data on megacryst-suite phases. Eggler et al. (1979) show small but distinct differences in absolute abundances and chondrite-normalized fractionation patterns for rare earth elements in the high-Cr and low-Cr megacryst suites in the Colorado/Wyoming kimberlites. Cr-rich diopside is slightly enriched in light REE/(La-Gd) and both Cr-rich diopside and garnet are depleted in heavy REE (Dy-Yb) relative to their Cr-poor counterparts; their patterns are similar to those for garnet and clinopyroxene in Lesotho garnet lherzolites (Shimizu 1975b). REE's in a Cr-poor clinopyroxene intergrown with ilmenite in a nodule from the Monastery Mine (Mitchell et al. 1973a) are considerably higher than those in the Cr-poor and high-Cr diopsides in the Colorado/Wyoming kimberlites, suggesting two types of low-Cr diopside megacrysts (one massive, the other in the lamellar intergrowths) analogous to the two distinctive types of ilmenite (above).

Shimizu (1975a) reports trace amounts of K (~ 185 ppm), Rb, Cs, Sr and Ba in a sub-calcic [Ca/(Ca + Mg) = 0.32] diopside megacryst from Thaba Putsoa; the rims are depleted in Rb, Cs and Ba relative to the core. Kramers (1979) also reports small amounts of K (125–301 ppm), and small amounts of Rb and Sr in four megacrysts from Bultfontein, Dutoitspan, Wesselton and De Beers Mine, with $^{87}Sr/^{86}Sr$ ratios tightly clustered from 0.70327–70399. These samples also have uniform and fairly radiogenic Pb isotope compositions, suggesting a common magmatic origin; however, in view of the large initial Pb isotopic variations in the host kimberlites, Kramers (1979) suggests that the megacrysts are not genetically linked with the kimberlites, a view contrary to the fractionation hypothesis outlined below.

Phlogopite megacrysts from the Monastery Mine contain Rb 620–880 ppm and Sr 38–402 ppm (Allsopp and Barrett 1975); they yield an isochron of 90 ± 4 m.y. and an initial $^{87}Sr/^{86}Sr$ ratio of 0.7046 which falls within the range of 0.7038–0.7048 for fresh kimberlites (Barrett and Berg 1975) and close to the Kimberley area clinopyroxene megacrysts (see above). It is apparent that, although the Pb isotopic evidence is contrary to a genetic relationship between the investigated megacrysts and kimberlites, the Sr isotopic evidence *does* support such a relationship.

The hydrogen and oxygen isotope data for phlogopite megacrysts plot in a distinct field on a δD vs. δO^{18} diagram (Fig. 22); as the field is relatively narrow (δD 40‰–70‰, $\delta^{18}O$ 5‰–8‰), and includes phlogopite megacrysts from S. Africa, U.S.A. and India, these values are interpreted as the isotopic values for deep-seated hydrogen and oxygen (Sheppard and Dawson 1975). The megacryst oxygen isotope range is very similar to that of carbonates and oxides in primary igneous carbonatites (Sheppard and Dawson 1973), and to that in peridotite and eclogite xenoliths from kimberlite (Garlick et al. 1971).

8.3 Conditions of Formation

As described earlier, some megacrysts occasionally contain small inclusions of another mineral of the megacryst suite, and the tie-lines between such co-existing garnets and clinopyroxenes or garnets and orthopyroxenes are in excellent accord with theoretical tie-lines calculated on the results of experimental systems (Akella and Boyd 1974). In addition, tie-lines between garnet and olivine suggest that the most subcalcic clinopyroxene and the most magnesian garnets, orthopyroxenes and olivines have formed in equilibrium with each other, and that gently rotating concordant tie-lines may be drawn for successively more iron-rich compositions. Boyd and Nixon (1975) and Gurney et al. (1979) broadly agree that the trends shown by the megacrysts represent a single differentiation sequence and that the megacrysts are cumulate minerals which precipitated from a magma which formed in a mantle-melting event. The earlierst, highest-temperature (~ 1315°) cumulus phase was highly magnesian orthopyroxene, followed by the magnesian olivines, magnesian garnets and the subcalcic clinopyroxenes which would have crystallized whilst the water content of the magma gradually increased; ilmenite also became a liquidus phase at about 1230° and crystallization of these co-existing phases was complete at approximately 1130°C. It is a possibility that the iron-rich olivines formed when the temperature dropped towards 950°C in a melt which was relatively rich in water compared with the initial CO_2-rich melt. The extremely wide range in temperature of crystallization and also the large size and (pegmatitic) aspect of the intergrowths may be due to the increasing water content in an essentially closed system. The temperature range is shown in Fig. 77.

It is apparent from studies on ilmenites (especially those of Haggerty et al. 1979) that they have had variable and sometimes complex histories. There are two major ilmenite compositional groups, one of the nodules and the other of those in ilmenite-pyroxene intergrowths, a feature originally recognized by Mitchell (1977) who proposes that there are two distinct types of ilmenite in kimberlite, one (the nodules) being phenocrystal and the other (lamellar intergrowths), being xenoliths derived from the low velocity zone. However, Haggerty et al. (1979) suggest there is sufficient overlap in the analytical data to indicate that both ilmenite types are derived from a common source region in the mantle which was ilmenite-saturated; the ilmenites in the intergrowths are more magnesian than the discrete nodule cores but the paths of equilibration result in terminal compositions for the nodules which are substantially more magnesian than for the intergrowth ilmenite composition. It is inferred that the growth of the nodules took place over a relatively long period of time and, based on the initial MgO contents, the discrete nodules possibly predate the intergrowths by a small amount. They suggest that the intergrowths formed at both lower pressure and temperature than the nodules and, having been encased within pyroxene, they were then armoured against interaction against the surrounding magma, whereas in the discrete nodules the interaction with the host kimberlite resulted in a trend towards the MgO-rich rims. The experimental data between 900° and 1200°C at a range of oxygen fugacity values between 10^{-14} and 10^{-8} atm suggest that the production of the exsolution lamellae of titano-haematite and spinel is related to both temperature and to the oxygen fugacity; the textures and composition of the subsolidus reduction assemblages lie within the stability fields demarcated by the buffers forsterite-magnetite-quartz at the upper limit and graphite-CO-CO_2 at the lower limit (Haggerty et al. 1979). Although Haggerty et al. (1979) attribute the whole of the ilmenite suite at Monastery (both discrete and lamellar intergrowth ilmenites) to a single episode of inter-related crystal-liquid reactions, and although the model outlined earlier for the megacryst-suite as a whole having formed during one igneous event appears to be realistic (Gurney et al. 1979), there is some evidence from the textural studies of Mitchell (1973b) that, prior to the kimberlite event certain nodules of ilmenite had a complex history which involved strain deformation with subsequent annealing; in addition, Pasteris et al. (1979) record that some of the ilmenite-pyroxene intergrowths from the Frank Smith mine are deformed. Both these observations suggest that in the cases cited the formation of ilmenite had taken place prior to the deformation associated with the kimberlite event. It is apparent that the chemical and textural data on the ilmenites (both in discrete nodules and in pyroxene-ilmenite intergrowths) may give variable and sometimes conflicting evidence from intrusion to intrusion.

Co-existing garnets and ilmenites have been synthesized at high pressure (21–40 kb) and temperatures between 900° and 1100°C from olivine basanite composition and a composition of pyrolite less 40% olivine (Green and Sobolev 1975). The compositions of the synthesized ilmenites and garnets are very similar to those in kimberlite from the point of view of Mg/Fe ratios and the amounts of Al_2O_3 and Cr_2O_3, and Green and Sobolev (1975) suggest that natural garnet/ilmenite pairs crystallized within the experimental temperature range, i.e. 900°–1100°C, but under varying pressure and fO_2 conditions.

The problems of megacryst formation are not unique to kimberlite, since megacryst of the composition of the megacryst phases described above are not unique to kimberlite. Some or all of the above phases have been found in nephelinite breccias at Kakanui, New Zealand; in basanite breccia from Elie Ness, Fife, Scotland; and in alnoite breccia from the island of Malaita, Solomon Islands (Colvine 1968; Dickey 1968; Dawson et al. 1978a; Nixon and Boyd 1979).

Chapter 9 The Sub-Continental Mantle and Crust – Evidence from Kimberlite Xenoliths

The examination of kimberlite xenolith populations can provide considerable insight into the nature of the mantle and lower crust beneath the cratons where kimberlites exist; the models proposed below cannot be extrapolated to continental margins or oceanic areas. The upper mantle is here regarded as the zone of the earth between (a) the seismically-determined Mohorovicic discontinuity at approximately 35–40 km beneath the cratons, and (b) the seismically-determined beginning of the transition zone at 300–400 km depth. Petrologically these two boundaries are tentatively interpreted as (a) the break between crustal basic granulites and upper-mantle harzburgite (reviewed by Dawson 1977), and (b) the onset of the inversion of upper-mantle orthorhombic olivine to its denser, cubic polymorph-ringwoodite. The fact that ringwoodite has not yet been identified in the kimberlites gives a maximum value of 300 km to the depth sampled by kimberlite.

Early models of the upper mantle were very simple, proposing that the upper mantle consists of either eclogite or garnet peridotite; these simple models were necessitated by the problems in interpreting the speed of seismic waves, or as a first approximation of upper-mantle compositions in connection with experimental production of basaltic melts. A more complex model, proposed by Clark and Ringwood (1964), comprised a lower zone composed of garnet lherzolite and an upper zone of barren harzburgite or dunite in which there were pods of eclogite – the products of partial melting of earlier garnet lherzolite. More recent studies of kimberlite xenolith suites have indicated that the upper mantle is a very heterogenous areas, much more complex than originally suspected and in many ways with as much diversity in rock type and processes as the earth's lower crust.

We may look at the upper mantle from the point of view of (a) the distribution of rock-types within the upper mantle, and (b) the processes operating there.

9.1 Distribution of Rock-Types Within the Upper Mantle

Most models for the upper mantle, ranging from the earliest model of Wagner (1914) to the most recent ones, assume a broad concentric arrangement of mineralogical different zones. The nature of these zones can be arrived at by study of the proportion of xenoliths within individual kimberlite xenolith suites, combined with estimates of the pressure and temperature conditions of origin of the rocks determined from experimentally-calibrated phase equilibria.

Judging from the proportion of xenoliths within most kimberlite xenolith suites, the most common upper-mantle rock types are harzburgite and garnet lherzolite; other rock types such as garnet harzburgite, garnet pyroxenite, lherzolite, Al-spinel-bearing lherzolite and the various types of eclogite are relatively uncommon, though we might note that eclogites are sometimes the dominant xenolith type in certain kimberlite intrusions. Regarding the two main rock types, density calculations indicate that the relatively light harzburgites should form the upper part of the upper mantle, whereas the denser garnet lherzolite should form the lower part; this is broadly consistent with the Clark and Ringwood model. However, within these two major zones it is apparent that there are numerous small-scale heterogeneities; for example, there are admittedly-rare examples of aluminium-spinel-bearing lherzolite which must be placed within the upper harzburgite zone, and conversely there are garnet-free lherzolites whose PT equilibration conditions place them within the garnet lherzolite zone; in the case of the latter, the absence of garnet may be a reflection of the bulk-rock composition. In addition, Jordan (1979) has calculated that more iron-rich, "fertile" garnet lherzolites will be denser than "depleted" lherzolites and should theoretically comprise the lower part of the garnet lherzolite zone. In the case of hydrous and potassic-bearing phases, experimental work (e.g. Mysen and Boettcher 1975) indicates that amphibole will be the potassic- and hydroxil-bearing phase at depths less than 75 km, whereas phlogopite will contain these components at greater depths; hence amphibole will be stable theoretically within the harzburgite zone and phlogopite within the garnet lherzolite zone, which is in good agreement with the observed petrographic distribution of these phases within the various peridotite types. (However, it will be appreciated that amphibole-bearing harzburgites cannot be regarded as restites left over from an earlier melting event unless it is assumed that the presence of amphibole is due to a later episode of metasomatism super-imposed on an earlier phase of melting.) In addition, experiments on appropriate CO_2-bearing systems indicate that the carbonates within the upper mantle, if present, will be calcite in the upper parts, giving way to dolomite and magnesite at increasing greater depths (reviewed by Wyllie 1979a,b); furthermore the vapour coexisting with silicate liquids should theoretically change with depth, varying from high-CO_2 vapour in the upper parts of the upper mantle to dominantly H_2O in the lower parts.

Regarding the distribution of individual elements, Sobolev (1977) has suggested that there are chemical variations within peridotites depending upon the depth of formation; he suggests that at greater depths there is an increased amount of chromium involved in the individual minerals and cites instances of high-chrome garnets, chrome kyanites and high-chrome pyroxenes as evidence for his hypothesis. In addition the overall mass balance of the earth suggests that iron should be an increasingly important element with increasing depth, and preliminary high-pressure Mössbauer spectra experiments (Burns et al. 1972) suggest that the iron in the upper mantle is considerably less oxidized than that in the crust. Furthermore, direct experimentation at very high pressures has shown that elements such as sodium and potassium may be incorporated within the crystal lattices of mineral species that would not otherwise accommodate them (reviewed by Ringwood 1975); this is corroborated by the presence of up to 0.5 wt% of K_2O in clinopyroxene inclusions in diamond (Sobolev 1977) and appreciable amounts of sodium and phosphorus in eclogite and garnet lherzolite garnets (Bishop et al. 1978).

In Chap. 7 it was shown that eclogites have formed over a wide temperature interval (though bearing in mind the uncertainties in calculation) and, assuming that these temperatures vary consistently with depth, over a wide depth interval, to depths as great as 200 km. On this assumption, eclogites (particularly the diamond-bearing ones) may exist down to depths of approximately 200 km, i.e. from high in the harzburgite zone down to well into the the garnet lherzolite zone. In the Clark and Ringwood model eclogites were confined to the upper parts of the upper mantle. However, Gurney and Harte (1980) suggest that the high temperatures indicated by some eclogites and eclogitic inclusions in diamond are relict temperatures associated with previous magmatic events in which the eclogite material was initially formed, and suggest that the xenoliths in kimberlite provide little direct evidence of rock types within the mantle below depths of approximately 175 km. This is in conflict with earlier interpretations of the origin of the high-temperature peridotites, referred to earlier in Chap. 7 (peridotite-pyroxenite suite, conditions of formation), and brings us on to the discussion of the inter-relationship between "fertile" and "depleted" garnet lherzolites within the major garnet lherzolite zone, together with the concept of palaeogeothermal gradients and ancient plate motions.

By using the pressure-temperature estimates from a group of kimberlite nodules, it is theoretically possible to determine a PT range which may be presumed to relate to the upper-mantle section sampled by the kimberlite. In a pioneering study, Boyd (1973a) interpreted these estimates for a suite of xenoliths from northern Lesotho as outlining a "fossil geotherm" for the late Cretaceous upper mantle and as indicating an inflexion to higher dT/dP in the deeper part of the mantle section. This technique was later extended to xenolith suites from other localities in southern Africa, Yakutia, the western United States and northern Canada by various other workers. In the case of the Thaba Putsoa xenolith suite, Boyd (1973a) showed that the higher-level, cooler limb of this palaeogeothermal gradient coincided very closely with the normal subcontinental geothermal gradient.

However the higher-temperature (interpreted as deeper) segment of the geothermal gradient was delineated by xenoliths equilibrated at higher temperatures than the normal subcontinental gradient (Fig. 82). In addition, this high-temperature segment of the geotherm was delineated by deformed peridotites (as opposed to undeformed peridotites forming the low-temperature segment of the geotherm) and in other xenolith suites, all high-T lherzolites consistently show evidence of

Fig. 82. Diagram showing differences and similarities between the theoretical sub-continental geotherm (C & R – Clark and Ringwood 1964), Boyd's N. Lesotho "kinked" geotherm, and the recalculated continental pyroxene geotherm (M & C) from Mercier and Carter (1975)

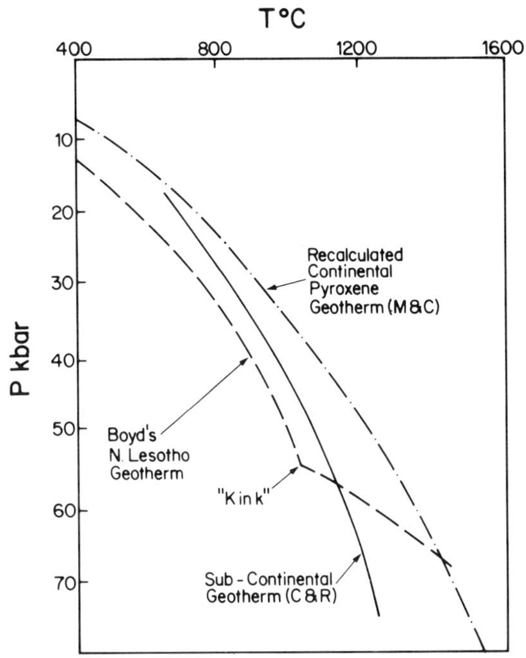

deformation; moreover, at Thaba Putsoa, the deformed peridotites contain higher amounts of elements such as titanium, iron and sodium and were interpreted as being relatively "fertile" garnet lherzolites. The inflection of the geothermal gradient caused by the high-T limb was interpreted by Nixon and Boyd (1973a) as due to shear heating at the base of the lithosphere as a result of plate motion during the break-up of Gondwanaland; deformation of the high-T xenoliths was interpreted as being due to movement of the base of the lithosphere, and the relatively fertile nature of these rocks as due to percolation of fluids from the low-velocity zone immediately below the base of the lithosphere.

This concept of palaeogeothermal gradients, implications for plate motions and migration of mantle fluids has given rise to one of the more interesting upper-mantle controversies of the 1970's. Whereas the theoretical application of current geothermometry and geobarometry to solving problems of palaeogeothermal gradients is undeniable, the possible extrapolation to major tectonic events has been questioned and has given rise to considerable debate. Various points have been raised to deny the original proposals. Some of these points are as follws:

1. A wide PT range of equilibration conditions, together with inferred inflections in the palaeogeothermal gradient, is not typical of all xenolith suites from kimberlite intrusions; for example, the xenolith suites at Matsoku (Cox et al. 1973) and at Bultfontein (Dawson et al. 1975) have a very restricted PT equilibration range compared with Thaba Putsoa (Fig. 59).

2. In the xenolith suite from Bultfontein some xenoliths are considerably more deformed than those at Thaba Putsoa; nonetheless, the PT equilibration conditions of the xenolith suite at Bultfontein are very restricted, lying on the low-temperature segment of the Thaba Putsoa geothermal gradient. Hence, there is no overall correlation between degree of deformation and PT conditions of equilibration.

3. Nixon and Boyd (1973a) drew attention to the apparent "fertility" (i.e. higher iron, titanium and sodium) of the deformed nodules at Thaba Putsoa compared with the coarse, undeformed peridotite nodules; this correlation has not been found within other kimberlite xenolith suites, leaving the question as to whether the correlation observed at Thaba Putsoa is a purely local, rather than widespread, phenomenon. In this context, in addition to noting that many deformed peridotites have been subjected to a certain amount of infiltration metasomatism, there also remains the possibility that peridotites have been subjected to changes in their bulk and individual phase chemistry during deformation and recrystallization. The concept of diffusive mass transfer, involving intercrystalline diffusive mass transport in response to stress gradients, is not new; it was first proposed over a century ago by Sorby (1863) and has been invoked to explain chemical changes arising during deformation in crustal rocks (e.g. Kerrich et al. 1977; McEwen 1978). In short, the question must be asked: are the high-iron and -titanium peridotites "fertile" or "fertilized"? with this feature being caused by metasomatism or diffusive mass transfer within narrow and localized upper-mantle deformation zones.

4. There is some evidence of isotopic disequilibrium in deformed garnet lherzolites (Allegre et al. 1978).

5. Boyd (1973a) proposed that the correlation between chemical composition and texture in the garnet lherzolite xenoliths is a feature of the base of the lithosphere, reflecting a sampling across the transition to a deeper, less-refractory low-velocity zone. However, a distinct relationship between xenolith texture and phase composition, similar to that reported in garnet lherzolites, has been reported in high-level spinel lherzolite xenoliths from California and Ninuvak Island, Alaska (Basu 1977; Francis 1978); the implication is that such correlations are not necessarily restricted to the base of the lithosphere but can be also related to processes operating at relatively high levels in the upper mantle.

6. Additional evidence for high-level deformation of peridotites is shown by disruption of pyroxene-spinel coronas around garnet in lherzolites from the Letseng satellite pipe, Lesotho (Lock and Dawson 1980); the interference is that the deformation took place after the lherzolite had moved out of the garnet stability field.

7. Lherzolite deformation is a very variable and sometimes very localized feature; in the xenoliths of the Bultfontein Mine, Dawson et al. (1975) record gradation from undeformed into highly deformed lherzolite over a distance of only 3–4 cm; this is hardly consistent with a major tectonic boundary such as the base of the lithosphere.

8. The presence of the deformation fabrics in deformed garnet lherzolite nodules is due to their preservation during the rapid ascent in kimberlite; Goetze (1975) has shown that such a fabric would be inconsistent with the thermal regime pertaining at the base of the continental lithosphere during plate motion.

9. On a theoretical level, Mercier and Carter (1975) claim that the pertubation of kinking in the geothermal gradient at Thaba Putsoa and other intrusions in an artefact due to the application of data from relatively simple experimental systems into complex natural systems and that the inflection disappears when correct thermodynamic corrections are made (Fig. 82).

Although there seems to be considerable evidence against the concept of shear heating and formation of the deformed garnet lherzolites at the base of the lithosphere, and also some scepticism as to the existence of "kinked" geotherms, the evidence for high-temperature peridotites still stands. Is there an alternative hypothesis, other than shear heating at the base of the lithosphere, for their generation? An implicit assumption in the Boyd and Nixon (1973b) model is that the higher temperature peridotites are derived from greater depths. An alternative explanation has been offered by Green and Geuguen (1974) who have reinterpreted the geothermal inflexion as due to the ascent of a mantle diapir. In effect, the high-temperature peridotites are envisaged in this alternative model as due to heating in the inner part of an upper-mantle thermal aureole around an ascending diapir. Movement along the diapir margin could account for the deformation features seen within these peridotites and the changes in chemistry may be attributed to the movement of fluids along the margins of the ascending diapir.

An additional alternative model has been proposed by Gurney and Harte (1980) which not only attempts to explain the high-temperature, metasomatised peridotites but also offers an explanation for the generation of the megacryst suite. They suggest that relatively localized melting occurs at depth within the upper mantle at temperatures a little above those indicated by the highest-temperature megacrysts (i.e. 1400°–1500°C); the magma then intrudes to a higher level, thus generating a horizontal temperature gradient at that level. The wide temperature and chemical range of the megacrysts results from cooling differentiation of the magma at this higher level, and there are possible variations in the cooling rate as a result of localized magma movement and injection; pegmatite bodies may form and later yield some of the megacrysts to kimberlite on disaggregation. The late-stage liquid generated by differentiation of the megacryst magma is kimberlitic and in some cases the kimberlite carrying the megacrysts and other nodules to the surface may originate from the same magma which initially precipitates the megacryst suite. In this model it is also suggested that, as in the Green and Geuguen (1974) model, the high-temperature peridotites form the immediate envelope to the intruding, megacryst-precipitating magma, and that they are formed from normal garnet lherzolites, their deformation and high-temperature features being due to the magma intrusion. Their distinctive chemical characteristics possibly result from metasomatic exchange between the magma and the thermally metamorphosed wall rock; the proposed wall-rock origin for these rocks may possibly explain the evidence for isotopic disequilibrium noted within them (Allegre et al. 1978) and the tendency of the PT estimates for high-temperature peridotites from specific intrusions to cluster tightly and separately from other peridotites. In the Gurney and Harte (1980) model the other peridotites and pyroxenites are believed to derive from the upper mantle outside the magma aureoles. In summary, in both the Green and Gueguen (1974) and the Gurney and Harte (1980) models the fabric of the high-temperature nodules, the high temperatures of the nodules themselves and the apparent inflexion in the calculated geothermal gradient are all attributed to vertical migration of hot, upper-mantle materials to a relatively high level in the upper mantle, rather than to horizontal movements at the base of the lithosphere. The differences in the three models are shown diagrammatically in Fig. 83.

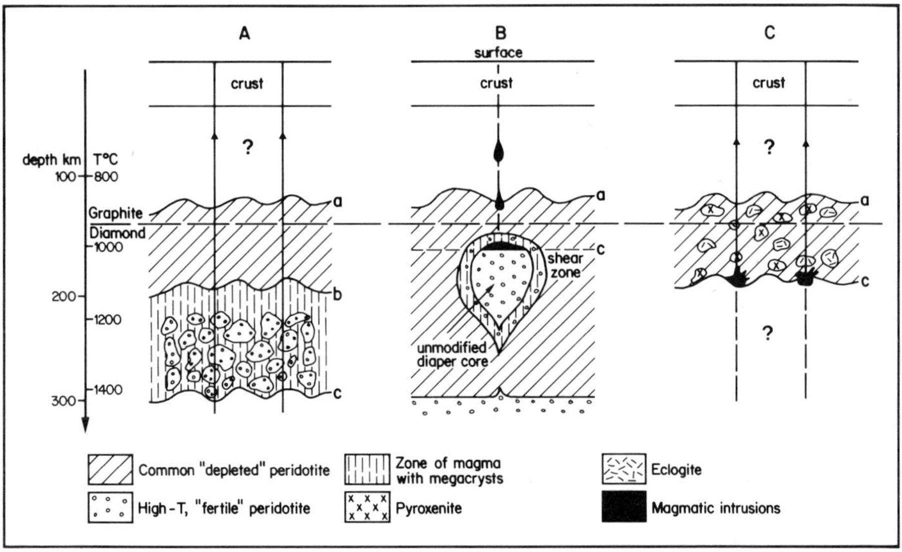

Fig. 83. Schematic vertical cross sections, illustrating three alternative models of the upper mantle sampled by kimberlite. *A* after Nixon et al. (1973) and Boyd and Nixon (1975); *B* after Mercier (1979); *C* after Gurney and Harte (1980). ***Vertical arrowed lines*** indicate ascent paths of kimberlite. Temperature scale is ambient temperature with increasing depth, ignoring perturbations by intrusions in models *B* and *C*. Boundaries are: *a* upper limit of extensive garnet peridotite sampling; *b* lower limit of sampled coarse peridotite and phlogopite, and top of low-velocity layer and geotherm "kink" in model *A*; *c* approximate lower limit of extensive mantle sampling; the boundaries are undulose to indicate uncertainties in temperature estimates and corresponding depths. Shear and metasomatic zones around magma intrusions in model *C* are not shown. The zone between boundary *a* and the base of the crust is probably dunite/harzburgite

In addition to the vertical variations noted in the models above, there is some slight evidence of lateral variation within the upper mantle in southern Africa across the margins of the Kaapvaal craton. The Al_2O_3 content of orthopyroxene in coarse garnet peridotite xenoliths is generally lowest for xenoliths in kimberlites in the centre of the craton; it increases towards the craton margins and is highest for xenoliths from kimberlites in the circum-cratonic fold belt (Gurney and Harte 1980). In addition diamondiferous kimberlites are confined to the craton whilst those in the circum cratonic fold belt are non-diamondiferous (Dawson 1970). These features are considered to relate mainly to differences in the thickness of the lithosphere and to the differences in heat flow between the cratons and the circum-cratonic fold belts.

Finally it may be stated that the inclusion suite and the included phases in diamonds from the Precambrian Premier kimberlite are essentially similar to those in the much younger Cretaceous kimberlites in South Africa (Danchin 1979; Tsai et al. 1979b). The inference is that there is no evidence of major variation in the lithology in the South African upper mantle over a period of approximately one thousand million years. However, Deines (in press) proposes that there is increasing evidence that the carbon isotopic composition of the mantle is not homogeneous, and shows a considerable range as indicated by the differences in mean $\delta^{13}C$ content of diamonds from different mines; diamonds from the Premier Mine contain less $\delta^{13}C$ (1.1‰) than diamonds from the Cretaceous Dan Carl Mine (Bellsbank area). This suggests that, whereas there are no gross lithological differences over the past 1000 m.y. in the mantle beneath South Africa, differences are apparent at the isotopic level.

9.2 Processes in the Upper Mantle

a) Evidence for Movement in the Upper Mantle. It has been inferred above from the fabrics within the nodules that there are upper mantle movements. The fabrics within the deformed xenoliths are the result of deformation within an anisotropic stress field and have been interpreted as the result of super-plastic flow within the upper mantle. These fabrics, which are now seen within the deformed peridotite xenoliths, would not have survived long under the temperatures pertaining within the upper mantle and hence their preservation is viewed as being due to deformation associated with the kimberlite event immediately prior to entrainment of the nodules within kimberlite and rapid transport to the surface (Bouiller and Nicolas 1975). Evidence for earlier deformation events within the upper mantle can be seen (a) in xenoliths which have a strong preferred orientation of the olivines and orthopyroxenes (the so-called tabular granuloblastic texture), and (b) in the rare coarse lherzolites in which large deformed orthopyroxene megacrysts display kink-bands and undulose extinction, the associated olivines having regrown from a presumed original porphyroclastic fabric. In the case of the garnet lherzolites of exsolution origin (in which there are still relict porphyroclasts of orthopyroxene exsolving garnet and clinopyroxene along kink-band junctions), the presence of the orthopyroxene megacrysts in their deformed state is testimony to an earlier period of strong deformation when the strain rate exceeded the ability of the orthopyroxene to relieve the stress by recrystallization (Dawson et al. 1980). Hence, within these examples we do have limited evidence of earlier episodes of deformation within the upper mantle which are unconnected with the kimberlite event.

There is also a limited amount of evidence for brittle fracture within the upper mantle. Some banded eclogites are of inequidimensional shape and very often the long axis of the nodule, delineated presumably by some type of fracture, is often at a high angle to the observed banding within the nodule; this may be attributed to a tightly spaced joint-system that was imposed upon the eclogite just prior to its incorporation within the kimberlite. Brittle fracture has been mentioned in the case of the orthopyroxenes above, and further examples of brittle fracture may be seen in the veined and metasomatised peridotites in which veinlets of mica, amphibole and diopside are injected along planar fractures within nodules (Dawson 1979); and planar fracture that acted as cooling surfaces for high-T liquids were inferred for the non-equilibrated nodule from Weltvreden (Rawlinson and Dawson 1979; see Chap. 7.5). In addition, if silicate-ilmenite intergrowths are interpreted as derived from upper-mantle pegmatites then we must infer a small amount of upper-mantle *dilation,* accompanying the brittle fracture, probably both being caused by the hydraulic pressures of the injecting liquids.

The existence of other movements in the mantle may be inferred from the textures within certain xenoliths and also phase chemistry. For example, the two-pyroxene-spinel coronas formed around garnets in garnet lherzolites result from the movement of the garnet lherzolites out of the garnet stability field and into that of aluminous spinel; theoretically this could have taken place by either a decrease in pressure or a rise in temperature (or both) but since there is no evidence in the coexisting phases for temperature increases, this reaction is taken as indicating a decrease in pressure attendant upon the slow uprise of the garnet lherzolite out of its stability field. There is evidence of prograde metamorphic reaction in the upper mantle from the generation of garnet lherzolites by pressure-induced exsolution of garnet and orthopyroxene from an earlier fertile harzburgite (Dawson et al. 1980), from formation of garnet harzburgites by pro-grade metamorphism of postulated chromite harzburgite cumulates (Dawson et al. 1978b), and from the PT evolution paths of some kyanite eclogites (Fig. 78). Whether these pro-grade reactions are connected with large-scale convective overturn or with ancient subduction zones is a matter for speculation.

b) Upper Mantle Metasomatism. There is increasing evidence for metasomatism within the upper mantle (see section on metasomatised peridotites, Chap. 7), and it is apparent that this is not solely confined to the parts of the mantle sampled by kimberlites since it has been recognized, for example, in xenolith suites in ankaramite from East Africa (Ridley and Dawson 1975), in xenoliths from the Eifel volcanic area, Germany (Lloyd and Bailey 1975) and in xenoliths from alkalic basalts in the western U.S.A. (Best 1974). As mentioned earlier, in the specific cases mentioned above, there has been considerable addition of elements such as K, Rb, Ba etc. to earlier peridotite and in the case of Lashaine, it was proposed that the metasomatic effects had been imposed on those of an earlier depletion event (Ridley and Dawson 1975). A more extreme view has been adopted by Boettcher et al. (1979) who propose that xenoliths in general have been subjected to some degree of metasomatism and hence are probably a poor representation of much of the mantle, and earth-models based on such evidence may portray an upper mantle richer in incompatible elements than is actually the case; this line of argument is directly contrary to the generally accepted concept that most xenoliths are depleted in these elements relative to pristine mantle lherzolite.

Another factor to be borne in mind is the question of scale. Metasomatism is directly observable in xenoliths on either side of small fractures infilled with micas, amphibole, rutile, ilmenite etc. Entire xenoliths which contain appreciable amounts of these phases have been interpreted as representing more-thoroughly metasomatised peridotites. However, whether these xenoliths represent fragments from a widespread upper-mantle region that has undergone a process akin to crustal granitization, as envisaged by Lloyd and Bailey (1975), or whether they are a product of a more localized metasomatic event within the aureole of an upper-mantle magma intrusion, is open to question. It will be important to determine the *timing* of the metasomatic events, relative to any earlier igneous activity.

9.3 The Deep Crust

The present consensus of opinion is that the deeper parts of the crust, as represented in the xenolith suites in kimberlitic and alkalic basalt diatremes, are composed mainly of granulite-facies rocks (summarized by Goldsmith 1976; Dawson 1977). It must be assumed that these rocks underlie the dominantly granite/greenstone-belt terrains which are often the most deeply exposed rocks in the cratonic areas. In many respects the granulite xenoliths resemble granulite-facies rocks observed in the circum-cratonic fold belts, although Griffin et al. (1979) have observed a higher jadeite component in the pyroxenes than is normal in granulite-facies rocks and it is possible that they may represent rocks transitional to crustal eclogites. Indeed, a complete transition, both mineralogically and compositionally, has been found between granulites and eclogites in the diatreme of northern Lesotho (Griffin et al. 1979) and hence it could be interpreted that the lower part of the crust under northern Lesotho consists of essentially basic rocks that have been metamorphosed under PT conditions varying from eclogite to granulite facies conditions.

An interesting facet of the granulites is that, despite their formation under relatively high PT conditions, they contain a certain amount of volatiles both in the form of fluid inclusions of CO_2 (Touret 1974) or as CO_3^{2-} and SO_4^{2-}radicle in scapolite (Goldsmith 1976).

The origins of deep crustal granulites have been a matter for debate. One theory is that they represent straightforward high-pressure equivalents of basaltic or gabbroic rocks. The other, favoured by the author, is that the basic granulites represent the residuum of rocks of more intermediate composition that have been subjected to an anatectic melting event during which the granitic constituents migrated into the upper crust (see summary by Dawson 1977).

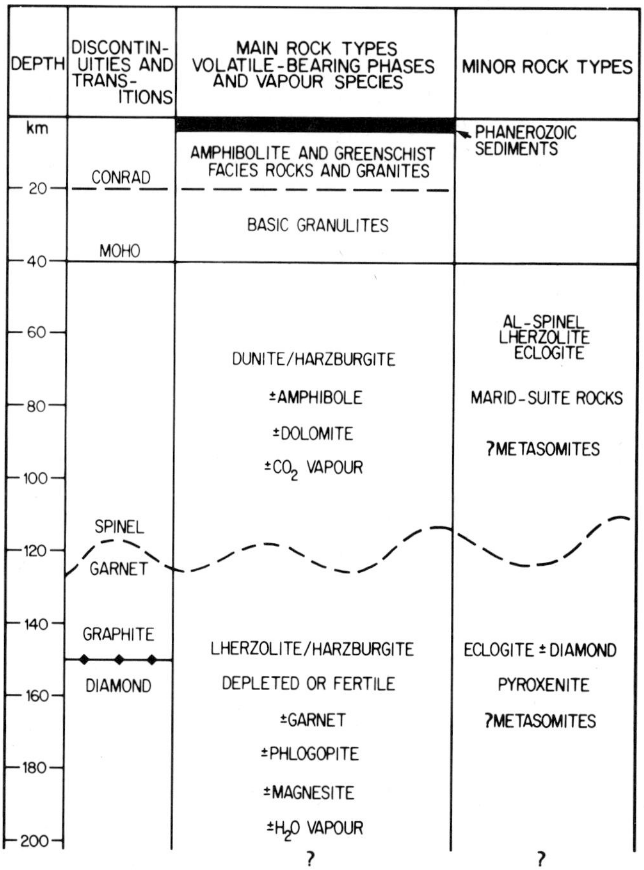

Fig. 84. Distribution of major and minor rock-types, and volatile-bearing and vapour phases, in the sub-cratonic lithosphere. The spinel-garnet transition is wavy to indicate experimental uncertainty. The distribution of theoretical carbonate species is based on experimental work (Wyllie 1979b). The depth of the Conrad discontinuity is based on seismic work on the S. African craton (Hales and Sacks 1959)

9.4 Summary

The earth's upper mantle is petrologically complex. Within the two main harzburgite and garnet lherzolite zones, themselves exhibiting large-scale and small-scale chemical and textural heterogeneities, are smaller bodies of less ultrabasic rock-types, such as pyroxenites, eclogites and MARID-suite rocks, that represent frozen liquids; these also show internal small-scale heterogeneities. Certain areas within the upper mantle have been subjected to magma intrusion, some of which was accompanied by shearing and metasomatism. The magma injection may have been accompanied by both brittle and plastic deformation, and the brittle deformation in some cases opened up channel-ways for injection of fluids that consolidated as upper-mantle veinlets and, possibly, pegmatites. In addition to zones of differential movement associated with upper mantle magma injection, there is a limited amount of evidence for movements transporting materials (? by convective overturn or ? subduction) from relatively low- to relatively high-pressure regimes. In short, there is a growing body of evidence that the vertical and lateral heterogeneities and type of processes which are known to operate within the earth's lower crust also are present within the upper mantle.

A tentative model for the distribution of major and minor rock-types, and possible additional solid and vapour phases, in the sub-cratonic lithosphere is given in Fig. 84.

Chapter 10 Kimberlite Genesis

10.1 The Three Hypotheses

In the earlier chapters on the petrology and mineralogy of kimberlite, it has been shown that kimberlite is a hybrid rock consisting of varying proportions of crystals derived from the fragmentation of upper mantle rock types together with crystals derived from the crystallization of an ultramafic magma, these being finally cemented by the late-crystallization products of the magma. Since many of the kimberlite phenocrysts are broadly similar in composition (with the exception of features such as high titanium and low chromium) to the similar species derived by the fragmentation of peridotite, it will be seen that the essential problem is the origin of the phenocrysts and the kimberlite matrix itself. In particular, it is the source and reasons for the concentration within the kimberlite matrix of the high amounts of incompatible elements (see Chap. 5, kimberlite geochemistry) that have to be explained.

Over the past few years three hypotheses have been proposed for the formation of kimberlite, these are:

1. The zone-refining hypothesis which was originally proposed by Harris (1957) and elaborated upon by Harris and Middlemost (1969). This hypothesis envisages that a liquid (the composition of which is not predicted) is generated at a depth of some 600 km. The liquid, being dynamically unstable in the earth's gravitational field, moves upwards by a process of solution stoping or a zone-refining process similar to that used in metallurgical refining. During the upward movement of such a magma body, the major components remain more or less in equilibrium between the liquid and the initial melt for a given pressure. However the incompatible elements (i.e. those elements that are unable to substitute readily in the major minerals of ultrabasic rocks) will undergo continuous enrichment in the liquids. As the melt ascends, it will eventually enter a lower temperature region where it partially crystallizes and decreases in volume, with the attendant fractionation and increasing concentration of the residual elements such as K, Rb, Cs, Ba, La, and Pb. The additional effect of concentration of volatiles (H_2O, CO_2, and F) is to depress the solidification temperature. This residuum is regarded by Harris and Middlemost (1969) as kimberlite, and since it contains large concentrations of volatiles is believed to be capable of being explosively injected from great depth.

The explosive injection of kimberlite from great depth is contrary to the intrusive features of the sub-diatreme dykes (see section on kimberlite emplacement, p. 41) but other objections can be made to the process outlined above. These are that: (a) this hypothesis takes no account of the physical two-component make-up of kimberlite, although acknowledging its hybrid nature from the chemical aspect; and (b) it is evident that the products of zone-refining of the mantle might be expected to be abundant and widespread, which is contrary to the insignificant amounts of kimberlite and other rare potassic lava types, whose genesis is likewise attributed by Harris and Middlemost to zone-refining

2. Another process, which may be referred to as the "residual process" was at one time very popular (e.g. O'Hara and Yoder 1967). Starting with a garnet peridotite parent (garnet + chrome diopside + enstatite + olivine), partial melting involving the garnet + diopside fraction at depths of 80–100 km is believed to produce a liquid with the composition of picrite basalt; the refractory olivine + enstatite proportion would remain as dunite or harzburgite. Fractional crystallization at high pressures of the picritic basalt is proposed to give rise to bimeneralic eclogite cumulates and a complementary series of silica-poor alkalic residual liquids that have the geochemical characteristics of kimberlite and other rare, ultrabasic potassic lavas. The association of xenoliths or garnet peridotite (a potential source rock that is widespread in the upper mantle) and xenoliths of eclogite (possible high-pressure cumulates from the partial melt liquid) within the kimberlite groundmass (the residual fluid of the proposed high-pressure fractionation) makes this an extremely attractive hypothesis. Two major objections are (a) that the assemblage garnet + clinopyroxene + enstatite + olivine is "barren" in that it contains insignificant amounts of oxides such as K_2O, TiO_2, CO_2 and P_2O_5 that are so obviously concentrated in the kimberlite matrix; hence it cannot be regarded as an ideal starting material; and (b) another problem is that, whereas the garnet peridotite in the above scheme may be regarded as an older material (and therefore xenolithic when incorporated into kimberlite), the eclogite is regarded as forming contemporaneously with the residual fluid, i.e. it is cognate; this is contrary to the geochronological evidence that eclogites are much older than their host kimberlites, and are not in isotopic equilibrium. However it must be acknowledged that fractionation of garnet and pyroxene (together with olivine, ilmenite and ?phlogopite) is believed to be instrumental in the formation of kimberlite but these garnet-pyroxene combinations of the megacryst suite [the co-called griquaites (see p. 92), are not eclogites in the sense as originally proposed by O'Hara and Yoder (1967)].

3. The very limited volume of kimberlite points to its being either the residual end-product of a long fractionation process or else the product of a very limited amount of partial melting. The arguments against the residual hypothesis have been outlined above, and we will discuss below the generation of kimberlite by incipient melting. Dawson (1971) argued that phlogopite-bearing garnet lherzolite was a potential parent with primary titaniferous phlogopite providing the source for many of the incompatible elements (K, Ti, Rb, Ba) that are concentrated

within kimberlite. It has subsequently been appreciated that carbonate, probably in the form of magnesite or calcite, must also be present within any source region for kimberlite and a phlogopite-carbonate-garnet lherzolite is the favoured parental material for the generation of kimberlite in the hypotheses reviewed by Wyllie (1979a) and Eggler and Wendlandt (1979).

Assuming that kimberlite does arise by partial melting of upper mantle materials, there are a considerable number of factors which will affect the composition of the melt, both initially and during its ascent and eventual consolidation. These factors are (a) the composition of the parental upper mantle materials; (b) the composition of the vapour phase present when melting takes place; (c) the depth at which melting takes place; (d) the degree of melting of the parental material, and whether this melting is partial or batch melting; (e) the mode of ascent, either stepwise or direct; (f) precipitation and fractionation of various phases during ascent; (g) loss of volatile and mobile elements during the emplacement process and, possibly, during interaction with wall rocks; (h) xenolith incorporation both in the mantle and at relatively high levels; (i) interaction with groundwater during intrusion; and (j) post-consolidation high-level percolation of groundwater. The interplay between these factors will be variable and often complex.

In the context of the composition of the melted upper mantle, it is apparent that the source area must contain materials containing the appropriate amounts of K, Ti, Ba, H_2O and other volatile elements (Dawson 1971) and phlogopite-carbonate garnet lherzolite is an ideal hypothetical starting material. A small thermal input at depths of approximately 300 km could yield a very small amount of partial melt liquid in which the materials yielded by melting of phlogopite would be a dominant constituent. The highly distinctive rare-earth element pattern of kimberlites is consistent with mica having been involved in their generation though this is also consistent with the involvement of diopside, a mineral whose involvement is strongly suggested also by the available strontium isotope and rare-earth data (Mitchell and Crockett 1971a; Mitchell and Brunfelt 1975). If, in addition, water was the main vapour phase, small amounts of hydrolitic weakening might be caused on a micro-scale in the wallrocks around the partial melt area. If by exploiting propagating microfractures, this melt could then ascent to depths of approximately 200 km, then additional melting would involve the melting of carbonate and evolution of CO_2 (Wyllie 1979b). At this point the liquids, of kimberlitic character, would become more charged with CO_2 vapour, and light REE elements would partition into the vapour in preference to the liquids and solidus phases (Wendlandt and Harrison 1979); the evolved CO_2 would change the X_vCO_2/H_2O, having a fundamental effect upon both the type of liquids that might be subsequently generated and certainly upon the species of the solidus phases precipitating out of any magma already generated. A further effect of evolution of CO_2 would be increased fracture propagation. As pointed out by Anderson (1979), a definite relationship exists between a crack and the availability of fluid because a crack tip cannot accelerate faster than the fluid within it can flow in the channle provided by the crack; the speed of the fluid is limited by its viscosity and volatile phases at the tip of the crack would allow the crack growth to accelerate. The compositional nature of the initial liquid is perhaps difficult to ascertain although its highly potassic and undersaturated nature appears to be beyond dispute.

Modification of the ascending kimberlite magma by crystal fractionation has been appealed to by many authors (e.g. Dawson 1971; Elthon and Ridley 1979; Gurney et al. 1979). The phases precipitated are olivine, Ti-enstatite, low-chrome diopside, ilmenite, titanian pyrope and, possibly, titaniferous phlogopite. The proportion of the olivine to enstatite precipitated will depend on the partial pressures of CO_2 and H_2O vapour and could presumably change depending upon the balance of the two vapours present. Moreover, the question must be asked: can these megacrysts be precipitated from a rapidly-ascending magma, or alternatively, does the magma have to be in an arrested stage of ascent? The composite nodules composed of species of the megacryst suite strongly suggest fractionation within a relatively static magma chamber during which precipitated phases could coalesce and intergrow. However, this is at variance with suggestions of rapid ascent, particularly from the aspect of high speed of ascent being necessary to suspend high-density megacrysts and xenoliths within the low-density magma. Nonetheless, is rapid ascent really necessary? It is only necessary if it is assumed that the ascending magma is a Newtonian fluid; however if it behaves like a Bingham fluid then it will have a very much higher shear velocity and could carry very high density nodules in suspension. In addition, a Bingham fluid could transmit stress, and transportation of nodules in such a fluid could account for the instances of deformation observed in some megacryst-suite minerals (see Chap. 8).

In most kimberlites, it must be assumed that the near-surface environment is one of complex chemical reaction. The reaction rims on ilmenites and spinel indicate rapidly changing fO_2 and also changes in the activity of Mg, Ti, Fe and Al (Haggerty 1975a; Boctor and Meyer 1979). In addition, the build up of CO_2-rich residual liquid is testified to by calcite-zircon-ilmenite interaction to give zirconolite (Raber and Haggerty 1979) and by the crystallization in the rapidly quenched kimberlite liquids of calcite and diopside which are present rather than melilite which would be unstable in the presence of large amounts of CO_2 (Yoder 1975). Finally, in the later stages of consolidation there will be numerous low-temperature effects such as serpentisation of olivine, serpentisation of previously formed calcite (Clement 1975), and the formation of heazlewoodite and bornite by combination of sulphur vapour with nickel and copper released during serpentisation (Boctor and Meyer 1979). In addition, the inter- and intra-grain chemical variations noted in groundmass phases such as micas (Smith et al. 1978) and oxides indicate small-scale disequilibrium, the attainment of equilibrium probably having been precluded by rapid quenching of the residual liquids during diatreme formation. Oxygen fugacities in the order of 10^{-20} b at the time of groundmass formation indicate that the kimberlite magmas are high reducing (Mitchell 1973a).

10.2 Relationship with Other Rock Types

In Chap. 1, it was demonstrated that kimberlite is associated chronologically with other minor rock types such as olivine melilitite, olivine nephelinite and carbonatite, though these other rock types are generally geographically separated from the kimberlite. Kimberlite is not unique in having been formed within the garnet lherzolite zone of the upper mantle, since garnet lherzolite xenoliths have been found in alnoite (Marchand 1970; Nixon and Boyd 1979), in olivine melilitite (Moore 1973), in ankaramite and carbonatite (Dawson et al. 1970) and basanite (Albert et al. 1967; Volyanyuk et al. 1976). However, kimberlites are unlike these other rock-types in that they are vitually confined to cratons; cratons are areas which have undergone many phases of earlier volcanic activity, i.e. they are areas of the earth's lithosphere where, since the early Precambrian, considerable amounts of upper mantle material have been transferred from the upper mantle to the crust. As such it may be inferred that a very thick layer of depleted mantle exists beneath the cratons, this zone of depletion extending to a greater depth than under circum-cratonic fold-belts, continental margins or oceanic areas. It is suggested that this zone of barren mantle is so thick that during the thermal event during which kimberlite is generated, melting can only take place at considerable depth, that depth being compatible with the generation of kimberlite rather than any of the closely related magma types. This is consistent with the evidence that certain members of the kimberlite xenolith suites (e.g. diamond-bearing xenoliths) are derived from deeper levels than most other xenolith suites, and is also consistent with current experimental evidence that nephelinitic, melilitic and basanitic melts can be generated experimentally from parental garnet lherzolite at pressures lower than those needed to form diamond (summarized by Ringwood 1975). In addition, it appears that, in addition to the depth factor, differences in the composition of mantle source areas and differing degrees of partial melting of source areas could account for the observed differences in composition between kimberlite and these other volumetric insignificant rock types; for example, it could be argued that the relatively low heat-flow under the cratons would produce very limited amounts of partial melting, thereby generating kimberlite liquids rich in K_2O, TiO_2, BaO and H_2O from melted phlogopite, whereas in the higher heat-flow regime of the circum-cratonic fold belts a higher degree of partial melting will be achieved and the phlogopite component of the melt will be diluted by the addition of Na, Ca and Al (together with Si, Mg, Fe and Cr) from melted diopside and garnet, producing the relatively high Na/K, Ca/Mg and Al_2O_3 of nephelinites and melilitites compared with kimberlite. In addition, it is evident that the modal proportions of the phases in the mantle source rock will be important. A phlogopite-free source could never produce kimberlite but could be parental to nephelinites or melilitites. In summary, the presence of phlogopite in parental garnet lherzolite, at depths consistent with diamond stability, together with low degrees of partial melting, are required for the formation of kimberlite; their confinement to old cratons may be due to earlier depletion and low heat-flow in the sub-cratonic upper mantle; and the presence on the cratons of distinct areas of kimberlite intrusions interspersed with areas of micaceous kimberlite intrusions (Chap. 4) points to varying proportions of phlogopite in the kimberlite source zone. This, in itself, may be some reflection of the sporadic distribution of earlier upper-mantle metasomatism.

10.3 The Kimberlite-Carbonatite Relationship

The hypothesis that kimberlite and carbonatites are intimately related is a long-standing one. It arose originally by the fact that carbonate-rich dykes of igneous origin cut the kimberlite of the Premier Mine, South Africa (Daly 1925); samples from these dykes were studied with other carbonatites in a geochemical study of carbonatites by Higazy (1954). Since then strontium isotope studies on kimberlite calcite have shown them to be isotopically similar to carbonatites (see Chap. 5), and calcite-rich kimberlites from the Bentfontein Sills, Kimberley, have Th/U ratios and light REE contents similar to other carbonatites (Dawson and Hawthorne 1973). In addition, Raber and Haggerty (1979) suggest that the carbonatitic fluids are involved in the interaction with kimberlite ilmenite and zircon to form secondary perovskite and zirconolite. Nonetheless, the following points may be argued against the carbonatite-kimberlite relationship. On a regional scale, true kimberlites are not found within carbonatite complexes and, with the exception of the Arkansas province and Pilaansberg province (the latter including the Premier kimberlite), kimberlites are found in tectonically different environments from carbonatite complexes, the latter being generally confined to rift valleys situated within the circum-cratonic fold belts (Chap. 1). On a more detailed scale, Mitchell (1979) points out that even the most carbonate-rich residues of kimberlite lack the mineralogical characteristics of carbonatites in not containing sodium-iron silicates and niobium-rich minerals; they contain primary serpentine and the spinel, where present, is generally magnetite by contrast with the carbonatites which contain titaniferous magnesio-ferrite-magnetites. On balance, there is little evidence for a *genetic* connection between kimberlite and carbonatite. However, it is fair to say that both kimberlites and the carbonatites found within alkalic complexes are both characterized by high-temperature calcites that are indistinguishable on their isotopic composition, trace-element content, distinctive morphology and formation of segregation bodies (probably by liquid immiscibility); the CO_2-charged kimberlite and nephelenitic parental magmas both originate in the upper mantle and the isotopic and trace element characteristics of their evolved carbonate segregations are indicative of their common mantle origin. Distinctive chemical features for some kimberlite and carbonatite groundmass minerals such as mica (Smith et al. 1978) and high-manganese ilmenite (McMahon and Haggerty 1979) probably reflect parallel but similar crystallization and reaction processes in the high-temperature carbonate segregations, despite their different parentage.

10.4 Kimberlite and Diamond

Although there is a scanty amount of evidence for daimond being derived from rocks other than kimberlite (summarized by Dawson 1979), kimberlite is by far the most abundant source of terrestrial diamond. The principal current debate is as to whether diamond is xenocrystal, having been derived by fragmentation of diamondiferous eclogite and peridotite, or whether it is a phenocrystal phase having precipitated from the kimberlite magma. Both viewpoints have their proponents. Features favouring xenocrystal origin are the common occurrence in some kimberlites of diamonds that have been plastically deformed and have had a long and complex history (see diamond, Chap. 6) and the fact that diamonds in Cretaceous kimberlites have yielded Precambrian ages (Kramers 1979). Moreover it has been argued (Robinson 1978) that the resemblance in chemical composition of mineral inclusions in diamond to those in eclogitic and peridotitic xenoliths points to their derivation by fragmentation of diamondiferous eclogites and garnet peridotites (Sobolev 1977). Proponents of the xenocrystal theory believe that kimberlite is merely a medium whereby diamond is transported from the mantle to the surface, and contend that the reason for diamonds being restricted to kimberlites amongst igneous rocks is that only kimberlite originates at sufficient depths to sample diamond-bearing mantle.

Proponents of the phenocrystal theory point out that certain features in diamonds coulc only have been derived by crystallization from a liquid. These features are (a) delicate growth features which could only have been achieved by crystallization of the diamonds from a liquid; (b) the presence in diamonds of submicroscopic liquid inclusions of carbonated picritic composition (Fesq et al. 1975a); and (c) that, although acknowledging the existence of diamondiferous eclogite and peridotite, an unusually large proportion of these diamondiferous xenoliths would have had to be preferentially fragmented compared with non-diamondiferous xenoliths in order to contribute the number of diamonds now found in kimberlite; and (d) the fact that certain garnets which are unusually low in calcium and unusually high in chromium, features which are normally usually found only in diamond inclusions, are found in the concentrates of the kimberlite of the Finsch Mine (Gurney and Switzer 1973). However it must be acknowledged that the delicate growth features in diamonds have been found also in diamonds from eclogite xenoliths (Robinson 1979) and are due to the growth of diamond from a much older liquid unconnected with the kimberlite event.

In this debate the significance of the inclusions in diamond is of considerable importance. (For a review of the inclusions parageneses, see diamond, Chap. 6.) The proponents of the xenocrystal hypothesis point to the similarity in composition between the eclogitic-suite inclusions and eclogite xenoliths and also between some peridotitic inclusions and the common garnet xenoliths; if the eclogites and the peridotites are regarded as xenoliths then it is argued, the diamonds containing phases of composition similar to those in the xenoliths must also be xenocrystal.

The high-chrome peridotite-inclusion suite is regarded as being a reflection of a particular high-chrome peridotite area within the upper mantle (Sobolev 1977). By contrast, it could be argued that the peridotitic suite of inclusions, generally low in diopside, could be interpreted as the solidus phases, in equilibrium with kimberlite magma, which could have been incorporated into diamond crystallizing from the kimberlite magma; in this context, Harte et al. (1980) have proposed that the low-calcium, high-chrome inclusion suite is formed during partial melting when strong preferential partitioning of calcium into the high-carbonate kimberlite liquid results in solidus phases with the low-calcium high-chromium characteristics of the olivine, garnet, orthopyroxene and chromite commonly found in diamond inclusions. However, any hypothesis proposing that diamond crystalizes from the kimberlite magma, must explain the absence of inclusions or intergrowths of high-Ti megacryst suite phases in diamonds; or, conversely, it must deny that the megacryst suite phases are connected with the kimberlite event.

Another hypothesis, that of Mitchell and Crockett (1971b), is an elaboration of the phenocrystal hypothesis; they propose that diamond seed nuclei form in the kimberlite magma within the diamond stability field but then this is followed by a stage of overgrowth of metastable diamond on the seed nuclei during the ascent. A variant of the nucleus/overgrowth theory is that small xenocrysts of diamond shed from fragmented diamondiferous peridotite (a potentially much more widespread source of xenocrystal diamond than eclogite) act as nuclei for diamond precipitating from the kimberlite magma (Dawson and Smith 1975b).

10.5 Relationship of Kimberlite Magmatism to Major Earth Movements

The deformation textures in some peridotite xenoliths testify to the fact that kimberlite magmatism was associated with movements within the upper mantle, and it is not unreasonable to assume that these upper-mantle movements had some expression at the earth's surface. However, what is not apparent is the sense of these upper-mantle movements. Were they vertical or were they horizontal? Strong horizontal movements, accompanying the break-up of Gondwanaland, were proposed by Boyd (1973a) and Nixon and Boyd (1973a) to explain the deformation fabrics in some peridotite xenoliths from the Thaba Putsoa pipe in Lesotho; and certainly the wide range of dates of South Africa kimberlites, from 150 to 80 m.y. B.P. (see Chap. 1) overlap the 130 m.y. B.P. initial date of opening of the South Atlantic (Francheteau 1973) and the early drift of South America away from the African continent. It is clear that this major event must have been accompanied and produced by major movements in the upper mantle. Nonetheless, we are still not certain as to the *absolute* amount of movement of the African Plate; and the absence of subduction zones around the African Plate during this period points to a relative immobility compared with the South American, Indian, Australian and Antarctic Plates where migration from the Gondwanaland spreading centre resulted in the formation of subduction zones and accreted margins along their leading edges. In addition, it may be noted that the African Plate remained reasonably stationary relative to the palaeomagnetic North Pole during the period 100 to 170 m.y. B.P. (McElhinny et al. 1968). In the context of subduction, Sharp (1974) proposes that the South African kimberlites result from melting at the deep end of a subducted oceanic plate; he suggests that an oceanic plate with entrained hydrous and carbonaceous material forms of diamond-bearing kimberlites upon melting and states that the emplacement of these kimberlites occurs deep under the continent parallel to the Cape Fold Belt. This hypothesis is contrary to the observed random distribution of kimberlites in Southern Africa (see Fig. 3) and Newton and Gurney (1975) point out that the concept of the Cape Fold Belt as representing a subduction zone is incorrect; the Belt resulted from the deformation of the Cape Karroo Trough, which is not a trench, and its deformation took place in the Middle Triassic which antedates the Cretaceous kimberlites by a considerable period of time; this refutes Sharp's hypothesis that requires the kimberlites to form during or immediately following the main deformation stage which produced the Cape Fold Belt. In general, the confinement of kimberlite magmatism to the cratonic centres of continents well away from any major fold belts, together with the very limited amounts of kimberlite magma, appear to preclude any connection with classical subduction zones. The Tertiary magmatic centres of the south-western United States, including the so-called "kimberlites", may be different since there is evidence of subducted oceanic crust beneath the south-western United States (Helmstaedt and Doig 1975); and Anderson and Perkins (1975) have proposed a model involving non-laminar asthenospheric flow with upwelling material from deep within the asthenosphere to account for the random patterns.

Turning to hypotheses in which major vertical upper-mantle movements are involved, the models of Green and Geuguen (1974) and Mercier (1979) have been mentioned in Chap. 9. In the specific case of South Africa, Du Toit (1954) has recorded ample evidence of strong vertical movements of the South African continent during the Cretaceous, resulting in peripheral faults downthrowing 6000 m towards the continguous ocean basins; and breaks in the offshore sedimentary record, apparently reflecting major marine regression associated with marginal upwarping of the southern Africa landmass, can be correlated with distinct episodes of alkalic vulcanism in the Cretaceous and early Tertiary, including the major episode of kimberlite activity (Moore 1976).

At the first sight, vertical models, particularly the proposed diapir model of Green and Geuguen (1974), might appear to be the continental equivalent of the oceanic "hot spots" with their attendant plumes, the presence of which has been predicted by Morgan (1971). Even bearing in mind the circular arguments postulating relationships between observed oceanic vulcanism and attendant spreading with inferred plumes, there are distinct differences between the ocean basin phenomena and the sub-cratonic mantle events connected with kimberlite magmatism. First, the major quantities of mantle-derived melts, produced on the hot spots and along spreading ridges, are generally considerably in excess of the products of the very limited kimberlite and nephelinite magmatism on the cratons. Second, although the upper-mantle forces attendant upon the kimberlite magmatism were sufficient to cause the minor dilation of the cratons necessary for the production of major fractures extending from the mantle to the surface, they were not sufficiently powerful to cause break-up of the cratonic areas; in this respect, there is a sharp contrast with the obviously major lateral forces operating at the spreading ridges, which must be orders of magnitude greater than those that operated beneath the cratons at the time of kimberlite emplacement. In Chap. 1, it was concluded that kimberlite magmatism was connected with a broad epeirogenic movement of the cratons. These movements give rise to continental crust swells in the order of 500 to 1000 km across which, in addition to the kimberlite magmatism, have been the sites of other types of alkalic magmatism throughout theMesozoic and Tertiary in Africa (Le Bas 1971). There is little firm evidence as yet as to the reason for these swells; whether they are due to phase transitions within the mantle, outgassing within the upper mantle, or due to purely physical buckling of the lithosphere, such as proposed in the membrane tectonics theory of Oxburgh and Turcotte (1974), must remain open to question.

An indirect connection between kimberlite magmatism and the development of oceanic basins has been proposed by Williams and Williams (1977) who suggest that the kimberlites of Sierra Leone and Liberia lie on a continental extension of the Vema, Guinea and Sierra Leone Fracture Zones in the south Atlantic. Similarly, Stracke et al. (1979) propose that the location of kimberlites in south-eastern Australia has been governed by fracture patterns initially developed prior to the separation of Australia from Antarctica; these fractures later became the sites of continental extension of transform faults associated with the Antarctic Ridge, during the formation of which kimberlite magmatism took place in the continental fracture zones.

10.6 Outstanding Problems

Although, in the author's opinion, the partial melting hypothesis outlined above explains the genesis of kimberlite more adequately than the residual hypothesis, it also contains some problems that cannot be adequately explained at present.

One of these problems is the source of carbon, phosphorus and nitrogen within the upper mantle. The source of other elements, including titanium and hydrogen, can be adequately explained by melting of known upper-mantle materials. Regarding carbon, moissanite is only a very rare accessory mineral in some peridotite xenoliths (Chap. 6) and although the presence of carbonate is invoked in most hypotheses for the genesis of kimberlite, only small amounts of carbonate, mainly calcite, have been found as inclusions in mantle olivines and garnets (see Chap. 6); it appears unlikely that the small observed amounts are adequate to explain the amount of carbon present as diamond and the carbonate in kimberlites and alkalic carbonatite complexes, and that released during basalt eruptions. Another apparent anomaly is that, whereas in the garnet-peridotite stability field either dolomite or magnesite should theoretically be the stable carbonate phase (Wyllie 1979a), the only observed carbonate inclusions in phases derived from these depths are calcite. In the context of a mantle source for carbon, Dawson (1972) suggested that carbon might be combined at depth with a metal to form a carbide, such as cohenite (Fe_3C); that the carbon may occur as inclusions or interstitial gas or liquid CO and CO_2; or that it may exist as a primary hydrocarbon vapour. The idea of an intergranular, methane-rich vapour in the upper mantle has been invoked by Eggler et al. (1979) and various types of hydrocarbon, including ethylene and butane, have been detected in diamonds from Africa, Arkansas and Brazil (see diamond, Chap. 6). In the case of phorphorus, small amounts are present in peridotite and eclogite garnets (Bishop et al. 1978) but again there appears to be insufficient in this particular source to satisfy the amounts which are known to exist in kimberlites and other igneous rock types. An additional source of phosphorus must be present in the upper mantle and the most obvious candidate must be apatite; however since other phosphorus-bearing species such as schreibersite, merrillite and farringtonite are known in meteorites, it is a possibility that these species may exist within the upper mantle but have so far been undetected. Nitrogen exists in platelets and in segregated nodes within diamond which is the only site of upper-mantle nitrogen known at present. Whereas some evidence, albeit scanty, exists for a source of carbon and phosphorus, the initial source of the nitrogen and other rare gases within diamond must remain an enigmal.

Another problem, and one which is of considerable economic importance, is why do some kimberlites contain diamond and others not? As pointed out by Dawson (1970), there is no reason why diamond should not exist in any part of the upper mantle providing there is carbon under reducing conditions at sufficiently high pressures to stabilize diamond. In addition to diamond forming it must also be preserved during the ascent of the transporting kimberlite to the surface. In this context, a method of evaluating the diamond-bearing potential of kimberlite has been proposed by Milashev (1965). This was based on the thesis that the diamond-bearing properties (DBP) of kimberlite are the result of two inter-related factors: (1) the chemical potential for diamond formation (CPD), and (2) the degree of preservation of the crystals (DPC). The CPD is expressed by the formula:

$$CPD = \frac{Fe : Ti}{\log (Fe + Ti) + 1/2 \log (Al + K + Na)} .$$

The argument behind this is that a magma with the highest Fe/Ti ratio would theoretically be the most favourable for crystallizing diamond, whereas Al, Na and K will be "diluting" factors. The degree of preservation of diamonds is held to be the result of two factors: (a) the PT conditions of crystallization of a particular body of kimberlite magma, which may be either favourable or unfavourable to diamond crystallization, and (b) the degree of resorption of the diamond in the magma when the initial PT conditions change. Both these factors are responsible for the resulting morphology of diamonds, and Milashev (1965) proposed that they may be estimated by the proportion of various morphological types of diamond within a kimberlite body. The DPC of a body is expressed by a formula:

$$\mathrm{DPC} \simeq \frac{o + 0.45\,i + 0.45\,d}{1 + \log d}$$

where o = the percentage of flat-faced octahedral crystals, d = percentage of rhombic-dodecahedra; and i = percentage of transitional forms (octahedral-dodecahedra). The diamond-bearing properties (DBP) are expressed by the formula DBP = CPD × DPC. Both factors are equally important since, even in a kimberlite with a high CPD, only the correct PT conditions will produce diamonds, whereas kimberlite crystallizing under the correct PT conditions will produce no diamond if the degree of resorption is too high. In short, Milashev (1965) is trying to express the complex interplay between the requisite chemical and physical conditions for diamond formation and the amount of resorption which is imposed upon the diamonds during their ascent. He also assumes the growth of diamond from the kimberlite magma.

Milashev's (1965) attempt to quantify the problem as to why some kimberlites contain diamonds and others do not highlights two fundamental problems in diamond geology. The first is the debate, already discussed above, as to whether diamonds are xenocrysts or phenocrysts. In this context, more isotopic and microchemical work is required, not only upon the individual growth layers which are so apparent within many diamonds, but also upon the rounded cores and regular overgrowths in diamonds exhibiting two stages of growth. The second point is that we do not really know the composition of kimberlite liquid; the complex problems attendant upon this have been outlined in Chap. 5 and all that can be said is that the actual nature of kimberlite liquid will only be arrived at by the stripping off of the chemical effects of high-level alteration, xenocrystal contamination and megacryst fractionation.

A particularly fascinating field for further work will be the search for more relicts of the original igneous rocks which existed prior to events that converted most of them into the metamorphic upper-mantle assemblages that comprise most kimberlite xenoliths. It must be confirmed whether the high-calcium, high-chrome garnets found as crystals within kimberlite (Clarke and Carswell 1977) are indeed derived from earlier cumulates, and whether the so-called fertile harzburgites (Dawson et al. 1980) are low-pressure cumulates which have been transported down into a higher-pressure regime. If more of these apparently-ancestral igneous rock-types can be found and investigated, and their igneous nature and chemistry confirmed, then we may be able to gain some further insight into the earlier stages of the earth's history.

Finally, there will be much to be learnt from isotope studies of upper mantle materials, particularly in the context of the nature and the timing of any upper mantle metasomatism. At present there are two main theories regarding the interplay between upper-mantle metasomatism and the generation of kimberlites. The first regards an earlier metasomatic event and the formation of incompatibles-enriched mantle as a pre-requisite for the generation of kimberlites, carbonatites and alkaline magmas (Bailey 1980); and in this context, recent neodymium and strontium isotope data have been interpreted as indicating an ancient, enriched and grossly-heterogeneous sub-continental mantle beneath southern Africa (Menzies and Murthy 1980). Conversely, samarium/neodynium isotope and neodynium/strontium isotope data on S. African kimberlites indicate that the kimberlite source is in mantle that is chondritic in rare earth element values and relatively primeval in composition (Basu and Tatsumoto 1979; Lock 1980). This controversy and, more importantiy, the timing of any metasomatic event relative to kimberlite magmatism will hopefully be resolved by further isotope studies on upper mantle xenoliths.

Note Added in Proof

The source of carbon in the upper mantle has been debated above (p. 222). Relevant to this important problem is the recent paper by Freund et al. (1980) who show that atomic carbon may be dissolved in solid solution in periclase and forsterite and that, upon exsolution, it combines with lattice oxygen to form CO_2. This has considerable implications for carbon storage in the mantle, for the evolution of CO_2 during magmatism, and also for the chemical evolution of the atmosphere and the origin of life.

Reference: Freund F, Kathrein H, Wengeler H, Knobel R, Heinen HJ (1980) Carbon in solid solution in forsterite – a key to the untractable nature of reduced carbon in terrestrial and cosmogenic rocks. Geochim Cosmochim Acta 44: 1319–1333

References

Åhman E (1950) Ett tynd av frisk olivin i kimberlit från kalix skargård. Geol Foeren Stockholm Foerh 72: 207–211

Ahrens LH, Cherry RD, Erlank AJ (1967) Observations on the Th-U relationship in zircons from granitic rocks and kimberlites. Geochim Cosmochim Acta 31: 2379–2387

Akella J, Boyd FR (1972) Partitioning of Ti and Al between pyroxenes, garnets and oxides. Carnegie Inst Wash Yearb 71: 378–384

Akella J, Boyd FR (1974) Petrogenic grid for garnet peridotites. Carnegie Inst Wash Yearb 73: 269–273

Akimov AP, Berzina IG, Gurvich MYu, Lutts BC (1968) Uranium content of eclogite inclusions from kimberlite pipes. Dokl Acad Sci Nauk USSR Earth Sci Sect 181: 208–210

Albert D, Albert A, Brousse R (1967) Enclaves de péridotites à pyrope chromifère dans les pipes de la région de Bédarieux (Hérault). Acad Sci 265: 657–659

Allègre CJ, Manhes G, Richard P, Rousseau D, Shimizu N (1978) Systematics of Sr, Nd and Pb isotopes in garnet lherzolite nodules in kimberlites. In: Zartman RE (ed) Short papers of the 4th Int Conf Geochron Cosmochron, pp 10–11. US Geol Surv Open File Rep, pp 78–701

Allen JB, Deans T (1965) Ultrabasic eruptives with alnöitic kimberlites affinities from Malaita, Solomon Islands. Min Mag 34 (Tilley Vol): 16–34

Allsopp HL, Barrett DR (1975) Rb-Sr age determinations on South African kimberlite pipes. Phys Chem Earth 9: 605–617

Allsopp HL, Kramers JD (1977) Rb-Sr and U-Pb age determinations on Southern African kimberlite pipes. 2nd Int Conf Kimb Santa Fe Ext Abst (unpaged)

Allsopp HL, Burger AJ, Van Zyl C (1967) A minimum age for the Premier kimberlite pipe yielded by biotite Rb-Sr measurements, with related galena isotopic data. Earth Planet Sci Lett 3: 161–166

Allsopp HL, Nicholaysen LO, Hahn-Weinheimer P (1969) Rb/K ratios and Sr-isotopic compositions of minerals in eclogitic and peridotitic rocks. Earth Planet Sci Lett 5: 231–244

Anderson OL (1979) The role of fracture dynamics in kimberlite pipe formation. In: Boyd FR, Meyer HOA (eds) Kimberlites, diatremes and diamonds: their geology, petrology and geochemistry. AGU, Washington, pp 344–353

Anderson OL, Grew PC (1977) Stress corrosion theory of crack propagation with applications to geophysics. Rev Geophys Space Phys 15: 77–104

Anderson OL, Perkins PC (1975) A plate tectonics model involving non-laminar asthenospheric flow to account for irregular patterns of magmatism in the southwestern United States. Phys Chem Earth 9: 113–122

Andrews JR, Emeleus CH (1975) Structural aspects of kimberlite dyke intrusion in southwest Greenland. Phys Chem Earth 9: 43–50

Andrews JR, Emeleus CH (1976) Kimberlites of West Greenland. In: Escher A, Watt WS (eds) Geology of Greenland. Geol Surv Greenland, Copenhagen, pp 575–581

Andrews-Jones DA (1966) Geology and mineral resources of the Northern Kambai schist belt and adjacent granulites. Geol Surv. Sierra Leone Bull 6: 100 pp

Angel F (1957) Einige ausgewählte Probleme eklogitischer Gesteinsgruppen des Österreichischen Ostalpen. Neues Jahrb Mineral Abh 91: 151–192

Aoki K (1974) Phlogopites and potassic richterites from mica nodules in South African kimberlites. Contrib Mineral Petrol 48: 1–7

Aoki K (1975) Origin of phlogopite and potassic richterite-bearing peridotite xenoliths from South Africa. Contrib Mineral Petrol 53: 145–156

Aoki K, Fodor RV, Keil K, Dowty E (1972) Tremolite with high richterite-molecular content in kimberlite from Buell Park, Colorado. Am Mineral 57: 1889–1893

Aoki K, Fujino K, Akaogi M (1976) Titanochondrodite and titanoclinohumite derived from the upper mantle in the Buell Park kimberlite, Arizona U.S.A. Contrib Mineral Petrol 56: 243–253

Arsenyev AA (1962) The laws of the distribution of kimberlites in the eastern part of the Siberian Platform. Dokl Akad Nauk USSR Earth Sci Sect 137: 355–357

Bailey DK (1980) Volatile flux, geotherms, and the generation of the kimberlite-carbonatite-alkaline magma spectrum. Min Mag 43: 695–699

Baldock JW, Hepworth JV, Marengwa BS (1976) Gold, base-metals and diamonds in Botswana. Econ Geol 71: 139–156

Banno S (1970) Classification of eclogite in terms of physical conditions of their origin. Phys Earth Planet Inter 3: 405–421

Barbosa A (1951) Contribucao a origem do diamante em Diamantina, Estado de Minas Ceraes. Bull Brasil Dep Nac Prod Minera 6 136

Bardet MG (1963) Controle géotectonique de la repartition des venues diamantifères dans la monde. Chron Mines Rech Minière 328-9: 67–89

Bardet M (1965) Les gisements de diamant d'U.R.S.S. Chron Mines Rech Minière 65: 1–40

Bardet MG (1973) Géologie du diamant. Mem Bur Rech Geol Minière 83: 235

Bardet MG, Vachette M (1966) Déterminations d'ages de kimberlites de l'Ouest Africain et essai d'interprétation des datations des diverses venues diamantifères dans le monde. Bur Rech Geol Minières Rep DS 66: 59

Barrett DR (1975) The genesis of kimberlite and associated rocks: strontium isotope evidence. Phys Chem Earth 9: 637–654

Barrett DR, Berg GW (1975) Complementary petrographic and strontium-isotope studies of South African kimberlite. Phys Chem Earth 9: 619–635

Basu AR (1977) Olivine-spinel equilibria in lherzolites from San Quintin, California. Earth Planet Sci Lett 33: 443–450

Basu AR, Tatsumoto M (1979) Samarium-neodymium systematics in kimberlites and in the minerals of garnet lherzolite inclusions. Science 205: 398–401

Beck R (1899) Neues von den afrikanischen Diamantlagerstätten. Z Prakt Geol 1899: 417–419

Berg GW (1968) Secondary alteration in eclogites from kimberlite pipes. Am Mineral 53: 1336–1346

Berg GW, Allsopp HL (1972) Low Sr^{87}/Sr^{86} ratios in fresh South African kimberlites. Earth Planet Sci Lett 16: 27–30

Besson M (1967) La teneure engeikielite des ilmenites des kimberlites. Bull Soc Fr Mineral Cristallogr 90: 192–201

Best MG (1974) Mantle-derived amphibole within inclusions in alkalic basaltic lavas. J Geophys Res 79: 2107–2113

Bilal A (1978) Les inclusion fluides des roches de la croûte continentale inférieure et du manteau supérieur. Unpubl Dr État Thesis, University of Paris

Bishop FC (1977) The distribution of Fe^{2+} and Mg between coexisting ilmenite and pyroxene with applications to geothermometry. Unpubl Ph D Thesis, Univ Chicago

Bishop FC, Smith JV, Dawson JB (1975) Pentlandite-magnetite intergrowth in De Beers spinel lherzolite: review of sulphides in nodules. Phys Chem Earth 9: 323–338

Bishop FC, Smith JV, Dawson JB (1978) Na, K, P and Ti in garnet, pyroxene and olivine from peridotite and eclogite xenoliths from African kimberlites. Lithos 11: 155–173

Blagulkina VA (1971) On the composition of ilmenites from kimberlites. Zap Vses Mineral Ova 1971(2): 194–198 (Russian)

Bobrievich AP, Bondarenko MN, Gnevushev MA, Kind NV, Koreshkov BYu, Kuryleva NA, Nefedoca ZD, Popugeva LA, Popova EZ, Skuiskii VD, Smirnov GI, Yurkevtsi RK, Fayunshteyeyuh GH, Shukin VI (1959) Almazy Sibirii (Diamonds of Siberia). Gozgeol teknizdat, Moscow (Russian)

Bobrievich AP, Lupin JP, Kozlov IT, Lebedev LI, Pankratov AA, Smirnov GI, Kharkiv AD (1964) Petrography and mineralogy of the kimberlite rocks at Yakutia. "Nedra" Izdatelstve, Moscow (Russian)

Boctor NZ, Meyer HOA (1979) Oxide and sulfide minerals in kimberlite from Green Mountain, Colorado. In: Boyd FR, Meyer HOA (eds) Kimberlites, diatremes and diamonds: their geology, petrology and geochemistry. AGU, Washington, pp 217–228

Boettcher AL (1970) The system CaO-Al_2O_3-SiO_2-H_2O at high pressures and temperatures. J Petrol 11: 337–379

Boettcher AL, O'Neill JR, Windom KR, Stewart DC, Wilshire HG (1979) Metasomatism of the upper mantle and the genesis of kimberlites and alkali basalts. In: Boyd FR, Meyer HOA (eds) The mantle sample: inclusions in kimberlites and other volcanics, AGU, Washington, pp 173–182

Bosch JL (1971) The petrology of some kimberlite occurrence in the Barkly West district, Cape Province. Trans Geol Soc S Afr 74: 75–101

Boullier AM, Nicolas A (1975) Classification of textures and fabrics of peridotite xenoliths from South African kimberlites. Phys Chem Earth 9: 467–476

Boyd FR (1968) Quantitative electron probe analysis of pyroxenes. Carnegie Inst Wash Yearb 66: 327–334

Boyd FR (1969) Electron-probe study of diopside inclusion in kimberlite. Am J Sci Schairer Vol 267-A: 50–69

Boyd FR (1970) Garnet peridotites and the system $CaSiO_3$-$MgSiO_3$-Al_2O_3. Mineral Soc Am Spec Pap 3: 63–75

Boyd FR (1971a) Pargasite-spinel peridotite xenolith from the Wesselton Mine. Carnegie Inst Wash Yearb 70: 138–142

Boyd FR (1971b) Enstatite-ilmenite and diopside-ilmenite intergrowths from the Monastery Mine. Carnegie Inst Wash Yearb 70: 134–138

Boyd FR (1973a) A pyroxene geotherm. Geochim Cosmochim Acta 37: 2533–2546

Boyd FR (1973b) Appendix of mineral analyses-Letseng-la-terae. In: Nixon PH (ed) Lesotho kimberlites. Lesotho Natl Dev Co, Maseru, pp 33–36

Boyd FR (1974) Ultramafic nodules from the Frank Smith kimberlite pipe, South Africa. Carnegie Inst Wash Yearb 73: 285–294

Boyd FR (1975) Stress heating and compositional variations in enstatites from sheared lherzolites. Carnegie Inst Wash Yearb 74: 525–528

Boyd FR, Danchin RN (1974) Discrete nodules from the Artur de Paiva kimberlite, Angola. Carnegie Inst Wash Yearb 73: 278–282

Boyd FR, Dawson JB (1972) Kimberlite garnets and pyroxene-ilmenite intergrowths. Carnegie Inst Wash Yearb 71: 373–378

Boyd FR, England JL (1960) The quartz-coesite transition. J Geophys Res 65: 749–756

Boyd FR, Finger LW (1975) Homogeneity of minerals in mantle rocks from Lesotho. Carnegie Inst Wash Yearb 74: 519–525

Boyd FR, Nixon PH (1970) Kimberlite diopsides. Carnegie Inst Wash Yearb 68: 324–329

Boyd FR, Nixon PH (1972) Ultramafic nodules from the Thaba Putsoa kimberlite pipe. Carnegie Inst Wash Yearb 71: 362–373

Boyd FR, Nixon H (1973a) Origin of the ilmenite-silicate nodules in kimberlites from Lesotho and South Africa. In: Nixon PH (ed) Lesotho kimberlites. Lesotho Natl Dev Co, Maseru, pp 254–268

Boyd FR, Nixon PH (1973b) Structure of the upper mantle beneath Lesotho. Carnegie Inst Wash Yearb 72: 431–445

Boyd FR, Nixon PH (1975) Origins of the ultramafic nodules from the kimberlites of northern Lesotho and the Monastery Mine, South Africa. Phys Chem Earth 9: 431–453

Boyd FR, Nixon PH (1978) Ultramafic nodules from the Kimberley pipes, South Africa. Geochim Cosmochim Acta 42: 1367–1382

Brooks CK (1970) The concentration of zirconium and hafnium in some igneous and metamorphic rocks and minerals. Geochim Cosmochim Acta 34: 411–416

Brookins DG (1967) The strontium geochemistry of carbonates in kimberlites and limestones from Riley County, Kansas. Earth Planet Sci Lett 2: 235–240

Brookins DG, McDermott VJ (1970) The mineralogy of the Randolph kimberlites, Riley County, Kansas. Trans Kans Acad Sci 73: 31–39

Brookins DG, Meyer HOA (1974) Crustal and upper mantle stratigraphy beneath eastern Kansas. Geophys Res Lett 1: 269–272

Brookins DG, Watson KD (1969) The strontium geochemistry of calcite associated with kimberlites at Bachelor Lake, Quebec. J Geol 77: 367–371

Brown DD, Bennet G, George PT (1967) The source of alluvial indicator minerals in the James Bay Lowlands. Ontario Dep Mines Misc Paper 7

Bureau d'Etudes Géologiques et Minière (1958) Sur le découvert de kimberlite dans la région de Kenieba (Soudain Français). Chron Mines Rech Minière 261: 76–79

Burkov VV, Podporina YeK (1965) Rare elements in kimberlite rocks. Dokl Akad Sci Nauk USSR Earth Sci Sect 163: 169–171

Burkov VV, Podporina YeK (1966) First data on rare earths in kimberlite. Dokl Akad Sci Nauk USSR Earth Sci Sect 171: 215–219

Burns RG (1975) Crystal field effects in chromium and its partitioning in the mantle. Geochim Cosmochim Acta 39: 857–864

Burns RG, Huggins F, Drickamer HG (1972) Applications of high-pressure Mössbauer spectroscopy to mantle mineralogy. Proc 24th Int Geol Congr Sect 14: 113–123

Butakova EL, Egorov LS (1962) The Meimecha-Kotui complex of alkaline and ultrabasic rocks. In: The petrology of eastern Siberia. Academy of Sciences, Moscow, pp 419–589 (Russian)

Camsell C (1911) A new diamond-bearing locality in the Tulameen District, British Columbia. Econ Geol 6: 604–611

Cann JR (1971) Major element variations in ocean-floor basalts. Phil Trans R Soc London Ser A 268: 495–505

Carstens H (1971) Plastic stress relaxation around solid inclusions in pyrope. Contrib Mineral Petrol 32: 289–294

Carswell DA (1975) Primary and secondary phlogopites in garnet lherzolite xenoliths. Phys Chem Earth 9: 417–430

Carswell DA, Dawson JB (1970) Garnet peridotite xenoliths in South African kimberlite pipes and their petrogenesis. Contrib Mineral Petrol 25: 163–184

Carswell DA, Griffin WL (1980) Calculation of equilibration conditions for garnet granulite and garnet websterite nodules in African kimberlite pipes. Trans R Soc Edinburgh Earth Sci Sect (in press)

Carswell DA, Clarke DB, Mitchell RH (1979) The petrology and geochemistry of ultramafic nodules from Pipe 200, northern Lesotho. In: Boyd FR, Meyer HOA (eds) The mantle sample: inclusions in kimberlites and other volcanics. Am Geophys Union, Washington, pp 127–144

Carter NL (1971) Static deformation of silica and silicates. J Geophys Res 76: 5514–5540

Chen Ju-Chin (1971) Petrology and chemistry of garnet lherzolite nodules in kimberlite from South Africa. Am Mineral 56: 2098–2110

Cherepanov VA (1967) Boron in the kimberlite and meimechite of Siberia. Dokl Akad Nauk USSR Earth Sci Sect 172: 187–190

Chinner GA, Cornell DH (1974) Evidence of kimberlite-grospydite reaction. Contrib Mineral Petrol 45: 153–160

Chinner GA, Boyd FR, England JL (1960) Physical properties of garnet solid solutions. Carnegie Inst Wash Yearb 59: 76–78

Choubert B (1946) Sur la présence du diamant au Gabon (A.E.F.) en rélation avec des kimberlites et des rocks carbonatés métamorphiques. CR Scéances Acad Sci 223: 638–640

Clark SP, Ringwood AR (1964) Density distribution and constitution of the mantle. Rev Geophys 2: 35–88

Clarke DB (1979) Synthesis of nickeloan djerfisherites and the origin of potassic sulphide at the Frank Smith Mine. In: Boyd FR, Meyer HOA (eds) The mantle sample: inclusions in kimberlites and other volcanics. AGU Washington, pp 300–308

Clarke DB, Carswell DA (1977) Green garnets from the Newlands kimberlite, Cape Province, South Africa. Earth Planet Sci Lett 34: 30–38

Clarke DB, Mitchell RG (1977) Mineralogy and petrology of the kimberlite from Somerset Island, N.W.T., Canada. Phys Chem Earth 9: 123–136

Clement CR (1973) Kimberlites from the Kao pipe, Lesotho. In: Nixon PH (ed) Lesotho kimberlites. Lesotho Natl Dev Co, Maseru, pp 110–121

Clement CR (1975) The emplacement of some diatreme-facies kimberlites. Phys Chem Earth 9: 51–59

Clement CR, Skinner EMW, Hawthorne JB, Kleinjan L, Allsopp HL (1979) Precambrian ultramafic rocks with kimberlite affinities in the Kimberley area. In: Boyd FR, Meyer HOA (eds) Kimberlites, diatremes and diamonds; their geology, petrology and geochemistry. AGU, Washington, pp 101–110

Cloos H (1941) Bau und Tätigkeit von Tuffschloten. Untersuchungen an den Schwäbischen Vulkanen. Geol Rundsch 32: 709–800

Coleman RG, Lee ED, Beatty LB, Brannock WW (1965) Eclogites and eclogites: their differences and similarities. Geol Soc Am Bull 76: 483–508

Collerson KD, Malpas J (1977) Partial melts in upper mantle nodules from Labrador kimberlites. 2nd Int Conf Kimb Santa Fe Ext Abst (unpaged)

Colvine RJL (1968) Pyrope from Elie, Fife. Scot J Geol 4: 283–286

Cooper AF, Gittins J (1974) Shortite in kimberlite from the Upper Canada Gold Mine, Ontario: a discussion. J Geol 82: 667–669

Cornelissen AK, Verwoerd WJ (1975) The Bushmanland kimberlites and related rocks. Phys Chem Earth 9: 71–80

Cox KG, Hornung G (1966) The petrology of the Karroo basalts of Basutoland. Am Mineral 51: 1414–1432

Cox KG, Gurney JJ, Harte B (1973) Xenoliths from the Matsoku pipe. In: Nixon PH (ed) Lesotho kimberlites. Lesotho Natl Dev Co, Maseru, pp 76–91

Craig H (1953) The geochemistry of stable carbon isotopes. Geochim Cosmochim Acta 3: 53–92

Crawford AR (1969) India, Ceylon and Pakistan: new age data and comparisons with Australia. Nature (London) 223: 380–384

Crawford AR, Compston W (1970) The age of the Vindhayan system of peninsular India. Q J Geol Soc London 125: 351–371

Crockett RN, Mason R (1968) Foci of mantle disturbance in southern Africa and their economic significance. Econ Geol 63: 532–540

Daly RA (1924) Carbonate dykes of the Premier diamond mine, Transvaal. J Geol 33: 659–684

Danchin RV (1979) Mineral and bulk chemistry of garnet lherzolite and garnet harzburgite xenoliths from the Premier Mine, South Africa. In: Boyd FR, Meyer HOA (eds) The mantle sample: inclusions in kimberlites and other volcanics. AGU, Washington, pp 104–126

Danchin RV, d'Orey F (1972) Chromian spinel exsolution in ilmenite from the Premier Mine, Transvaal, South Africa. Contrib Mineral Petrol 35: 43–49

Danchin RV, Ferguson J, McIver JR, Nixon PH (1975) The composition of late-stage kimberlite liquids as revealed by nucleated autholiths. Phys Chem Earth 9: 235–245

David TWE (1950) The geology of the Commonwealth of Australia. Edward Arnold, London

Davidson CF (1964) On diamantiferous diatremes. Econ Geol 59: 1368–1380

Davidson CF (1967) The kimberlites of the U.S.S.R. In: Wyllie PJ (ed) Ultramafic and related rocks. Wiley, New York, pp 251–256

Davis GL (1977) The ages and uranium contents of zircons from kimberlites and associated rocks. 2nd Int Conf Kimberlite, Santa Fe, Ext Abst (unpaged)

Davis BTC, Boyd FR (1966) The join $Mg_2Si_2O_6$-$CaMgSi_2O_6$ at 30 kilobars pressure and its application to pyroxenes from kimberlites. J Geophys Res 71: 3367–3576

Dawson JB (1960) A comparative study of the geology and petrography of the kimberlites of the Basutoland province. Unpubl. Ph D Thesis, University Leeds

Dawson JB (1962) Basutoland kimberlites. Geol Soc Am Bull 73: 545–560

Dawson JB (1964) Carbonate tuff cones in northern Tanganika. Geol Mag 101: 129–137

Dawson JB (1967a) A review of the geology of kimberlite. In: Wyllie PJ (ed) Ultramafic and related rocks. Wiley, New York, pp 241–251

Dawson JB (1967b) Geochemistry and origin of kimberlite. In: Wyllie PJ (ed) Ultramafic and related rocks. Wiley, New York, pp 269–278

Dawson JB (1968) Recent researches on kimberlite and diamond geology. Econ Geol 63: 504–511

Dawson JB (1970) The structural setting of African kimberlite magmatism. In: Clifford TN, Gass IG (eds) African magmatism and tectonics. Oliver and Boyd, Edinburgh, pp 321–335

Dawson JB (1971) Advances in kimberlite geology. Earth Sci Rev 7: 187–214

Dawson JB (1972) Kimberlites and their relationship to the upper mantle. Philos Trans R Soc London Ser A 271: 297–311

Dawson JB (1977) Sub-cratonic crust and upper-mantle models based on xenolith suites in kimberlite and nephelinitic diatremes. J Geol Soc 134: 173–184

Dawson JB (1979) Veined peridotites from the Bultfontein Mine. 2nd Kimb Symp, Cambridge Abstr (unpaged)

Dawson JB, Hawthorne JB (1970) Intrusion features of some hypabassal South African kimberlites. Bull Volcanol 34: 740–757

Dawson JB, Hawthorne JB (1973) Magmatic sedimentation and carbonatitic differentiation in kimberlite sills at Benfontein, South Africa. J Geol Soc London 129: 61–85

Dawson JB, Reid AM (1970) A pyroxene-ilmenite intergrowth from the Monastery Mine, South Africa. Contrib Mineral Petrol 26: 296–301

Dawson JB, Smith JV (1973) Garnet exsolution from stressed orthopyroxene in garnet lherzolite from the Monastery Mine. 1st Int Conf Kimb, Cape Town, Ext Abstr, pp 81–82

Dawson JB, Smith JV (1975a) Chromite-silicate intergrowths in upper-mantle peridotites. Phys Chem Earth 9: 339–350

Dawson JB, Smith JV (1975b) Occurrence of diamond in a mica-garnet lherzolite from kimberlite. Nature (London) 254: 580–581

Dawson JB, Smith JV (1975c) Chemistry and origin of phlogopite megacrysts in kimberlite. Nature (London) 253: 336–338

Dawson JB, Smith JV (1977) The MARID (mica-amphibole-rutile-ilmenite-diopside) suite of xenoliths in kimberlite. Geochim Cosmochim Acta 41: 309–323

Dawson JB, Stephens WE (1975) Statistical analysis of garnets from kimberlites and associated xenoliths. J Geol 83: 589–607

Dawson JB, Stephens WE (1976) Statistical analysis of garnets from kimberlites and associated xenoliths: addendum. J Geol 84: 495–496

Dawson JB, Powell DG, Reid AM (1970) Ultrabasic lavas and xenoliths from the Lashaine volcano, Tanzania. J Petrol 11: 519–548
Dawson JB, Gurney JJ, Lawless PJ (1975) Palaeogeothermal gradients derived from xenoliths in kimberlite. Nature (London) 257: 299–300
Dawson JB, Smith JV, Hervig RL (1977) Late-stage diopside in kimberlite matrix. Neues Jahrb Mineral Mitt 1977: 529–553
Dawson JB, Delaney JS, Smith JV (1978a) Aspects of the mineralogy of alnöitic breccia, Malaita, Solomon Islands: comparison with continental kimberlites. Contrib Mineral Petrol 67: 189–193
Dawson JB, Smith JV, Delaney JS (1978b) Multiple spinel-garnet peridotite transitions in upper mantle: evidence from a harzburgite xenolith. Nature (London) 273: 741–743
Dawson JB, Smith JV, Hervig RL (1980) Heterogeneity in upper-mantle lherzolites and harzburgites. Philos Trans R Soc Lond Ser A 297: 323–331
De Sousa Machado FJ (1958) The volcanic belt of Angola and its carbonatites. Comm Tech Coop Africa South of the Sahara, Reg Comm Geol Publ 44: 309–317
Deines P (1980) The carbon isotopic composition of diamonds; relationship to diamond shape, colour, occurrence and vapour composition. Geochim Cosmochim Acta (in press)
Deines P, Gold DP (1973) The isotopic composition of carbonatite and kimberlite carbonates and their bearing on the isotopic composition of deep-seated carbon. Geochim Cosmochim Acta 37: 1709–1733
Delaney JS, Smith JV, Carswell DA, Dawson JB (1980) Chemistry of micas from kimberlites and xenoliths – II. Primary- and secondary-textured micas from peridotite xenoliths. Geochim Cosmochim Acta 44: 857–872
Desborough GA, Czamanske GK (1973) Sulfides in eclogite nodules from a kimberlite pipe, South Africa, with comments on violarite stoichimetry. Am Mineral 58: 195–202
Dickey JS (1968) Eclogite and other inclusions in the Mineral Breccia Member of the Deborah Volcanic Formation at Kakanui, New Zealand. Am Mineral 53: 1304–1319
Dimroth E (1970) Meimechites and carbonatites of the Castignon Lake complex, New Quebec. Neues Jahrb Mineral Abh 112: 239–278
Dingle E (1973) Mesozoic palaeogeography of the southern Cape, South Africa. Palaeogeogr Palaeoclimatol Palaeoecol 13: 203–213
Donaldson CF, Reid AM (1977) Emplacement and crystallisation of a kimberlite dyke from the De Beers Mine, Kimberley, South Africa. 2nd Int Conf Kimb Santa Fe, Ext Abstr (unpaged)
Draper D (1911) The diamond-bearing deposits of Bagagem and Agua Suja in the State of Minas Geraes, Brazil. Geol Soc S Afr Trans 14: 18–20
Draper D (1923) Additional evidence regarding the origin of the high-level diamond-bearing breccias of Diamantina, Brazil. Geol Soc S Afr Trans 26: 7–13
Drever HI, Johnston R (1967) Picritic minor intrusions. In: Wyllie PJ (ed) Ultramafic and related rocks. Wiley, New York, pp 71–82
Dubey VS, Merh S (1954) Diamondiferous plug of Majgawan, Panna State, central India. Q J Geol Min Metal Soc India 21: 1–10
Du Toit AL (1954) The geology of South Africa. Oliver and Boyd, Edinburgh
Dunin-Barkovskaya EA, Ikhanov AV (1974) Bismuth in the kimberlites of Yakutia. Geochem Int 11: 1215–1219
Dyer RG, Brookins DG (1970) Petrography and geochemistry of the Leonardville kimberlite, Riley County, Kansas. Trans Kans Acad Sci 73: 460–480
Eckermann H von (1948) The alkaline district of Alnö Island. Sver Geol Unders Ser C 36
Eckerman H von (1967) A comparison of Swedish, African and Russian kimberlites. In: Wyllie PJ (ed) Ultramafic and related rocks. Wiley, New York, pp 302–312
Edwards CB, Howkins JB (1966) Kimberlites in Tanganyika with special reference to the Mwadui occurrence. Econ Geol 61: 537–554
Eggler DH, McCallum ME (1976) A geotherm from megacrysts in the Sloan kimberlite pipes, Colorado. Carnegie Inst Wash Yearb 75: 538–541

Eggler DH, Wendlandt RF (1979) Experimental studies on the relationship between kimberlite magmas and partial melting of peridotite. In: Boyd FR, Meyer HOA (eds) Kimberlites, diatremes and diamonds: their geology, petrology and geochemistry. AGU Washington, pp 308–338

Eggler DH, McCallum ME, Smith CB (1979a) Megacryst assemblages in kimberlites from northern Colorado and southern Wyoming: petrology, geothermometry, barometry, and areal distribution. In: Boyd FR, Meyer HOA (eds) The mantle sample: inclusions in kimberlites and other volcanics. AGU, Washington, pp 213–216

Eggler DH, Wendlandt RF, Olafsson M (1979b) Volatiles and the generation of kimberlitic liquids: experimental studies. 2nd Kimb Symp Cambridge 1979 (unpaged)

Egorov LS (1970) Carbonatites and ultrabasic-alkaline rocks of the Maimecha-Kotui region, N. Siberia. Lithos 3: 341–359

Ellis DJ, Green DH (1979) An experimental study of the effect of Ca upon garnet-clinopyroxene exchange equilibria. Contrib Mineral Petrol 71: 13–22

Elthon D, Ridley WI (1979) The oxide and silicate mineral chemistry of a kimberlite from the Premier Mine: implications for the evolution of kimberlite magmas. In: Boyd FR, Meyer HOA (eds) Kimberlites, diatremes and diamonds: their geology, petrology and geochemistry. AGU, Washington, pp 206–216

Emeleus CH, Andrews JR (1975) Mineralogy and petrology of kimberlite dykes and sheet intrusions and included peridotite xenoliths from South-west Greenland. Phys Chem Earth 9: 179–198

Erlank AJ (1970) Distribution of potassium in mafic and ultramafic nodules. Carnegie Inst Wash Yearb 68: 433–439

Erlank AJ (1973) Kimberlite potassium richterite and the distribution of potassium in the upper mantle. Int Conf Kimb, Cape Town, Ext Abstr, pp 103–106

Erlank AJ, Finger LW (1970) The occurrence of potassic richterite in a mica nodule from the Wesselton kimberlite, South Africa. Carnegie Inst Wash Yearb 68: 442–443

Erlank AJ, Kushiro I (1970) Potassium content of synthetic pyroxenes at high temperatures and pressures. Carnegie Inst Wash Yearb 68: 233–236

Erlank AJ, Rickard RS (1977) Potassic richterite-bearing peridotites from kimberlite and the evidence they provide for upper-mantle metasomatism. 2nd Int Conf Kimb, Santa Fe, Ext Abstr (unpaged)

Erlank AJ, Shimizu N (1977) Strontium and strontium isotope distributions in some kimberlite nodules and minerals. 2nd Int Conf Kimb, Santa Fe, Ext Abstr (unpaged)

Fairbairn PE, Robertson RMS (1966) Stages in the tropical weathering of kimberlite. Clay Minerals 6: 351–370

Ferguson J, Sheraton JW (1979) Petrogenesis of kimberlitic rocks and associated xenoliths in southeastern Australia. In: Boyd FR, Meyer HOA (eds) Kimberlites diatremes and diamonds: their geology, petrology and geochemistry. AGU, Washington, pp 140–160

Ferguson J, Danchin RV, Nixon PH (1973a) Fenitisation associated with kimberlite magmas. In: Nixon PH (ed) Lesotho kimberlites. Lesotho Natl Dev Co, Maseru, pp 207–213

Ferguson J, Danchin RV, Nixon PH (1973b) Petrochemistry of kimberlite autoliths. In: Nixon PH (ed) Lesotho kimberlites. Lesotho Natl Dev Co, Maseru, pp 285–293

Ferguson J, Martin H, Nicolaysen LO, Danchin RV (1975) Gross Brukkaros: a kimberlite/carbonatite volcano. Phys Chem Earth 9: 219–234

Fesq HW, Bibby DM, Erasmus CS, Kable EJD, Sellschop JPF (1975a) A comparative trace-element study of diamonds from Premier, Jagersfontein and Finsch mines, South Africa. Phys Chem Earth 9: 817–836

Fesq HW, Kable EJD, Gurney JJ (1975b) Aspects of the geochemistry of kimberlites from the Premier Mine and other South African occurrences, with particular reference to the rare earth elements. Phys Chem Earth 9: 687–707

Fièremans C (1966) Contribution à l'étude petrographique de la brèche kimberlitique de Bakwanga. Mem Inst Geol Univ Louvain 24: 1–92

Forbes RB (1965) The comparative chemical composition of eclogite and basalt. J Geophys Res 70: 1515–1521

Fourie CP (1958) Die diamantvoorkomste in die omgewing van Swartruggens, Transvaal. Bull Geol Surv S Afr 26: pp 16 (Afrikaans)

Francheteau J (1973) Plate tectonics model of the opening of the Atlantic Ocean south of the Azores. In: Tarling DH, Runcorn SK (eds) Implications of continental drift to the earth sciences. Academic Press, London New York, pp 197–202

Francis DM (1978) The implications of the compositional dependence of texture in spinel lherzolite xenoliths. J Geol 86: 473–485

Frank FC (1969) Diamonds and deep fluids in the mantle. In: Runcorn SK (ed) The application of modern physics to earth and planetary interiors. Wiley, New York, pp 247–250

Frankel JJ (1956) An inclusion-bearing olivine melilitite from Mukorob, South West Africa. Trans R Soc S Afr 35: 115–123

Frantsesson EV (1968) The petrology of kimberlite. Nedra, Moscow (Russian)

Frantsesson EV (1969) Conjugation of carbonatite and kimberlite on the Siberian platform. Dokl Acad Sci USSR, Earth Sci Sect 183: 127–128

Frantsesson EV (1970) The petrology of kimberlite. Aust Univ Press, Canberra (Translation by D.A. Brown)

Frantsesson EV, Kaminski FV (1974) Carbonado – a diamond variety of non-kimberlite genesis. Dokl Akad Nauk USSR 219: 187–189 (Russian)

Frantsesson EV, Lutts BC (1970) On the discovery of graphite-bearing pyrope peridotite in the kimberlite pipe Mir. Dokl Akad Sci Nauk USSR, Earth Sci Sect 191: 159–161

Frey FA, Ferguson J, Chappell BW (1977) Petrogenesis of South African and Australian kimberlite suites. 2nd Int Conf Kimb, Santa Fe, Ext Abstr (unpaged)

Frick C (1972) The garnets in kimberlite and in the associated griquaite and ultramafic nodules. Contrib Mineral Petrol 35: 63–76

Frick C (1973a) Kimberlitic ilmenites. Trans Geol Soc S Afr 76: 85–94

Frick C (1973b) The sulphides in griquaite and garnet-peridotite xenoliths in kimberlite. Contrib. Mineral Petrol 39: 1–16

Frick C (1974) Ten kimberlites from Pofadder, Namaqualand, South Africa. Trans Geol Soc S Afr 77: 31–36

Frisch T (1970) A note on "The petrological environment of magnesium ilmenite". Lovering JF, Widdowson JR. Earth Planet Sci Lett 8: 329–330

Galimov EM, Prokhorov DV, Fedoseyev DV, Varnin VP (1973) Heterogeneous carbon isotope effects in synthesis of diamond and graphite from gas. Geokhimiya 1973: 416–424 (Russian)

Garlick GD, MacGregor ID, Vogel DE (1971) Oxygen isotope ratios in eclogites from kimberlites. Science 172: 1025–1027

Geijer P (1928) Alnöitic dykes from the coast of Lulea and Kalix in northern Sweden. Fennia 50: No 11

Giardini AA, Hurst VJ, Melton CE, Stormer JC (1974) Biotite as a primary inclusion in diamond: its nature and significance. Am Mineral 59: 783–789

Gittins J, Hewins RH, Laurin AF (1975) Kimberlite-carbonatite dykes of the Saguenay valley, Quebec, Canada. Phys Chem Earth 9: 137–148

Gnevushev MA, Kratsov YuM (1960) On the composition of impurities in Ural and Yakutian diamonds. Dokl Akad Nauk SSSR Geol Ser 130: 154–156 (Russian)

Goetze C (1975) Sheared lherzolites from the point of view of rock mechanics. Geology 3: 172–173

Goldsmith JR (1976) Scapolites, granulites, and volatiles in the lower crust. Bull Geol Soc Am 87: 161–168

Gonshakova VI, Ruzhitskiy VO, Boychuk MD, Zaritskiy AI, Strekozov NF (1969) Volcanic pipes and dikes of kimberlite rocks in the Russian platform. Int Geol Rev 11: 60–73
Grantham DR (1964) The diamond deposits of Panna, Central India. Ind Diamond Rev 24: 30–35
Grantham DR (1967) The diamond host-rocks of Ghana. Abstr 4th Symp Afr Geol, Sheffield 1967 (unpaged)
Grantham DR, Allen JB (1960) Kimberlites in Sierra Leone. Overseas Geol Miner Resour 8:5–25
Green DH, Ringwood AE (1967a) An experimental investigation of the gabbro to eclogite transition and its petrological applications. Geochim Cosmochim Acta 31: 767–833
Green DH, Ringwood AR (1967b) The stability fields of aluminous pyroxene peridotite and garnet peridotite and their relevance in upper mantle structure. Earth Planet Sci Lett 3: 151–160
Green DH, Sobolev NV (1975) Co-existing garnets and ilmenites synthesised at high pressures from pyrolite and olivine basanite and their significance for kimberlite assemblages. Contrib Mineral Petrol 50: 217–229
Green HW, Geuguen Y (1974) Origin of kimberlite pipes by diapiric upwelling in the upper mantle. Nature (London) 249: 617–620
Green TH (1967) An experimental investigation of subsolidus assemblages formed at high pressure in high alumina basalt, kyanite eclogite and grosspydite compositions. Contrib Mineral Petrol 16: 84–114
Griffin WL, Murthy VR (1968) Abundance of K, Rb, Sr, and Ba in some ultramafic rocks and minerals. Earth Planet Sci Lett 4: 497–501
Griffin WL, Jensen BB, Misra SN (1971) Anomalously elongated rutile in eclogite-facies pyroxene and garnet. Nor Geol Tidsskr 51: 177–185
Griffin WL, Carswell DA, Nixon PH (1979) Lower crustal granulites and eclogites from Lesotho, southern Africa. In: Boyd FR, Meyer HOA (eds) The mantle sample: inclusions in kimberlites and other volcanics. AGU, Washington, pp 59–86
Grinenko LN, Ukhanov AV (1977) Sulfur levels and isotopic compositions in upper-mantle xenoliths from the Obnazhennaya kimberlite pipe. Geochem Int 1977: 169–171
Gurney JJ, Ahrens LH (1973) The zinc content of some ultramafic and mafic rocks. Trans Geol Soc S Afr 76: 301–308
Gurney JJ, Berg GW (1968) Potassium, sodium and caesium in South African kimberlites and their peridotite xenoliths. Geol Soc S Afr Spec Publ 2: 417–427
Gurney JJ, Ebrahim S (1973) Chemical composition of Lesotho kimberlites. In: Nixon PH (ed) Lesotho kimberlites. Lesotho Natl Dev Co, Maseru, pp 280–284
Gurney JJ, Harte B (1980) Chemical variations in upper mantle nodules from Southern African kimberlites. Philos Trans R Soc London 297: 273–293
Gurney JJ, Hobbs JBM (1973) Potassium, thorium and uranium in some kimberlite from South Africa. Int Conf Kimb, Cape Town Ext Abstr 143–145
Gurney JJ, Switzer GS (1973) The discovery of garnets closely related to diamonds in the Finsch pipe, South Africa. Contrib Mineral Petrol 39: 103–116
Gurney JJ, Berg GW, Ahrens LH (1966) Observations on caesium enrichment and the potassium/rubidium relationships in eclogites from the Roberts Victor Mine, South Africa. Nature (London) 210: 1025–1027
Gurney JJ, Siebert JC, Whitfield GG (1968) A diamondiferous eclogite from the Roberts Victor Mine. Geol Soc S Afr Spec Publ 2: 352–357
Gurney JJ, Mathias M, Siebert C, Moseley G (1971) Kyanite eclogites from the Rietfontein kimberlite pipe, Mier Coloured Reserve, Gordonia, Cape Province, South Africa. Contrib Mineral Petrol 30: 46–52
Gurney JJ, Fesq HW, Kable EJD (1973) Clinopyroxene-ilmenite intergrowths from kimberlite: a re-appraisal. In: Nixon PH (ed) Lesotho kimberlites. Lesotho Natl Dev Co, Maseru, pp 238–253

Gurney JJ, Dawson JB, Harte B, Lawless PJ (1975) The bulk chemical composition of peridotite facies rocks from the Matsoku and Bultfontein pipes. Kimb Symp Cambridge 1975, Ext Abstr 3–5

Gurney JJ, Harte B, Cox KG (1976) Mantle xenoliths in the Matsoku kimberlite pipe. Phys Chem Earth 9: 507–524

Haggerty SE (1973) Spinels of unique composition associated with ilmenite reactions in the Liqhobong kimberlite pipe, Lesotho. In: Nixon PH (ed) Lesotho kimberlites. Lesotho Natl Dev Co, Maseru, pp 149–158

Haggerty SE (1975a) The chemistry and genesis of opaque minerals in kimberlite. Phys Chem Earth 9: 295–307

Haggerty SE (1975b) An analysis of correlative and non-correlative oxide parameters for spinels and ilmenite in groundmass kimberlite. Kimb Symp Cambridge 1975 Abstr (unpaged)

Haggerty SE, Boyd FR (1975) Kimberlite inclusions in an olivine megacryst from Monastery. Kimb Symp Cambridge 1975 Abstr (unpaged)

Haggerty SE, Hardie RB, McMahon BM (1979) The mineral chemistry of ilmenite nodule associations from the Monastery diatreme. In: Boyd FR, Meyer HOA (eds) The mantle sample, inclusions in kimberlites and other volcanics. AGU, Washington, pp 249–256

Hales AL, Sacks IS (1959) Evidence for an intermediate layer from crustal structure studies in the eastern Transvaal. Geophys J R Astron Soc 2: 15–33

Harger HS (1911) Note on the occurrence of oriental ruby in kimberlites. Trans Geol Soc S Afr 14: 64–70

Harris JW (1968) The recognition of diamond inclusions. I. Supergene inclusions. Ind Diamond Rev 28: 402–410

Harris JW (1972) Black material on mineral inclusions and in internal fracture planes in diamond. Contrib Mineral Petrol 35: 22–33

Harris JW, Gurney JJ (1979) Inclusions in diamond. In: Field JJ (ed) The properties of diamond. Academic Press, London New York; pp 555–591

Harris JW, Hawthorne JB, Oosterveld MM (1979) Regional and local variations in the characteristics of diamonds from some southern African kimberlites. In: Boyd FR, Meyer HOA (eds) Kimberlites, diatremes and diamonds: their geology, petrology and geochemistry. AGU, Washington, pp 27–41

Harris PG (1957) Zone refining and the origin of potassic basalts. Geochim Cosmochim Acta 12: 195–208

Harris PG, Middlemost EAK (1969) The evolution of kimberlites. Lithos 3: 77–88

Harte B (1977) Rock nomenclature with particular relation to deformation and recrystallisation textures in olivine-bearing xenoliths. J Geol 85: 279–288

Harte B (1978) Kimberlite nodules, upper mantle petrology and geotherms. Philos Trans R Soc London Ser A 288: 487–500

Harte B, Gurney JJ (1975a) Evolution of clinopyroxene and garnet in an eclogite nodule from the Roberts Victor kimberlite pipe, South Africa. Phys Chem Earth 9: 367–388

Harte B, Gurney JJ (1975b) Ore mineral and phlogopite mineralisation within ultramafic nodules from the Matsoku kimberlite pipe, Lesotho. Carnegie Inst Wash Yearb 74: 528–536

Harte B, Cox KG, Gurney JJ (1975) Petrography and geological history of upper mantle xenoliths from the Matsoku kimberlite pipe. Phys Chem Earth 9: 477–506

Harte B, Gurney JJ, Cox KC (1977) Clinopyroxene-rich sheets in garnet peridotite xenolith specimens from the Matsoku kimberlite pipe, Lesotho. 2nd Int Conf Kimb, Santa Fe, Ext Abstr (unpaged)

Harte B, Gurney JJ, Harris JW (1980) The formation of peridotitic suite inclusions in diamond. Contrib Mineral Petrol 72: 181–190

Haskin LA, Frey FA, Schmitt RA, Smith RH (1966) Meteoritic, solar and terrestrial rare-earth distribution. Phys Chem Earth 7: 167–321

Haskin LA, Haskin MA, Frey FA, Wildeman R (1968) Relative and absolute terrestrial abundance of the rare earths. In: Ahrens LH (ed) Origin and distribution of the elements. Pergamon, London, pp 889–912

Hatton CJ (1978) Geochemistry and origin of eclogite xenoliths from the Roberts Victor Mine. Unpubl Ph D Thesis, University of Cape Town

Hatton CJ, Gurney GG (1977a) Igneous fractionation trends in Roberts Victor eclogites. 2nd Int Conf Kimb, Santa Fe, Ext Abstr (unpaged)

Hatton CJ, Gurney JJ (1977b) Kyanite eclogites from the Roberts Victor Mine. 2nd Int Conf Kimb, Santa Fe, Ext Abstr (unpaged)

Haüy RJ (1822) Traite de minéralogie. Bachelier, Paris

Hawthorne JB (1968) Kimberlite sills. Trans Geol Soc S Afr 71: 291–311

Hawthorne JB (1975) Model of a kimberlite pipe. Phys Chem Earth 9: 1–15

Hawthorne JB, Carrington AJ, Clement CR, Skinner EMW (1979) Geology of the Dokolwayo kimberlite and associated palaeo-alluvial diamond deposits. In: Boyd FR, Meyer HOA (eds) Kimberlites, diatremes and diamonds: their geology, petrology and geochemistry. AGU, Washington, pp 59–70

Hearn BC (1968) Diatremes with kimberlitic affinities in north-central Montana. Science 159: 522–625

Heath DC, Toerien DK (1962) A cryptovolcanic structure on Hatzium II28, South West Africa. Ann Geol Surv S Afr 1: 81–85

Heier KS (1963) Uranium, thorium and potassium in eclogitic rocks. Geochim Cosmochim Acta 27: 849–860

Heier KS, Compston W (1966) K/Rb ratios of eclogites. Earth Planet Sci Lett 1: 293–294

Helmstaedt H, Doig R (1975) Eclogite nodules from kimberlite pipes of the Colorado Plateau-samples of subducted Franciscan-type oceanic lithosphere. Phys Chem Earth 9: 95–112

Higazy RA (1954) Trace elements of volcanic ultrabasic potassic rocks of south-western Uganda and adjoining part of the Belgian Congo. Bull Geol Soc Am 65: 39–70

Holmes A (1936) A contribution to the petrology of kimberlite and its inclusions. Trans Geol Soc S Afr 39: 379–428

Hornung G, Nixon PH (1973) Chemical variations in the knorringite-rich garnets. In: Nixon PH (ed) Lesotho kimberlites. Lesotho Natl Dev Co, Maseru, pp 122–127

Howells S, O'Hara MJ (1978) Low solubility of alumina in enstatite and uncertainties in estimated palaeogeotherms. Philos Trans R Soc London Ser A 288: 471–486

Howells S, Begg C, O'Hara MJ (1975) Crystallisation of some natural eclogites and garnetiferous ultrabasic rocks at high temperature and pressure. Phys Chem Earth 9: 895–902

Huckenholz HG, Knittel D (1975) Stability of grossularite-uvarovite solid solutions. Contrib Mineral Petrol 49: 211–232

Ilupin IP (1962) Distribution of some hydrothermal and supergene minerals in the kimberlites of Yakutia. Sov Geol 3: 152–156 (Russian)

Ilupin IP (1972) On successive stages in kimberlite intrusion. Geol Geofiz 1972: 151–157 (Russian)

Ilupin IP, Kozlov IT (1970) Zircon in kimberlite. In: Kovalski VV, Yurkevich RK (eds) Geology, petrography and mineralogy of magmatic formations of the north-eastern part of the Siberian Platform. Acad Sci SSSR, Moscow, pp 254–266 (Russian)

Ilupin IP, Levshev PP (1970) Application of X-ray analysis in the study of the groundmass of kimberlite. In: Kovalski VV, Yurkevich RK (eds) Geology, petrography and mineralogy of magmatic formations of the north-east part of the Siberian Platform. Acad Sci SSSR, Moscow, pp 301–307 (Russian)

Ilupin IP, Lutts BG (1971) The chemical composition of kimberlite questions on the origin of kimberlite magma. Sov Geol 6: 61–73 (Russian)

Ilupin IP, Rovsha VS (1972) On the mineralogy of rhombic pyroxenes in kimberlite. Dokl Akad Nauk SSSR 205: 181–184 (Russian)

Ilupin IP, Khomiakov AP, Balashov IA (1971) Rare earths in accessory minerals of Yakutian kimberlites. Dokl Akad Sci Nauk USSR, Earth Sci Sect 171: 1214–1217

Ilupin IP, Milashev VA, Tomanovskaya YuI, Evdomikov AN (1974) Ilmenites in kimberlite intrusions. In: Milashev VA (ed) Mineralogy, geochemistry and prediction of diamond deposits. NIIGA, Leningrad, pp 5–29 (Russian)

Irving AJ (1976) On the validity of palaeogeotherms determined from xenoliths suites in basalts and kimberlites. Am Mineral 61: 638–642

Ito K, Kennedy GC (1967) The melting and phase relations in a natural peridotite to 40 kilobars. Am J Sci 265: 519–538

Ivanov VV, Ilupin IP, Staroshitskaya MI, Grosse YuI (1977) Copper, gold and silver distributions in deep-seated inclusions and in kimberlites of Yakutia. Geokhimiya 1977: 1038–1051 (Russian)

Jackson PM, Harte B (1977) The nature and conditions of formation of granulite facies xenoliths from the Matsoku kimberlite pipe. 2nd Int Conf Kimb, Santa Fe, Ext Abstr (unpaged)

Jakob WRO (1977) Geochemical aspects of the megacryst suite from the Monastery kimberlite pipe. Unpubl M Sci Thesis, University of Cape Town

Janse AJA (1969) Gross Brukkaros, a probable carbonatite volcano in the Nama Plateau of South West Africa. Geol Soc Am Bull 83: 573–586

Janse AJA (1971) Monticellite-bearing porphyritic peridotite from Gross Brukkaros, South-West Africa. Trans Geol Soc S Afr 74: 45–55

Janse AJA (1975) Kimberlites and related rocks from the Nama Plateau of South West Africa. Phys Chem Earth 9: 81–94

Jordan TH (1979) Mineralogies, densities and seismic velocities of garnet lherzolites and their geophysical implications. In: Boyd FR, Meyer HOA (eds) The mantle sample: inclusions in kimberlites and other volcanics. AGU, Washington, pp 1–14

Junner NR (1943) The diamond deposits of the Gold Coast. Bull Geol Surv Gold Coast 12: pp 23

Kable EJD, Fesq HW, Gurney JJ (1975) The significance of the inter-element relationships of some minor and trace elements in South African kimberlites. Phys Chem Earth 9: 709–734

Kaminski FV (1972) Factors controlling the location of kimberlites (heterofacial) and related rocks in the Siberian platform. Dokl Akad Nauk SSSR 204: 1187–1190 (Russian)

Kaminski FV, Frantsesson EV, Khostova VP (1974) First information on metals of the platinum group (Pt, Pd, Rh, Ir, Ru, Os) in kimberlitic rocks. Dokl Akad Nauk SSSR 219: 204–207 (Russian)

Kaneoka I, Aoki K (1978) $^{40}Ar/^{39}Ar$ analyses of phlogopite nodules and phlogopite-bearing peridotites in South African kimberlites. Earth Planet Sci Lett 40: 119–129

Kaneoka I, Takaoka N, Aoki K (1977) Rare gases in a phlogopite nodule and a phlogopite-bearing peridotite in South African kimberlites. Earth Planet Sci Lett 36: 181–186

Kennedy CS, Kennedy GC (1976) The equilibrium boundary between graphite and diamond. J Geophys Res 81: 2467–2470

Kennedy GC, Nordlie BE (1968) The genesis of diamond deposits. Econ Geol 63: 495–503

Kerrich R, Fyfe WS, Gorman BE, Allison I (1977) Local modification of rock chemistry by deformation. Contrib Mineral Petrol 65: 183–190

Keyser A (1940) A kimberlite occurrence in the Phillipstown District. Trans Geol Soc S Afr 43: 69–83

Kharkiv AD (1967) Instances of high temperature metamorphism connected with kimberlite magmas. Geol Geofiz 1967: 124–126 (Russian)

Kharkiv AD (1975) Kimberlite veins interlinked with pipes as an independent phase of kimberlite magmatism Dokl Akad Nauk SSSR 224: 190–193 (Russian)

Kharkiv AD, Sobolev NV, Tchumirin KG (1972) Chrome diopside inclusions in zircon from kimberlite rocks of the Malo-Botuobinsky region. Zap Vses Mineral Ova 1972: 431–433 (Russian)

Khitarov NI, Slutski AB, Pugin VA, Revin NI, Soldatov IA (1971) High-alumina basalt and the mechanism of its melting and crystallisation at great and small depths. Geochem Int 8: 631–642

Khomyakov AP, Frantsesson YeV (1971) Composition of apatite from carbonatite of the Yakut diamond province and distribution of rare-earth elements. Geol Geofiz 1971 (12): 121–125 (Russian)

Kidwell AL (1951) Mesozoic igneous activity in the northern Gulf Coastal Plain. Trans Gulf Coast Assoc Geol Soc 1951: 182–199

Knopf D (1970) Les kimberlites et les roches apparentés de Côte d'Ivoire. Sodemi, Abidjan

Kobelski BJ, Gold DP, Deines P (1979) Variations in stable isotope compositions for carbon and oxygen in some South African and Lesothan kimberlites. In: Boyd FR, Meyer HOA (eds) Kimberlite, diatremes and diamonds: their geology, petrology and geochemistry. AGU, Washington, pp 252–271

Komar PD (1972) Mechanical interactions of phenocrysts and flow differentiation of igneous dykes and sills. Geol Soc Am Bull 83: 973–988

Kopecky L (1960) Diamond prospects in the Czeck Massif. Izv Akad Nauk SSSR Ser Geol 1960: 46–55 (Russian)

Kopecky L, Pisova J, Pokorny L (1967) Pyrope-bearing diatremes of the Ceske Stredhori mountains. Sb Geol Ved Geol 12: 81–130 (Czeck)

Koptil VI, Lazko EE, Serenko VP (1976) Diamond-bearing kyanite eclogite from the kimberlite of the Sitikanskaya pipe – first find in the U.S.S.R. Dokl Akad Nauk SSSR 225: 924–927 (Russian)

Kovalski VV (1963) The kimberlitic rocks of Yakutia. Acad Sci SSSR, Moscow (Russian)

Kovalski VV, Cherski NV (1972) The carbon isotope composition of diamonds and possible sources of carbon for natural diamond formation. Geol Geofiz 1972: 10–15 (Russian)

Kozlov IT (1966) On the geology and petrography of kimberlites in Guinea. Sov Geol 9: 113–125 (Russian)

Kozlov IT, Levshev PP (1962) Amakinite, a new mineral of the brucite-pyrochroite group. Zap Vses Mineral Ova 91: 72–77 (Russian)

Kramers J (1977) Lead and strontium isotopes in Cretaceous kimberlites and mantle-derived xenoliths from southern Africa. Earth Planet Sci Lett 34: 419–431

Kramers JD (1979) Lead, uranium, strontium, potassium and rubidium in inclusion-bearing diamonds and mantle-derived xenoliths from southern Africa. Earth Planet Sci Lett 42: 58–70

Kresten P (1973a) The coating on kimberlite zircons: a preliminary study. In: Nixon PH (ed) Lesotho kimberlites. Lesotho Natl Dev Co, Maseru, pp 220–223

Kresten P (1973b) Differential thermal analysis of kimberlites. In: Nixon PH (ed) Lesotho kimberlites. Lesotho Natl Dev Co, Maseru, pp 269–279

Kresten P (1973c) Kimberlite zircons. Int Conf Kimb Cape Town, Ext Abstr 191–194

Kresten P, Fels P, Berggren G (1975) Kimberlite zircons – a possible aid in prospecting for kimberlites. Miner Deposita 10: 47–56

Kröner, A (1973) Comments on "Is the African plate stationary?". Nature (London) 243: 29–30

Kropotova OI, Fedorenko BV (1970) Carbon isotope composition of diamond and graphite from eclogite. Geokhimiya 1970: 1279 (Russian)

Kropotova OI, Grinenko VA, Bezrukov GN (1967) Possible isotope effects during crystallisation of diamonds. Geochem Int 4: 819–820

Kukharenko AA, Mikhailov BM, Orlova MT (1971) On the mineralogy of kimberlites of the Liberian shield (West Africa). Sov Geol 1971: 91–103 (Russian)

Kuno H, Aoki K (1970) Chemistry of ultramafic nodules and their bearings on the origin of basaltic magmas. Phys Earth Planet Inter 3: 273–301

Kuroda Y, Suzuoki T, Matsuo S, Aoki K (1975) D/H ratios of co-existing phlogopite and richterite from mica nodules and a peridotite in South African kimberlites. Contrib Mineral Petrol 52: 315–318

Kushiro I, Aoki K (1968) Origin of some eclogite inclusions in kimberlite. Am Mineral 53: 1347–1367

Lambert IB, Heier KS (1968) Geochemical investigations of deep-seated rocks in the Australian Shield. Lithos 1: 30–53

Lambert IB, Wyllie PJ (1968) Stability of hornblende and a model for the low-velocity zone. Nature (London) 219: 1240–1241

Lappin MA (1978) The evolution of a grospydite from the Roberts Victor Mine, South Africa. Contrib Mineral Petrol 66: 229–241

Lappin MA, Dawson JB (1975) Two Roberts Victor cumulate eclogites and their re-equilibration. Phys Chem Earth 9: 351–366

Larsen ES (1940) Petrographic province of central Montana. Geol Soc Am Bull 51: 887–948

Lawless PJ (1974) Some aspects of the geochemistry of kimberlite xenocrysts. Unpubl M Sci Thesis, University Cape Town

Lawless PJ, Gurney JJ, Dawson JB (1979) Polymict peridotites from the Bultfontein and De Beers Mines, Kimberley, South Africa. In: Boyd FR, Meyer HOA (eds) The mantle sample: inclusion in kimberlites and other volcanics. AGU, Washington, pp 145–155

Lazko EE, Afanasev VP (1974) Mineral inclusions in olivines from the Udachanya kimberlite pipe. Zap Vses Mineral Ova 103: 77–84 (Russian)

Le Bas MJ (1971) Per-alkaline volcanism, crustal swelling and rifting. Nature (London) 230: 85–87

Lebedev AP (1964) Kimberlites of north-eastern U.S.S.R. and allied problems. Geol J 4:87–104

Lee HA, Lawrence DE (1968) A new occurrence of kimberlite in Gauthier Township, Ontario. Geol Surv Can Pap 68-22: pp 16

Lewis HC (1887) On a diamantiferous peridotite and the genesis of diamond. Geol Mag New Ser 5: 22–24

Lewis HC (1888) The matrix of the diamond. Geol Mag New Ser 5: 129–131

Lindsley DH, Dixon SA (1977) Diopside-enstatite equilibria at 850° to 1400°C, 5 to 35 kb. Am J Sci 276: 1285–1301

Lloyd FE, Bailey DK (1975) Light element metasomatism of the continental mantle: the evidence and the consequences. Phys Chem Earth 9: 389–416

Lock NP (1980) The geology of the Letseng kimberlites, Lesotho. Unpubl Ph D Thesis, University of Sheffield

Lock NP, Dawson JB (1980) Garnet-olivine reaction in the upper mantle: evidence from peridotite xenoliths in the Letseng le Terae kimberlite, Lesotho. Trans R Soc Edinburgh Earth Sci 71: 47–53

Lorenz V (1979) Phreatomagmatic origin of the olivine melilitite diatremes of the Schwabian Alb, Germany. In: Boyd FR, Meyer HOA (eds) Kimberlites, diatremes and diamonds, their geology, petrology and geochemistry. AGU, Washington, pp 354–363

Loubet M, Bernat M, Javoy M, Allègre CJ (1972) Rare earth contents in carbonatites. Earth Planet Sci Lett 14: 226–232

Lovering JF, Tatsumoto M (1968) Lead isotopes and the origin of granulite and eclogite inclusions in deep-seated pipes. Earth Planet Sci Lett 4: 350–356

Lutts BG, Gladkikh VS, Pyatenko IK, Viktorova ME (1974) Distribution of niobium and tantalum in ultrabasic inclusions from kimberlites and basalts. Sov Geol 60-71 (Russian)

Maaløe S, Aoki K (1977) The major element composition of the upper mantle estimated from the composition of lherzolites. Contrib Mineral Petrol 63: 161–173

MacDonald GA, Katsura T (1964) Chemical composition of Hawaiian lavas. J Petrol 5:82–133

MacGregor AM (1921) The geology of the diamond-bearing grounds of the Somabula Forest. Bull Geol Surv S Rhodesia 8: pp 44
MacGregor AM, Ferguson JC, Amm FL (1937) The geology of the country around the Queen's Mine, Bulawayo district. Bull Geol Surv S Rhodesia 30: pp 175
MacGregor ID (1979) Mafic and ultramafic xenoliths from the Kao kimberlite pipe. In: Boyd FR, Meyer HOA (eds) The mantle samples; inclusions in kimberlites and other volcanics. AGU, Washington, pp 156–172
MacGregor ID, Carter JL (1970) The chemistry of clinopyroxenes and garnets of eclogite and peridotite xenoliths from the Roberts Victor Mine, South Africa. Phys Earth Planet Inter 3: 391–397
Macintyre RM, Dawson JB (1976) Age and significance of some South African kimberlites. 4th Eur Colloq Geochron Cosmochron Isotope Geol Amsterdam, Abstr 66
Makhilaev LV, Surina NP (1965) Maymecha-Kotui province of ultrabasic and alkalic rocks – new region of kimberlite magmatism. Dokl Akad Sci Nauk SSSR, Earth Sci Sect 153: 178–180
Malde HE (1954) Serpentine pipes at Garnet Ridge, Arizona. Science 119: 618
Malkov BA (1972) On petrological differences of kimberlites and meimechites. Dokl Akad Nauk SSSR 206: 951–954 (Russian)
Malkov BA (1974) Brucite in kimberlite. Dokl Akad Sci Nauk SSSR, Earth Sci Sect 215: 157–160
Malkov BA (1975) Carbonatite kimberlites – a new type of diamond-bearing rock. Dokl Akad Sci Nauk SSSR 221: 1170–1173 (Russian)
Mannard G (1962) The Singida kimberlite pipes. Unpubl Ph D Thesis, McGill University
Manton WI, Tatsumoto M (1971) Some Pb and Sr isotopic measurements in eclogites from the Roberts Victor Mine, South Africa. Earth Planet Sci Lett 10: 217–226
Marchand M (1970) Ultramafic nodules from Île Bizard, Quebec. Unpubl M Sci Thesis, McGill University
Markovski BA, Rotman VK (1975) Geochemistry of rocks of the ultrabasic volcanic province of Kamchatka. Sov Geol 1975: 70–82 (Russian)
Marsh JS (1973) Relationships between transform directions and alkaline igneous rock lineaments in Africa and South America. Earth Planet Sci Lett 18: 317–323
Marshintsev VK (1970a) Carbonatitic explosion breccias from the eastern part of the Anabar massif. In: Kovalski VV, Yurkevich RK (eds) Geology, petrography and mineralogy of magmatic formations of the north-eastern part of the Siberian Platform. Acad Sci SSSR, Moscow, pp 129–169
Marshintsev VK (1970b) Discovery of baddelyite in kimberlitic rocks of Yakutia. In: Kovalski VV, Yurkevich RK (eds) Geology, petrography and mineralogy of magmatic formations of the north-eastern part of the Siberian Platform. Acad Sci SSSR, Moscow, pp 247–253
Marshintsev VK, Sukneva LS (1970) Scandium in minerals and rocks from Yakutian kimberlite pipes. Geokhimiya 1970: 1519–1522 (Russian)
Mathias M, Rickwood PC (1969) Ultramafic xenoliths in the Matsoku kimberlite pipe. Geol Soc S Afr Spec Publ 2: 359–369
Mathias M, Siebert JC, Rickwood PC (1970) Some aspects of the mineralogy and petrology of ultramafic xenoliths in kimberlite. Contrib Mineral Petrol 26: 75–123
Mathur SM (1962) Geology of the Panna diamond deposits. Rec Geol Surv India 87: 787–818
McCallum ME, Eggler DH (1971) Mineralogy of the Sloan diatreme, a kimberlite diatreme in northern Larrimer County, Colorado. Am Mineral 56: 1735–1749
McCallum ME, Eggler DH (1976) Diamonds in an upper mantle peridotite nodule from kimberlite in southern Wyoming. Science 192: 253–256
McCallum ME, Eggler DH, Burns LK (1975) Kimberlite diatremes in northern Colorado and southern Wyoming. Phys Chem Earth 9: 149–162

McElhinny MW, Briden JC, Jones JL, Brock AB (1968) Geological and geophysical implications of palaeomagnetic results from Africa. Rev Geophys 6: 201–238

McEwen TJ (1978) Diffusional mass transfer processes in pitted pebble conglomerates. Contrib Mineral Petrol 67: 405–415

McGetchin TR (1968) The Moses Rock dyke: petrology and mode of emplacement of kimberlite-bearing Breccia dyke, San Juan County, Utah. Unpubl Ph D Thesis, Calif Inst Tech

McGetchin TR, Besancon JR (1973) Carbonate inclusions in mantle-derived pyropes. Earth Planet Sci Lett 18: 408–410

McGetchin TR, Silver LT (1970) Compositional relations in minerals from kimberlite and related rocks in the Moses Rock dykes, San Juan County, Utah. Am Mineral 55: 1738–1771

McGetchin TR, Silver LT, Chodos AA (1970) Titanoclinohumite: a possible mineralogical site for water in the upper mantle. J Geophys Res 75: 255–259

McGetchin TR, Nihan YS, Chodos AA (1973) Carbonatite-kimberlite relations in the Cane Valley diatreme, San Juan County, Utah. J Geophys Res 78: 1854–1869

McIver JR, Ferguson J (1979) Kimberlitic, melilitic, trachytic and carbonatite eruptives at Saltpetre Kop, Sutherland, South Africa. In: Boyd FR, Meyer HOA (eds) Kimberlite, diatremes and diamonds: their geology, petrology and geochemistry. AGU, Washington, pp 111–128

McKinlay ACM (1955) Kimberlite intrusions cutting Karroo sediments in the Ruhuhu depression of south-west Tanganyika. Rec Geol Surv Tanganyika V: 63–80

McMahon BM, Haggerty SE (1979) The Oka carbonatite complex: magnetite compositions and the related role of titanium in pyrochlore. In: Boyd FR, Meyer HOA (eds) Kimberlites, diatremes and diamonds: their geology, petrology and geochemistry. AGU, Washington, pp 382–392

Melton CE, Giardini AA (1974) The composition and significance of gas released from natural diamonds from Africa and Brazil. Am Mineral 59: 775–782

Melton CE, Giardini AA (1975) Experimental results and a theoretical interpretation of gaseous inclusions found in Arkansas natural diamonds. Am Mineral 60: 413–417

Menzles M, Murthy VR (1980) Enriched mantle: Nd and Sr isotopes in diopsides from kimberlite nodules. Nature 283: 634–636

Mercier JCC (1979) Peridotite xenoliths and the dynamics of kimberlite intrusion. In: Boyd FR, Meyer HOA (eds) The mantle sample: inclusions in kimberlites and other volcanics. AGU, Washington, pp 197–212

Mercier JCC, Carter NL (1975) Pyroxene geotherms. J Geophys Res 80: 3349–3362

Merh S (1952) Further study of Majgawan diamond mine, Panna State, central India. Q J Geol Min Soc India 24: 125–132

Meyer HOA (1976) Kimberlites of the continental United States: a review. J Geol 84: 377–403

Meyer HOA, Boctor NZ (1975) Sulfide-oxide minerals in eclogite from Stockdale kimberlite, Kansas. Contrib Mineral Petrol 52: 57–68

Meyer HOA, Boyd FR (1972) Composition and origin of crystalline inclusions in natural diamonds. Geochim Cosmochim Acta 36: 1255–1273

Meyer HOA, Brookins DG (1971) Eclogite xenoliths from Stockdale kimberlite, Kansas. Contrib Mineral Petrol 34: 60–72

Meyer HOA, Brookins HOA (1976) Sapphirine, sillimanite and garnet in granulite xenoliths from Stockdale kimberlite, Kansas. Am Mineral 61: 1194–1202

Meyer HOA, Kridel Baugh SJ (1977) Green Mountain kimberlite, Colorado: mineralogy and petrology. 2nd Int Conf Kimb, Santa Fe, Ext Abstr (unpaged)

Meyer HOA, Svizero DP (1975) Mineral inclusions in Brazilian diamonds. Phys Chem Earth 9: 785–796

Meyer HOA, Tsai H, Gurney JJ (1979) A unique enstatite megacryst with co-existing Cr-poor and Cr-rich garnet, Weltvreden floors, South Africa. In: Boyd FR, Meyer HOA (eds) The mantle sample: inclusions in kimberlite and other volcanics. AGU Washington, pp 279–291

Meyer de Stadelhofen (1963) Les brèches kimberlitiques du territoire de Bakwanga (Congo). Arch Sci 16: 87–143
Mikheyenko VI (1968) Mechanism of the formation of banded flow texture in kimberlite. Dokl Akad Nauk SSSR 179: 194–197 (Russian)
Mikheyenko VI (1972) On kimberlite serpentinisation. Dokl Akad Sci Nauk SSSR, Earth Sci Sect 202: 167–169
Mikheyenko VI, Nenashev NI (1962) Absolute age of the kimberlites of Yakutia. Int Geol Rev 4: 916–924
Mikheyenko VI, Vladimirov BM, Nenashev NI, Seldisheva YB (1970) Cobble of diamond-bearing eclogite from the kimberlite of the Mir pipe. Dokl Akad Sci Nauk SSSR Earth Sci Sct 190: 179–181
Milashev VA (1965) Petrochemistry of the kimberlites of Yakutia and some factors in diamond formation. Nedra, Moscow (Russian)
Milashev VA, Krutoyarski MA, Rabhkin MI, Erlich EN (1963) Kimberlite rocks and picritic porphyries of the north-eastern part of the Siberian Platform. Tr Nauchno Issled Inst Geol Arktiki 126 (Russian)
Mitchell RH (1972) Composition of perovskite in kimberlite. Am Mineral 57: 1748–1753
Mitchell RH (1973a) Composition of olivine, silica activity and oxygen fugacity in kimberlite. Lithos 6: 65–82
Mitchell RH (1973b) Magnesian ilmenite and its role in kimberlite petrogenesis. J Geol 81: 301–311
Mitchell RH (1975) Theoretical aspects of gaseous and isotopic equilibria in the system C-H-O-S with application to kimberlite. Phys Chem Earth 9: 903–916
Mitchell RH (1977) Geochemistry of magnesian ilmenites from kimberlites in South Africa and Lesotho. Lithos 10: 29–37
Mitchell RH (1978) Composition of spinels in micaceous kimberlites from the Upper Canada Mine, Kirkland Lake, Ontario. Can Mineral 16: 591–595
Mitchell RH (1979) The alleged kimberlite-carbonatite relationship: additional contrary mineralogical evidence. Am J Sci 279: 570–589
Mitchell RH, Brunfelt AO (1975) Rare earth element geochemistry of kimberlite. Phys Chem Earth 9: 671–686
Mitchell RH, Carswell DA (1976) Lanthanum, samarium and ytterbium abundances in some South Africa garnet lherzolites. Earth Planet Sci Lett 31: 175–178
Mitchell RH, Crockett JH (1971a) The isotopic composition of strontium in some South African kimberlites. Contrib Mineral Petrol 30: 277–290
Mitchell RH, Crockett JH (1971b) Diamond genesis – a synthesis of opposing views, Miner Deposita 6: 392–403
Mitchell RH, Fritz P (1973) Kimberlite from Somerset Island, District of Franklin, N.W.T. Can J Earth Sci 10: 384–393
Mitchell RH, Brunfelt AO, Nixon PH (1973a) Trace elements in magnesian ilmenites from Lesotho kimberlites. In: Nixon PH (ed) Lesotho kimberlite. Lesotho Natl Dev Co, Maseru, pp 230–235
Mitchell RH, Carswell DA, Brunfelt AO (1973b) Mineralogy and rare-earth geochemistry of an ilmenite-clinopyroxene xenolith from the Monastery Mine. In: Nixon PH (ed) Lesotho kimberlites. Lesotho Natl Dev Co, Maseru, pp 224–229
Moor GG, Sheinmann J (1946) Meimechite, a new rock from the northern border of the Siberian platform. Dokl Akad Nauk SSSR 51: 145–148 (Russian)
Moor GG (1941) Micaceous kimberlites in the north of Central Siberia. C R Acad Sci URSS 31: 363–365 (Russian)
Moore AE (1973) The olivine melilitite-kimberlite association of Namaqualand. 1st Int Conf Kimb, Cape Town, Ext Abstr 239–242

Moore AE (1976) Controls of post-Gondwanaland alkaline volcanism in southern Africa. Earth Planet Sci Lett 31: 291–296

Morgan J (1971) Convection plumes in the lower mantle. Nature (London) 230: 42–43

Mori T, Green DH (1978) Laboratory duplication of phase equilibria observed in natural garnet lherzolites. J Geol 86: 83–97

Muramatsu Y (1977) Geochemische Untersuchungen an Kimberliten, einem Granatperidotit- und einem Eklogit-Einschluß aus Kimberley, Südafrika. Dr Dissertation, University of Göttingen

Mysen BO, Boettcher AL (1975) Melting of a hydrous mantle: I. Phase relations of natural peridotites at high pressures and temperatures with controlled activities of water, carbon dioxide and hydrogen. J Petrol 16: 520–548

Naeser CW (1971) Geochronology of the Navajo-Hopi diatremes, Four Corners area. J Geophys Res 76: 4878–4985

Nekrasova RA, Gamyanina VV (1968) Composition of rare earths in kimberlite minerals. Dokl Akad Sci Nauk USSR, Earth Sci Sect 182: 195–198

Newton AR, Gurney JJ (1975) Discussion of "A plate tectonics origin for diamond-bearing kimberlites". Earth Planet Sci Lett 27: 356–358

Nikishov KN, Gamyanina VV (1970) Trace elements in perovskite from intrusive kimberlites. In: Kovalski VV, Yurkevich RK (eds) Geology, petrography and mineralogy of magmatic formations of the north-eastern part of the Siberian Platform. Acad Sci SSSR, Moscow, pp 281–287 (Russian)

Nikishov KN, Nikishova V (1966) The possibility of reaction relations between olivine and monticellite. Geochem Int 3: 1200–1206

Nikishov KN, Kovalski VV, Marshintsev VK (1972) Alkaline-ultrabasic rocks (alnöites, kimberlite and carbonatite) of the north-east Siberian platform. Proc 24th Int Geol Cong Part 14: 51–55

Nixon PH (1973a) Kimberlitic volcanoes in East Africa. Overseas Geol Mineral Resour 41: 119–130

Nixon PH (ed) (1973b) Lesotho kimberlites. Lesotho Natl Dev Co, Maseru

Nixon PH, Boyd FR (1973a) Petrogenesis of the granular and sheared ultrabasic nodule suite in kimberlite. In: Nixon PH (ed) Lesotho kimberlites. Lesotho Natl Dev Co, Maseru, pp 48–65

Nixon PH, Boyd FR (1973b) The discrete nodule (megacryst) association in kimberlites from northern Lesotho. In: Nixon PH (ed) Lesotho kimberlites. Lesotho Natl Dev Co, Maseru, pp 67–75

Nixon PH, Boyd FR (1979) Garnet-bearing lherzolites and discrete nodule suites from the Malaita alnöite, Solomon Islands, S.W. Pacific, and their bearing on oceanic mantle composition and geotherm. In: Boyd FR, Meyer HOA (eds) The mantle sample: inlcusions in kimberlites and other volcanics. AGU, Washington, pp 400–423

Nixon PH, Hornung G (1968) A new chromium garnet end-member, knorringite, from kimberlite. Am Mineral 53: 1833–1840

Nixon PH, Kresten P (1973) Chromium and nickel in kimberlite ilmenites. In: Nixon PH (ed) Lesotho kimberlites. Lesotho Natl Dev Co, Maseru, pp 235–237

Nixon PH, Knorring O von, Rooke JM (1963) Kimberlites and associated inclusions of Basutoland: a mineralogical and geochemical study. Am Mineral 48: 1090–1132

Nixon PH, Boyd FR, Bouiller AM (1973) The evidence of kimberlite and its inclusions on the constitution of the outer part of the earth. In: Nixon PH (ed) Lesotho kimberlites, Lesotho Natl Dev Co, Maseru, pp 312–318

Nixon PH, Chapman NA, Gurney JJ (1978) Pyrope-spinel (alkremite) xenoliths from kimberlites. Contrib Mineral Petrol 65: 341–346

O'Daniel H (1930) Ein chromhaltiger Pyroxen von Jagersfontein (Südafrika). Z Kristallogr 75: 575

O'Hara MJ, Mercy ELP (1966) Eclogites, peridotite and pyrope from the Navajo country, Arizona and New Mexico. Am Mineral 51: 336–352

O'Hara MJ, Yoder HS (1967) Formation and fractionation of basic magmas at high pressure. Scott J Geol 3: 67–117

O'Hara MJ, Richardson SW, Wilson G (1971) Garnet peridotite stability and occurrence in crust and mantle. Contrib Mineral Petrol 32: 48–68

O'Hara MJ, Saunders M, Mercy ELP (1975) Garnet peridotite, possible ultrabasic magmas and eclogite: interpretation of upper mantle processes in kimberlite. Phys Chem Earth 9: 681–713

Orlov YuL (1977) The mineralogy of the diamond. Wiley, New York

Oxburgh ER, Turcotte DL (1974) Membrane tectonics and the East African Rift. Earth Planet Sci Lett 22: 133–140

Parry J, Wright FE (1925) Afwillite, a new hydrous calcium silicate from the Dutoitspan Mine, Kimberley, South Africa. Min Mag 20: 277–286

Parry J, Williams AF, Wright FE (1932) On bultfonteinite, a new fluorine-bearing hydrous calcium silicate from South Africa. Min Mag 23: 145–162

Pasteris JD, Boyd FR, Nixon PH (1979) The ilmenite association at the Frank Smith mine, R.S.A. In: Boyd FR, Meyer HOA (eds) The mantle sample: inclusions in kimberlites and other volcanics. AGU, Washington, pp 257–264

Paul DK (1979) Isotopic composition of strontium in Indian kimberlites. Geochim Cosmochim Acta 43: 389–394

Paul DK, Potts PJ (1976) Rare-earth abundances in kimberlites from Greenland and Zambia. Chem Geol 18: 161–167

Paul DK, Potts PJ, Gibson IL, Harris PG (1975a) Rare-earth abundances in Indian kimberlite. Earth Planet Sci Lett 25: 151–158

Paul DK, Rex DC, Harris PG (1975b) Chemical characteristics and K-Ar ages of Indian kimberlite. Geol Soc Am Bull 86: 364–366

Paul DK, Crocket JH, Nixon PH (1979) Abundances of palladium, indium and gold in kimberlites and associated nodules. In: Boyd FR, Meyer HOA (eds) Kimberlites, diatremes and diamonds: their geology, petrology and geochemistry. AGU, Washington, pp 272–279

Pavlov DI (1969) Role of chlorine in the endogenic mineralisation of kimberlite. Geol Geofiz 1969: 133–134 (Russian)

Pavlov DI, Ilupin IP (1973) Halite in Yakutian kimberlite, its relation to serpentine and the source of its parent solutions. Dokl Akad Sci Nauk USSR, Earth Sci Sect 213: 178–180

Pecherski VP (1965) On the discovery of micaceous diamondiferous kimberlites in the foothills of eastern Sayan. Sov Geol 4: 131–133 (Russian)

Philpotts JA, Schnetzler GC, Thomas HM (1972) Petrogenetic implications of some new geochemical data on eclogitic and ultrabasic inclusions. Geochim Cosmochim Acta 36: 1131–1166

Pokhilenko NP, Sobolev NV, Lavrentyev YuG (1977) Xenoliths of diamondiferous ultramafic rocks from Yakutian kimberlites. 2nd Int Conf Kimb, Santa Fe, Ext Abstr (unpaged)

Pollard ER, Dixon CG, Dujardin RJ (1957) Diamond resources of British Guiana. Bull Geol Surv Br Guiana 28: pp 45

Ponomarenko AI (1975) Alkremite – a new variety of aluminous hyperbasite xenoliths from the kimberlites of the Udachnaya pipe. Dokl Akad Nauk SSSR 225: 928–931 (Russian)

Ponomarenko AI, Spetsius ZV (1978) Coesite of eclogites from the Udachnaya kimberlite pipe. XI. Gen Meet Int Mineral Assoc, Novosibirsk, Abstr Vol 2, p 23

Ponomarenko AI, Serenko VP, Lazko EE (1973) First find of diamond-bearing eclogites in the kimberlite pipe "Udachnaya". Dokl Akad Nauk SSSR 209: 188–189 (Russian)

Ponomarenko AI, Sobolev NV, Pokhilenko NP, Lavrentyev YuG, Sobolev VS (1976) Diamond-bearing kyanite eclogite from the Udachanya pipe, Yakutia. Dokl Akad Sci Nauk USSR, Earth Sci Sect 226: 158–161

Popov AS (1967) Palaeozoic volcanism of the Kola Peninsula. Dokl Akad Sci Nauk USSR, Earth Sci Sect 174: 32–34

Prinz M, Manson DV, Hlava PF, Keil K (1975) Inclusions in diamonds: garnet lherzolite and eclogite assemblages. Phys Chem Earth 9: 797–816

Raber E, Haggerty S (1979) Zircon-oxide reactions in diamond-bearing kimberlites. In: Boyd FR, Meyer HOA (eds) Kimberlites„diatremes and diamonds: their geology. petrology and geochemistry. AGU, Washington, pp 229–240

Rabhkin MI, Krutoyarskii MA, Milashev VA (1962) Classification and nomenclature of Yakutian kimberlites. Tr Issled Inst Geol Arktiki 121: 154–164 (Russian)

Råheim A, Green DH (1974) Experimental determination of the temperature and pressure dependence of the Fe-Mg partition coefficient for co-existing garnet and clinopyroxene. Contrib Mineral Petrol 48: 179–203

Ramberg IB, Barth TFW (1966) Eocambrian volcanism in southern Norway. Nor Geol Tidskr 46: 219–234

Rao PS, Phadtare PH (1966) Kimberlite pipe rocks of the Wajrakurar area – Anantpur District, Andhra Pradesh. J Geol Soc India 7: 110–117

Rawlinson PJ, Dawson JB (1979) A quench pyroxene-ilmenite xenolith from kimberlite: implications for pyroxene-ilmenite intergrowths. In: Boyd FR, Meyer HOA (eds) The mantle sample: inclusions in kimberlites and other volcanics. AGU, Washington, pp 292–299

Real F (1959) Intrusoes kimberliticas da Lunda. Serv Geol Port Mem 5

Reay A, Wood CP (1974) Ilmenites from Kakanui, New Zealand. Min Mag 39: 721–723

Reid AM, Dawson JB (1972) Olivine-garnet reaction in peridotites from Tanzania. Lithos 5: 115–124

Reid AM, Hanor JS (1970) Pyrope in kimberlites. Am Mineral 55: 1374–1379

Reid AM, Donaldson CH, Ridley WI, Brown RW, Dawson JB (1975a) The Igwisi Hills extrusive 'kimberlites'. Phys Chem Earth 9: 199–218

Reid AM, Donaldson CH, Brown RW, Ridley WI, Dawson JB (1975b) Mineral chemistry of peridotite xenoliths from the Lashaine volcano, Tanzania. Phys Chem Earth 9: 525–544

Reid AM, Brown RW, Dawson JB, Whitfield GG, Siebert JC (1976) Garnet and pyroxene compositions in some diamondiferous eclogites. Contrib Mineral Petrol 58: 203–220

Reis B (1972) Preliminary note on the distribution and tectonic control of kimberlites in Angola. Proc 24th Int Geol Congr Sect 4: 276–281

Rickwood PC, Mathias M (1970) Diamondiferous eclogite xenoliths in kimberlite. Lithos 3: 223–235

Rickwood PC, Mathias M, Siebert JC (1968) A study of garnets from eclogite and peridotite xenoliths found in kimberlite. Contrib Mineral Petrol 19: 271–301

Rickwood PC, Gurney JJ, White-Cooper DR (1969) The nature and occurrence of eclogite xenoliths in the kimberlites of southern Africa. Geol Soc S Afr Spec Publ 2: 371–393

Ridley WI, Dawson JB (1975) Lithophile trace element data bearing on the origin of peridotite xenoliths, ankaramite and carbonatite from Lashaine volcano, Tanzania. Phys Chem Earth 9: 559–570

Rimann E (1915) Über Kimberlit und Alnöit in Brasilien. Tschermaks Mineral Petrogr Mitt 33: 244–262

Rimann E (1931) Über das Muttergestein der Diamanten von Minas Geraes, Brasilien. Fortschr Mineral Kristallogr Petrogr 16: 93–96

Rimsaite J (1971) Distribution of major and minor constituents between mica and host ultrabasic rocks and between zoned mica and zoned spinel. Contrib Mineral Petrol 33: 259–272

Ringwood AE (1975) Composition and petrology of the earth's mantle. McGraw-Hill, New York

Ringwood AE, Green DH (1966) An experimental investigation of the gabbro-eclogite transformation and some geophysical implications. Tectonophysics 3: 383–427

Ringwood AE, Lovering JF (1978) Significance of pyroxene-ilmenite intergrowths among kimberlite xenoliths. Earth Planet Sci Lett 7: 371–375

Robinson DN (1975) Magnetite-serpentine-calcite dykes at Premier Mine and aspects of their relationship to kimberlite and to the carbonatite of alkalic carbonatite complexes. Phys Chem Earth 9: 61–70

Robinson DN (1978) The characteristics of natural diamond and their interpretation. Miner Sci Eng 10: 55–71

Robinson DN (1979) Diamond and graphite in eclogite xenoliths from kimberlite. In: Boyd FR, Meyer HOA (eds) The mantle sample: inclusions in kimberlites and other volcanics. AGU, Washington, pp 50–58

Roedder E (1965) Liquid CO_2 inclusions in olivine-bearing nodules and phenocrysts from basalts. Am Mineral 50: 1746–1782

Rogers NW (1977) Granulite xenoliths from Lesotho kimberlites and the composition of the lower continental crust. Nature (London) 270: 681–684

Rovsha VS, Ilupin IP (1970) Chrome spinels in kimberlite of Yakutia. Geol Geofiz 1970: 47–56 (Russian)

Ruotsala AP (1975) Alteration of the Finsch kimberlite pipe, South Africa. Econ Geol 70: 587–590

Sampson DN (1953) The volcanic hills at Igwisi. Rec Geol Surv Tanganyika 3: 48–53

Schmid R, Wood BJ (1976) Phase relationships in granulitic metapelites from the Ivrea-Verbano zone (Italy). Contrib Mineral Petrol 54: 255–279

Schutte CEG (1966) The chemical composition and hafnium/zirconium ratios of South Africa zircons. Bull Geol Surv S Afr 46: pp 71

Schutte FJP (1967) Xenoliete en xenotriste in kimberlit van Barkly-Wes. Unpubl M Sci Thesis, University of Potchefstroom (Afrikaans)

Scott BH (1979) Petrogenesis of kimberlites and associated potassic lamprophyres from Central West Greenland. In: Boyd FR, Meyer HOA (eds) Kimberlites, diatremes and diamonds: their geology, petrology and geochemistry. AGU, Washington, pp 190–205

Sellschop JPF, Mingay DW, Bibby DM, Erasmus SA (1974) Determination of impurities in diamond by nuclear methods. Diamond Res 1974: 43–50

Sen SN, Rao CN (1967) Igneous activity in Cuddapah Basin and adjacent areas and suggestions on the palaeogeography of the Basin. Proc Int Symp Indian Upper Mantle Project, Hyderabad, pp 261–285

Sen SN, Rao CN (1970) Chelima dykes. Proc 2nd Symp Indian Upper Mantle Project, Hyderabad, pp 435–439

Sergeyeva NYe (1971) Microscopic study of micro-morphology of picroilmenites from the kimberlite pipes of Yakutia. Int Geol Rev 13: 1242–1246

Shand SJ (1922) The alkaline rocks of the Franspoort Line, Pretoria District. Trans Geol Soc S Afr 25: 81–101

Sharp WE (1966) Pyrrhotite: a common inclusion in South African diamonds. Nature (London) 211: 402–403

Sharp WE (1974) A plate tectonic origin for diamond-bearing kimberlites. Earth Planet Sci Lett 21: 351–354

Shchelkova SG, Brovkin AA (1969) Titanolivine from Siberian kimberlites. Zap Vses Mineral Ova 98: 246–247 (Russian)

Shee SR, Gurney JJ (1979) The mineralogy of xenoliths from Orapa, Botswana. In: Boyd FR, Meyer HOA (eds) The mantle sample: inclusions in kimberlites and other volcanics. AGU, Washington, pp 37–49

Sheppard SMF, Dawson JB (1973) $^{13}C/^{12}C$ and D/H isotope variations in "primary" igneous carbonatites. Fortschr Mineral 50: 128–129

Sheppard SMF, Dawson JB (1975) Hydrogen, carbon and oxygen isotope studies of megacrysts and matrix minerals from Lesothan and South African kimberlites. Phys Chem Earth 9: 747–763

Shimizu N (1975a) Geochemistry of ultramafic inclusions from Salt Lake Crater, Hawaii, and from southern African kimberlites. Phys Chem Earth 9: 655–670

Shimizu N (1975b) Rare earth elements in garnets and clinopyroxenes from garnet lherzolite nodules in kimberlites. Earth Planet Sci Lett 25: 26–32

Singewald JT, Milton C (1930) An alnöite pipe, its contact phenomena and ore deposition near Avon, Missouri. J Geol 38: 54–66

Skinner EMW, Clement CR (1979) Mineralogical classification of southern African kimberlites. In: Boyd FR, Meyer HOA (eds) Kimberlites, diatremes and diamonds: their geology, petrology and geochemistry. AGU, Washington, pp 129–139

Smirnov GI (1959) Mineralogy of Siberian kimberlites. Int Geol Rev 1: 21–39

Smith D, Levy S (1976) Petrology of the Green Knobs diatreme and implications for the upper mantle below the Colorado Plateau. Earth Planet Sci Lett 29: 107–125

Smith JV, Dawson JB (1975) Chemistry of Ti-poor spinels, ilmenites and rutiles from peridotite and eclogite xenoliths. Phys Chem Earth 9: 309–322

Smith JV, Brennesholtz R, Dawson JB (1978) Chemistry of micas from kimberlites and xenoliths. I. Micaceous kimberlites. Geochim Cosmochim Acta 42: 959–971

Smith JV, Hervig RL, Ackermand D, Dawson JB (1979) K, Rb, and Ba in micas from kimberlite and peridotite xenoliths and implications for origin of basaltic rocks. In: Boyd FR, Meyer HOA (eds) Kimberlites, diatremes and diamonds: their geology, petrology and mineralogy. AGU, Washington, pp 241–251

Smulikowski K (1972) Classification of eclogites and allied rocks. Krystallinikum 9: 107–130

Smyth JR (1977) Quartz pseudomorphs after coesite. Am Mineral 62: 828–830

Smyth JR, Hatton CJ (1977) A coesite-sanidine grospydite from the Roberts Victor kimberlite. Earth Planet Sci Lett 34: 284–290

Snyder FG, Gerdemann PE (1965) Explosive igneous activity along an Illinois-Missouri-Kansas axis. Am J Sci 263: 465–493

Sobolev NV (1970) Eclogites and pyrope peridotites from the kimberlites of Yakutia. Phys Earth Planet Inter 3: 398–404

Sobolev NV (1977) Deep-seated inclusions in kimberlites and the problems of the composition of the upper mantle. Am Geophys Union, Washington

Sobolev NV, Kuznetsova IK (1965) More facts on the mineralogy of eclogite from Yakutian kimberlite pipes. Dokl Akad Sci Nauk USSR, Earth Sci Sect 163: 137–140

Sobolev NV, Lavrentyev YuG (1972) Isomorphic sodium admixture in garnets formed at high pressure. Contrib Mineral Petrol 31: 1–12

Sobolev NV, Botkunov AI, Kuznetsova IK (1970) A diamond-bearing eclogite with calcium-rich garnet from the "Mir" pipe, Yakutia. Int Geol Rev 12: 1357–1358

Sobolev NV, Kuznetsova IK, Zyuzin NI (1968) The petrology of grospydite xenoliths from the Zagadochnaya kimberlite pipe in Yakutia. J Petrol 9: 253–280

Sobolev NV, Pustinshev VI, Kuznetsova IK, Kharkiv AD (1969) New data on the mineralogy of diamond-bearing eclogites from the "Mir" pipe, Yakutia. Geol Geofiz 1969: 113–116 (Russian)

Sobolev NV, Botkunov AI, Lavrentev YuG, Pospelova LN (1971) Compositional characteristics of minerals associated with diamonds in the Mir pipe, Yakutia. Zap Vses Mineral Ova 100: 558–564 (Russian)

Sobolev SF, Lapin AV, Nazarenko II, Kislova IV, Ilupin IP (1975) Selenium in the kimberlites and plutonic peridotite xenoliths of Yakutia. Geochem Int 12: 40–45

Sobolev VS, Sobolev NV (1964) Xenoliths in kimberlite of northern Yakutia and the structure of the mantle. Dokl Akad Sci Nauk USSR, Earth Sci Sect 158: 22–26

Sobolev VS, Kharkiv AD, Sobolev NV, Lavrentyev YuG, Pospelova LN (1972) Zoned garnet in kimberlite from the Mir pipe. Dokl Akad Nauk SSSR 207: 421–424 (Russian)

Sobolev VS, Nay BS, Sobolev NV, Lavrentyev YuG, Pospelova LN (1969) Xenoliths of diamond-bearing pyrope serpentinite from the Aykhal pipe, Yakutia. Dokl Akad Sci Nauk USSR, Earth Sci Sect 188: 168–170

Sorby HC (1863) On the direct correlation of mechanical and chemical forces. Proc R Soc 12: 538–550

Starkov NP (1968) Picrite porphyry associates of the western side of the Urals. Dokl Akad Sci Nauk USSR, Earth Sci Sect 177: 146–149

Stavrov OD, Ukhanov AV (1971) Alkali elements and fluorine in rocks and minerals of the mantle. Geochem Int 8: 168–178

Steele KF, Wagner GH (1979) Relationship of the Murfreesboro kimberlite and other igneous rocks of Arkansas, U.S.A. In: Boyd FR, Meyer HOA (eds) Kimberlites, diatremes and diamonds: their geology, petrology and geochemistry. AGU, Washington, pp 393–399

Stephens WE, Dawson JB (1977) Statistical comparison between pyroxenes from kimberlites and their associated xenoliths. J Geol 85: 433–449

Stockley GM (1948) Report on the geology of Basutoland. Basutoland Gov, Maseru

Stracke KJ, Ferguson J, Black LP (1979) Structural setting of kimberlite in south-eastern Australia. In: Boyd FR, Meyer HOA (eds) Kimberlites, diatremes and diamonds: their geology, petrology and geochemistry. AGU, Washington, pp 71–91

Stueber AM, Goles GG (1967) Abundances of Na, Mn, Cr, Sc and Co in ultramafic rocks. Geochim Cosmochim Acta 31: 75–93

Stueber AM, Murthy VR (1966) Strontium isotope and alkali element abundances in ultramafic rocks. Geochim Cosmochim Acta 30: 1243–1259

Sutton JR (1928) Diamond: a descriptive treatise. Murby, London

Suwa K, Oana S, Wada H, Osaki S (1975) Isotope geochemistry and petrology of African carbonatites. Phys Chem Earth 9: 735–745

Svisero DP, Meyer HOA, Tsai H (1979) Kimberlites in Brazil: an initial report. In: Boyd FR, Meyer HOA (eds) Kimberlites, diatremes and diamonds: their geology, petrology and geochemistry. AGU, Washington, pp 92–100

Switzer G, Melson WG (1969) Partially melted kyanite eclogite from the Roberts Victor mine, South Africa. Smithson Contrib Earth Sci 1: pp 9

Taljaard MS (1936) South African melilite basalts and their relations. Trans Geol Soc S Afr 39: 281–316

Taylor HP (1968) The oxygen isotope geochemistry of igneous rocks. Contrib Mineral Petrol 19: 1–71

Teale EO (1932) The kimberlites and associated occurrences of the Iramba plateau. Geol Surv Tanganyika, Short Pap 10: pp 10

Thode HG, Monster J, Dunford HB (1961) Sulfur isotope geochemistry. Geochim Cosmochim Acta 5: 286–298

Thompson G, Melson WG (1972) The petrology of oceanic crust across fracture zones in the Atlantic Ocean: evidence of a new kind of sea-floor spreading. J Geol 80: 526–538

Thompson LF (1928) The upland diamond deposits of the Diamantina district, Minas Geraes, Brazil. Econ Geol 23: 705–723

Thorsteinsson R, Tozer ET (1969) Geology of the Arctic archipeligo. In: Douglas RJW (ed) Geology and economic minerals of Canada. Geol Surv Canada, Ottawa, pp 548–590

Tolansky S (1955) The microstructures on diamond surfaces. Natl Assoc Gemmol, London

Tolman C, Landes KK (1937) Igneous rocks of the Missouri lead-zinc deposits. Geol Soc Am Spec Pap 24: 71–103

Touret J (1974) Facies granulite et inclusions carbonique. Ann Soc Geol Belg, Vol P Michot: 267–287

Tsai HM, Meyer HOA, Moreau J, Milledge HJ (1979a) Mineral inclusions in diamond: Premier, Jagersfontein and Finsch kimberlites, South Africa, and Williamson Mine, Tanzania. In: Boyd FR, Meyer HOA (eds) Kimberlites, diatremes and diamonds: their geology, petrology and geochemistry. AGU, Washington, pp 16–26

Tsai HM, Shieh Y, Meyer HOA (1979b) Mineralogy and S^{34}/S^{32} ratios of sulfides associated with kimberlite, xenoliths and diamonds. In: Boyd FR, Meyer HOA (eds) The mantle sample: inclusions in kimberlites and other volcanics. AGU, Washington, pp 87–103

Ukhanov AV (1965) Olivine melilitite from the diamond bearing diatremes in Anabar. Dokl Akad Sci Nauk USSR, Earth Sci Sect 153: 176–178

Ukhanov AV (1968) Nickel in ultrabasic unclusions from kimberlite pipes of north Yakutia. Geochem Int 5: 1190–1198

Ukhanov AV, Pchelintseva NF (1972) Gold content in peridotites and eclogites inclusions from kimberlite of the Obnazhennaya pipe. Geokhimiya 1972: 247

Van Bemmeln RW (1949) The geology of Indonesia. Netherlands Gov Print Office, The Hague

Vanuxem L (1937) Geological Survey of New York. 2nd Annu Rep Geol Surv Third Dt

Velde D (1975) Armalcolite -Ti phlogopite-diopside-analcite lamproites from Smoky Butte, Garfield County, Montana. Am Mineral 60: 566–573

Verhoogen J (1938) Les pipes de kimberlites de Katanga. Com Spec Katanga Ann Serv Mines 9: 1–50

Verwoerd WJ (1967) The carbonatites of South Africa and South West Africa. Geol Surv S Afr Handb 6: pp 452

Viljoen RP, Viljoen MJ (1969) Evidence for the composition of the primitive mantle and its products of partial melting from a study of the mafic and ultramafic rocks of the Barberton mountain land. Geol Soc S Afr Spec Publ 2: 275–304

Villar HM (1969) El complejo ultrabasico de Novilla Muerto, Cordillera Frontal, Provincia de Mendoza, Republica Argentina. Rev Assoc Geol Argent 24: 223–238 (Spanish)

Vinogradov AP (1962) Average contents of chemical elements in the principal types of igneous rocks of the earth's crust. Geochemistry 1962: 641–664

Vinogradov AP, Kropotova O, Orlov YuL, Grinenko VA (1966) Isotopic compositions of diamond and carbonado crystals. Geochem Int 3: 1123–1125

Volyanyuk NYa, Semenova VG, Yelizaryeva TI, Bondareva GV (1976) Inclusions of pyrope and pyrope-spinel lherzolite in basanite of the Vitim plateau. Dokl Akad Sci Nauk USSR, Earth Sci Sect 228: 159–162

Voskrenskaya VB, Kovalskii VV, Nikishov KN, Parinova ZF (1965) Discovery of titan-olivine in Siberian kimberlites. Zap Vses Mineral Ova 94: 600–603 (Russian)

Vredevoogd JJ, Forbes WC (1975) The system diopside-ureyite at 20 kb. Contrib Mineral Petrol 52: 147–156

Wagner PA (1911) Petrographical notes on the kimberlite occurrences in the Pretoria district. Trans Geol Soc S Afr 14: 43–63

Wagner PA (1914) The diamond fields of southern Africa. Transvaal Leader, Johannesburg

Wagner PA (1916) Graphite-bearing xenoliths from the Jagersfontein diamond mine. Trans Geol Soc S Afr 19: 54–56

Watkinson DH, Chao GY (1973) Shortite in kimberlite from the Upper Canada gold mine, Ontario. J Geol 81: 229–233

Watson KD (1967a) Kimberlites of eastern North America. In: Wyllie PJ (ed) Ultramafic and related rocks. Wiley, New York, pp 312–323

Watson KD (1967b) Kimberlite pipes of north-eastern Arizona. In: Wyllie PJ (ed) Ultramafic and related rocks. Wiley, New York, pp 261–269

Watson KD (1973) Kimberlites of the Superior Province, Canadian Shield. Int Conf Kimb Cape Town Ext Abstr 305–308

Wedepohl KH (1975) The contribution of chemical data to assumptions about the origin of magmas from the mantle. Fortschr Mineral 52: 141–172

Wedepohl KH, Muramatsu Y (1979) The chemical composition of kimberlites compared with the average composition of three basaltic magma types. In: Boyd FR, Meyer HOA (eds) Kimberlites, diatremes and diamonds: their geology, petrology and geochemistry. AGU, Washington, pp 300–312

Wendlandt RH, Harrison WJ (1979) Rare earth partitioning between immiscible carbonate and silicate liquids and CO_2 vapour: results and implications for the formation of light rare earth-enriched rocks. Contrib Mineral Petrol 69: 409–419

White RW, Leo GW (1969) Geological reconnaissance in western Liberia. Liberia Geol Surv Spec Pap 1

White RW, Leo GW (1970) Geological summary of age provinces of Liberia. Bull Geol Min Metal Soc Liberia 4: 96–106

Whitfield GG (1971) A petrological and mineralogical study of peridotite and eclogite xenoliths from certain kimberlite pipes. Unpubl M Sci Thesis, Rhodes University

Wickmann FE (1956) The cycle of carbon and stable carbon isotopes. Geochim Cosmochim Acta 9: 136–153

Williams AF (1932) The genesis of the diamond. Ernest Benn, London

Williams GJ (1939) The kimberlite province and associated diamond deposits of Tanganyika Territory. Bull Tanganyika Geol Surv 12: pp 41

Williams HR, Williams RA (1977) Kimberlites and plate tectonics in West Africa. Nature (London) 270: 507–508

Wood BJ (1974) The solubility of alumina in orthopyroxene coexisting with garnet. Contrib Mineral Petrol 46: 1–15

Woolsey TS, McCallum ME, Schumm SA (1975) Modelling of diatreme emplacement by fluidization. Phys Chem Earth 9: 29–42

Wyatt BA (1977) The melting and crystallization behaviour of a natural clinopyroxene-ilmenite intergrowth. Contrib Mineral Petrol 61: 1–9

Wyatt B (1979) Kimberlite chromian picro-ilmenites with intergrowths of titanian chromite and rutile. In: Boyd FR, Meyer HOA (eds) The mantle sample: inclusions in kimberlites and other volcanics. AGU, Washington, pp 257–264

Wyllie PJ (1979a) Kimberlite magmas from the system peridotite-CO_2-H_2O. In: Boyd FR, Meyer HOA (eds) Kimberlites, diatremes and diamonds: their geology, petrology and geochemistry. AGU, Washington, pp 319–329

Wyllie PJ (1979b) Magmas and volatile components. Am Mineral 64: 469–500

Yoder HS (1975) Relationship of melilite-bearing rocks to kimberlite: a preliminary report on the system akermanite-CO_2. Phys Chem Earth 9: 883–894

Yoder HS, Tilley CE (1962) Origin of basalt magmas: an experimental study of natural and synthetic rock systems. J Petrol 3: 342–532

Zartman RE, Brock MR, Heyl AV, Thomas HH (1967) K-Ar and Rb-Sr ages of some alkalic intrusive rocks from central and eastern United States. Am J Sci 265: 848–870

Subject Index

Volume 10: J. T. Wasson

Meteorites

Classification and Properties

1974. 70 figures. X, 316 pages
ISBN 3-540-06744-2

Contents: Introduction: Meteorites as Probes of Processes Occurring Very Early in the History of the Solar System. - Classification of Meteorites. - The Study of Meteorites: Sources, Bibliographies, and History. - Bulk Composition. - Mineralogy and Phase Composition. - Petrology. - Trace Elements. - Stable Isotopes: Elements Other than Rare Gases. - Stable Isotopes of the Rare-Gas Elements and Related Particle-Track Studies. - Primordial Radionuclides and Associated Chronologies. - Breakup and Accretional History of Parent Bodies, Meteoroid Shape and Erosion, and Terrestrial Ages aof Meteorites. - Orbits. - Fall and Recovery. - Morphology and Macrostructure. - Organic Matter. - Magnetic Properties. - Miscellaneous Physical Studies. - An Example of the Interpretation of Meteorite Properties: The Origin of Ordinary Chondrites. - Appendix I: Glossary. - Appendix II: Lists of Classified Meteorites.

"... The book is written by a scientist, who has been one of the leading specialists in meteorite research during the past decade. He is responsible for a good deal of progess in meteorite science. His survey book on meteorites will be used by experienced specialists as a good reference book and by starting specialists and students as an excellent introduction into meteorite research. Moreover, it will be a handbook for everybody, who wants to classify new or unknown meteorite samples."
Journ. of the Brit. Interplanetary Soc.

Volume 12: R. G. Coleman

Ophiolites

Ancient Oceanic Lithosphere?

1977. 72 figures, 18 tables. IX, 229 pages
ISBN 3-540-08276-X

Contents: What is an Ophiolite? - Plate Tectonics and Ophiolites. - Igneous Petrology. - Metamorphic Petrology. – Ore Deposits in Ophiolites. - Geologic Character. – Emplacement Tectonics. - Geologic, Tectonic, and Petrologic Nature of Four Ophiolites.

"... Petrologists and structural geologists will find this book a valuable perspective in that while presenting evidence in a systematic way, it very responsibly exposes the gaps and declines to reach conclusions. On the hand by pointing up the fragmentary nature of much of the global sampling it stimulates interest in filling out the model."
Esrisat

Volume 13: M. S. Paterson

Experimental Rock Deformation - The Brittle Field

1978. 56 figures, 3 tables. XII, 254 pages
ISBN 3-540-08835-0

Contents: Experimental Procedures. - Experimental Studies on the Brittle Fracture Stress. - Theories of Brittle Failure. - The Role of Pore Fluid Pressure. - Friction and Sliding Phenomena. - Pre-Failure, Post-Failure, and Mechanistic Aspects of Brittle Behaviour. - Brittle-Ductile Transition. - Appendix.

"... This well-written and comprehensive book, by one of the major investigators in the field, provides an excellent summary of the sophisticated and extensive knowledge that has been accumulated over the past few decades on brittle rock deformation."
Of Geophysics

Volume 14: A. K. Gutpa, K. Yagi

Petrology and Genesis of Leucite-Bearing Rocks

1980. 99 figures, 43 tables. XV, 252 pages
ISBN 3-540-09864-X

Contents: Introduction. - Nomenclature and Petrography of Leucite-Bearing Rocks. - Mineralogical Composition of Leucite-Bearing Rocks. - Minor and Trace Element Geochemistry, Initial $^{87}SR/^{86}SR$ Ratios and Oxygen Isotopic Ratios of Leucite-Bearing Rocks. - Distribution of Leucite-Bearing Rocks. - Conditions of Formation of Leucite-Bearing Mafic and Ultramafic Rocks. - Leucite-Bearing Ternary Joins and Systems. - Leucite-Albite Incompatibility. - Leucite- and Feldspar-Bearing Quaternary Joins and Systems. - The System Forsterite-Diopside-Akermanite-Leucite. - The System Diopside-Nepheline-Akermanite-Leucite. - Solubility of $KFe^{3+}Si_2O_6$ in Leucite. - Survival of Leucite. - Study of Leucite-Bearing Systems Under Different Pressures and the P_{H_2O}-T Stabilities of Phlogopite, Potassium-Rich Richterite, and Kaersutite. Study of Leucite-Bearing Rocks in Air and Under Various P-T Conditions. -Leucite Bearing Rocks and Their Relation to Kimberlites. - Structural and Tectonic Control of Alkali Magnetism with Special Reference to Leucite-Bearing Lavas. - Generation of Potassium-Rich Mafic and Ultramafic Magma Capable of Producing Leucite-Bearing Rocks. - References. - Subject Index.

Springer-Verlag
Berlin Heidelberg NewYork